TO MY FORMER STUDENTS,
who participated in the development
of this text and the underlying notions

Scott Nolan
7/79

INTERACTIVE COMPUTER GRAPHICS

Data Structures, Algorithms, Languages

Wolfgang K. Giloi

Technical University of Berlin
and University of Minnesota

PRENTICE-HALL, INC., Englewood Cliffs, New Jersey 07632

Library of Congress Cataloging in Publication Data

Giloi, Wolfgang.
Interactive computer graphics.

Bibliography: p.
Includes index.
1.–Computer graphics. 2.–Interactive computer systems. 3.–Algorithms. 4.–Programming languages (Electronic computers)—Computer graphics. 5.–Data structures (Computer science) I.–Title.
T385.G54 001.6′443 78-5425
ISBN 0-13-469189-X

Printed in the United States of America

10 9 8 7 6 5 4 3 2

PRENTICE-HALL INTERNATIONAL, INC., *London*
PRENTICE-HALL OF AUSTRALIA PTY. LIMITED, *Sydney*
PRENTICE-HALL OF CANADA, LTD., *Toronto*
PRENTICE-HALL OF INDIA PRIVATE LIMITED, *New Delhi*
PRENTICE-HALL OF JAPAN, INC., *Tokyo*
PRENTICE-HALL OF SOUTHEAST ASIA PTE. LTD., *Singapore*
WHITEHALL BOOKS LIMITED, *Wellington, New Zealand*

Contents

Preface

Stimulated by Nikolaus Wirth's definition[†]: ALGORITHMS + DATA STRUCTURES = PROGRAMS, we may present a similar formula for computer graphics:

COMPUTER GRAPHICS ⇒ DATA STRUCTURES + GRAPHIC ALGORITHMS + LANGUAGES.[‡]

Consequently, the emphasis of this book is on data structures suitable for computer graphics, on algorithms suitable for picture generation and transformation, and on the appropriate language constructs (in the widest sense) for the generation of graphic objects. The book consists of two major parts, Part I: Data Structures and Algorithms, and Part II: Languages and Their Interpreters.

Part I begins with an introductory chapter in which basic definitions are given and the topics of the book are outlined. Chapter 2 is devoted to the complex of data structures, their representation in a computer, and the construction principles for data management systems. For the application programmer who uses computer graphics as a tool, the importance of being acquainted with data structures and file management techniques cannot be overemphasized. This causes a dilemma for the writer of a text on computer graphics. On one hand, he can hardly ignore the topics of data structures and file management. On the other hand, a more-than-superficial discussion of these topics must unavoidably lead to a tome in its own right. As a matter of fact, one such tome exists.[§] In this book, we

[†]N. Wirth, *Algorithms + Data Structures = Programs* (Prentice-Hall, Englewood Cliffs, N.J., 1975).

[‡]Note that our formula expresses a valid implication but not an identity.

[§]D. E. Knuth, *The Art of Computer Programming*, Vols. 1, 2, and 3 (Addison-Wesley, Reading, Mass., 1968).

try to solve this problem by describing data structures and their properties in a very concise and simple algebraic form, as a set of data together with a set of relations on the data. The various structures can then be identified by specifying certain general constraints on the "generating relations." The memory representation of the given structures is also concisely but sufficiently discussed, as is the topic of data models used as construction principles for data bases. The practical aspects of data base management, the algorithms and programming techniques, cannot, of course, be covered. However, to provide some guidelines for the casual programmer of a display system, we discuss some simple file-handling techniques which are adequate for display programming (but which, of course, may not be adequate for constructing a data base for the underlying application program).

A picture can be defined as a set of graphic primitives together with a set of relationships, such as "to be connected with," "to be the successor of" (with respect to a given ordering), or "to belong to the same object," between them. We recognize the identity of the preceding definition with the general definition of data structures; and, in fact, picture structures can be equated with the structure of the graphic data of the picture. Graphic data are tuples of coordinate values and are thus constrained to the data type *real* or even to the data type *integer*. In Chapter 3 we introduce a specific picture structure which (1) is easily managed and (2) can be represented directly by arrays, a structure type common in high-level programming languages. The various picture transformations and their effect on the introduced picture structure are discussed in detail.

The problem one is confronted with when writing about curve and surface interpolation and approximation is similar to the one encountered in the discussion of data structures. Any profound discussion would require a book in its own right. Again, only a survey is given, in which we outline the general principles and techniques which are particularly relevant for graphics applications (Chapter 4). In the chapter on the hidden-surface problem, we are primarily concerned with presenting a formalization of the problem so that the classes of operations which form the common denominator of the many existing methods can be isolated and taxonomized. Chapter 5 is based to a wide extent on the outstanding diploma thesis of Walter Klos, a former student of the author's.

Part II: Languages and Their Interpreters, is concerned first with the topics of interaction handling (Chapter 6). Chapter 7 discusses the hardware aspects of computer display systems to the extent to which the application programmer and even the end user of an application system should familiarize himself or herself with the organization and functioning of the hardware. Special attention is given to a development which may in the foreseeable future strongly affect the area of interactive computer graphics, the advent of low-cost TV raster displays equipped with microprocessors. Chapter 8 discusses implementational aspects of display files and picture files and, hence, supplements the discussion of the organizational aspects given in Chapter 7.

Chapter 9 is devoted to a discussion of the language aspects of computer graphics in general. Kulsrud's graphic language concept is presented as an early guide. The dichotomies of *language extensions* versus *procedure packages* and of the *prefabricated structure* versus the *building-block* concept are discussed and illustrated by cases in point. The importance of an intermediate language for device independence and program portability is recognized, and the construction rules for intermediate languages are illustrated. Chapter 10 presents a conclusion in the form of a model language, GRIP, which is based on the previously established interrelationships among data structures, picture structures, and language structures. GRIP represents our philosophy of structured display programming. Some of the most commonly used high-level programming languages are examined with respect to their suitability for an implementation of the GRIP concept, and a model implementation is outlined in the Appendix.

The more computer graphics becomes a basic tool in such important application fields as, for example, computer-aided design, computer-aided training, and computer-aided delivery of health care, the more the aspects of display programming will be intertwined with other important aspects which are idiosyncratic of the respective applications. To cover all these aspects in one book would be an impossible task. Therefore, this book is intended to be primarily a textbook rather than a comprehensive reference book in which all the answers can be found.

In general, the early phase of a new field is characterized by a prevalence of ad hoc methods and solutions. Computer graphics is no exception to this rule. Over the years, numerous proposals have been published concerning special file management systems and languages or language extensions for graphics. This creates a potential danger, that a textbook might turn out to be an eclectic conglomerate of such proposed solutions. We tried to avoid this danger by presenting general concepts rather than special solutions, logical rather than more or less arbitrary choices. Whenever implementations are discussed, the purpose is to exemplify the underlying concepts.

During the final phase of writing this book, the author became familiar with the first draft of the CORE recommendations of the ACM/SIGGRAPH Graphic Standards Planning Committee.† CORE is the concept of a universal graphic programming system that encompasses the whole range from plotting to interactive graphics. This system, well designed by a group of renowned authorities in the field, will undoubtedly become *the* standard graphic programming system. We pay our reverence to this development by devoting a section of Chapter 3 to the unique picture-generation process of CORE. On the other hand, we shall introduce a picture structure somehow different from that of CORE for two reasons: first, we consider the three-level tree structure introduced in this text to be more efficient for interactive graphics (the plotting aspects, however important, do not concern us) and better matched to the data structures and language constructs of high-level programming languages; and second, the CORE structure is included in our picture structure as a special case anyway. It should be emphasized, however, that it is not our intention to propose different standards—or any standards at all—but to explore the richness of possible concepts, certainly a valid objective for a text book.

†*Computer Graphics 11*, 3, Fall 1977.

The material presented in this book was taught for several years in a two-quarter sequence at the University of Minnesota. To a considerable extent it also reflects research work carried out by the author and his cooperators over the past decade at the Technical University of Berlin, the University of the Saarland, and the University of Minnesota. The end-of-chapter questions precipitated directly from this experience. The questions are supplementary to the text and should help the reader to reinforce learning. For the instructor they provide a reservoir of questions for assignments and examinations.

Of the many students who participated in the design and realization of various hardware and software systems in the realm of computer graphics, only a few names can be mentioned here. Particular credit must be given to Messrs. Borgendale, Brüders, Eckert, Dr. Encarnacao, Grosskopf, Günther, Dr. Kestner, Klos, Messer, Reismann, Savitt, Dr. Strasser, Dr. Troeller, and Zech. The author wants to express his particular appreciation to Helmut Berg, who coauthored Chapter 2; to José Encarnacao, who was very helpful with many discussions and suggestions; to Wolfgang Strasser, for stimulating discussions and for providing most of the illustrations of Chapter 4; and to Steven L. Savitt, who must be credited for his leading role in the implementation of the GRIP concept, for contributing the Appendix, for many valuable discussions and suggestions, and for a critical reading of the text.

E. W. Dijkstra contrasts the craftsman and the pure scientist as the representatives of two extreme attitudes and calls for that blend which contains the endeavor of the scientist to make knowledge explicit with the skill of the craftsman. He also warns of "a disastrous blending, viz. that of the technology of the craftsman with the pretence of the scientist. The craftsman has no conscious, formal grip on his subject matter, he just 'knows' how to use his tools. If this is combined with the scientist's approach of making one's knowledge explicit, he will describe what he knows explicitly, i.e., his tools, instead of describing how to use them!" It is our endeavor to avoid this "disastrous blending" and to help the reader obtain a "conscious, formal grip" on the subject matter, culminating in high-level, structured display programs in which the potential of existing programming languages is fully exploited and matched to the structure of the objects with which we are concerned.

Naturally, not all topics of interactive computer graphics are equally relevant to a reader, depending on the reader's orientation. The same holds for the use of this text in teaching. In the following, we present a table that may provide guidance for the reader of this book or for the instructor who uses it as a textbook.

Orientation of the reader	Most significant chapters
Graphics application programmer	1, 3, 5, 6, 10
Graphics system programmer	1, 2, 3, 6, 8, 9, 10, Appendix
Graphics system designer	1, 3, 6, 7, 8, 10
CAD application programmer	1, 2, 3, 4, 5, 6, 10

W. K. Giloi

ACKNOWLEDGEMENTS FOR ILLUSTRATIONS

We wish to express our gratitude to the following persons or institutions who provided illustrations for this book: ADAGE INC., Boston, Mass. (Fig. 7.11); Dr. Encarnacao, Darmstadt, Germany (Fig. 2.11); EVANS & SUTHERLAND COMPUTER CORP., Salt Lake City, Utah (Fig. 7.10(A) and Fig. 5.16); Dr. Fu, Purdue University, Lafayette, Ind. (Fig. 3.3); INFORMATION DISPLAYS INC., Elmsford, N.Y. (Fig. 7.8); Dr. Strasser, Berlin, Germany (Fig. 3.5, 4.3, 4.4, 4.5, 4.6, 4.7, 4.8, 4.11, 4.12, 7.7); and IEEE COMPUTER SOCIETY PUBLICATION OFFICE (Fig. 7.11).

PART I

DATA STRUCTURES AND ALGORITHMS

1

Introduction

1.1 WHAT IS COMPUTER GRAPHICS?

Computer graphics (CG) involves the generation, representation, manipulation, processing, or evaluation of graphic objects by a computer as well as the association of graphic objects with related nongraphic information residing in computer files. Graphic objects may be photographic images, or they may be created with the aid of a computer in the form of alphanumeric characters, special symbols, line drawings, or gray-shaded areas. Such artificially created objects may be rendered in black and white or in color.

Figure 1.1 is an attempt to classify computer graphics into subareas, dividing it into three main categories:

1. Generative graphics.
2. Image analysis.
3. Cognitive graphics.

Generative graphics involves artificially created graphic objects, usually in the form of line drawings. The main tasks of generative graphics are:

1. Model (object) construction and picture generation.
2. Model and picture transformation.
3. Object identification and information retrieval.

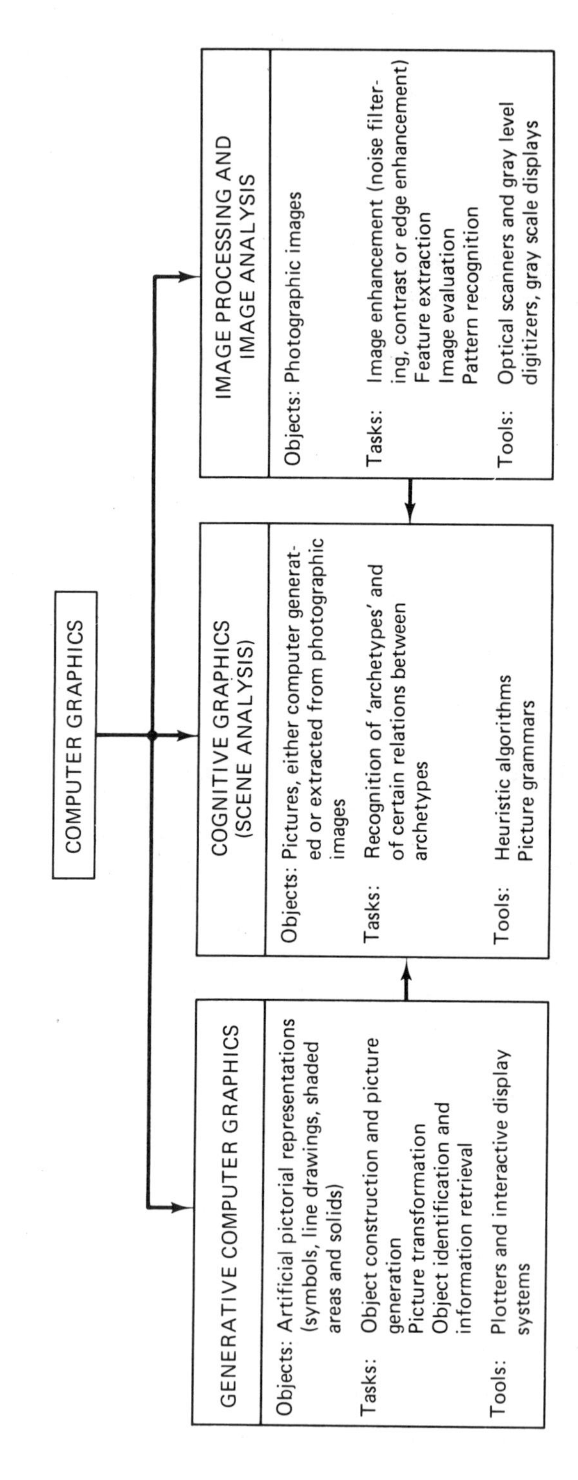

Figure 1.1 Various fields of computer graphics.

In compliance with the ACM/SIGGRAPH† glossary [1], we use the term *model* for an abstract description of a graphic object that can be understood by a computer and transformed into a corresponding picture on a *display surface*. Hence, a *picture* or *image* is the concrete visualization of a model. The tools for generative graphics are, in addition to the standard computer system, plotters or interactive display systems.

Plotters are strictly output devices for graphics; i.e., they offer no direct interaction between the system and its user.

Interactive display systems not only make it possible for their users to have information rendered in graphical form but also to interact directly with the system for the purpose of creating, manipulating, and designing graphic objects. To this end, such systems include a certain user–machine interface, called *display console*, as a medium through which a user–machine dialog can be carried out. The display surface in an interactive display system is in most cases the surface of a cathode ray tube (CRT), but other instruments, such as plasma displays, are also in use. Therefore, it is more to the point to call the display surface in an interactive display system the *display screen*.

Image processing and analysis involves photographic images or, more precisely, discretized representations of images given in the form of an array of numbers representing the gray-scale values of the corresponding picture elements. Therefore, the image must first be scanned and sampled; i.e., its gray levels at certain points are measured and the measurements are converted into numbers. Hence, image processing is the application of numerical algorithms to such arrays of numbers. Special gray-scale displays reconvert the result of an image-processing procedure into a photographic image. The main tasks of image processing and analysis are to perform:

1. Image enhancement (contrast enhancement, background-noise suppression).

2. Image evaluation (evaluation of size, shape, and location of certain objects in the image).

3. Pattern recognition (feature extraction and classification).

Cognitive graphics or "scene analysis" is concerned with abstract models of graphic objects and the relationships between them. An "abstract model" is the idea of an object regardless of its instantaneous appearance (e.g., a triangle is a triangle regardless of the length of its sides, its location and orientation in a particular space, etc.). Cognitive graphics may involve artificial graphic objects as well as certain objects in photographic images. In the latter case, however, these objects must first be "extracted" from the image. Thus, as in the case of pattern recognition in images, the first step of a scene analysis is to extract certain features

†ACM/SIGGRAPH, Special-Interest Group on Computer Graphics of ACM.

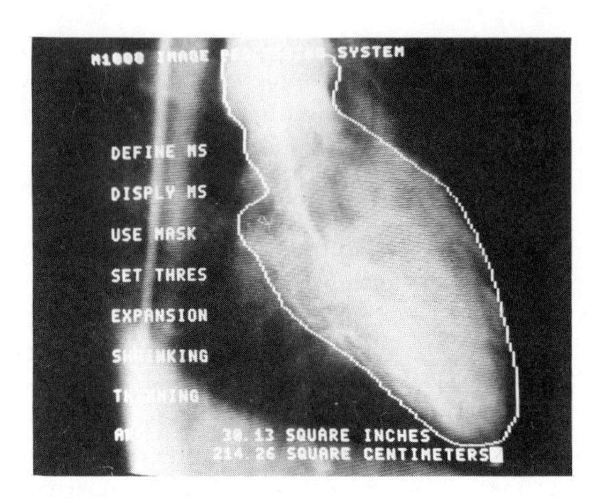

Figure 1.2 Coronary angiogram (X-ray of the heart) and the heart contour outlined manually with a lightpen.

as a basis for the subsequent conversion into graphic objects. Hence, cognitive graphics is the area of computer graphics where the generative and the image-processing fields are joined. As an example, Figure 1.2 shows an X-ray image and a graphic object extracted from that image, both superimposed on one CRT screen [48].

In this text we shall restrict ourselves to generative graphics, and we shall focus our attention on special computer systems called *interactive display systems*. Furthermore, we shall be concerned with basic tools rather than with particular applications.

In interactive display systems, graphic objects are represented on the display screen. Although the size of such a display screen typically is that of a large TV screen, such a display terminal is not capable of generating as precise a picture as a mechanical plotter or a film recorder. The more precise plotter, on the other hand, operates at considerably lower speed because of its mechanical nature. Thus, in the case of a display, we trade precision for speed. The main advantage of a CRT display system, however, is that it can be a highly interactive instrument, enabling the user to carry out a dialog with a computer, in which graphic representations are the communication medium. This does not necessarily mean that all CRT displays are interactive to such a high degree. In fact, sometimes CRT displays are only used as a plotter (with a rather restricted resolution).

Hence, the distinction between plotters and interactive display systems is not the distinction between different technical solutions for the task of drawing pictures, nor is it the distinction between a durable representation (as produced by a plotter) and a volatile one (as produced on a CRT screen), but it is a distinction in the mode of operation. As we shall see, the interactive aspects are the crucial topic in generative graphics, of which the plotting aspects form only a part.

1.2 THE MOTIVATION FOR INTERACTIVE COMPUTER GRAPHICS

One of the most renowned scholars in the computer graphics area, Ivan Sutherland, wrote in a very remarkable paper on computer displays in *Scientific American* (June 1970) [132]:

> Whereas a microscope enables us to examine the structure of a subminiature world and a telescope reveals the structure of the universe at large, a computer display enables us to examine the structure of a man-made mathematical world simulated entirely within an electronic mechanism. I think of a computer display as a window on Alice's Wonderland in which a programmer can depict either objects that obey well-known natural laws or purely imaginary objects that follow laws he has written into his program. Through computer displays I have landed an airplane on the deck of a moving carrier, observed a nuclear particle hit a potential well, flown in a rocket at nearly the speed of light and watched a computer reveal its innermost workings.
>
> Computer displays have become of major importance to two groups of people. One group has a pictorial problem in the workaday world for which it would like computer help. These users, for example, may want to shape a metal part of a computer-controlled machine tool; they begin by describing the part to a general purpose computer, which draws a picture of the part and verifies that the description is accurate. Other users employ computers to produce the intricate high resolution photographic masks required for making integrated electronic circuits. Similar pictorial problems in which computers can help arise in highway planning, automobile and aircraft design, topographical mapping, architecture, the layout of publications and the production of clothing patterns. In these and many more areas, written language is far from adequate.
>
> The other group using computer displays is interested in gaining insight into complex natural or mathematical phenomena. These users simulate physical situations of various kinds in the computer and use display devices to present the result of the simulation. For example, an organic chemist may want to synthesize a particular molecule; he creates a picture of the molecule on a display screen and then initiates a program by which the computer presents a selection of simpler molecules from which the desired substance can be synthesized. An engineer designing a communication circuit asks the computer for a graph showing how circuit response varies with frequency. A physician studying how blood flows through the arteries obtains a plot that reveals high vorticity at exactly the locations where the lesions of arteriosclerosis are most common. A physicist programs a computer to illustrate how elementary particles interact with their own electric fields to give his students some feeling for quantum-mechanical behavior. A circuit designer draws a circuit and asks the computer to simulate its operation and to plot its performance in a graph of voltage and current. A feedback theorist describes the location of poles

and zeros on a complex plane and watches as the computer plots the root locus. A mathematician enters the equations for conformal mappings and observes the maps produced by each equation. A pilot practices takeoffs and landings on a simulated airfield that can assume any orientation on the display screen as he operates "controls" for engine power and aircraft altitude. All these people, interested in educating themselves or others, use computer displays as one of the many tools for gaining deeper understanding of a problem.

Generally speaking, we may state that computer graphics has become a valuable tool if not a necessary prerequisite in such fields as *computer-aided design* (CAD)—be it in architecture, mechanical engineering, electronics, civil engineering, chemical engineering, free-form design in automotive or aerospace engineering, or many other fields—in *computer-aided instruction* (CAI) and training, in medical report generation and diagnosis, in simulation of physical and technical systems, in mathematical problem solving, and even in artificial intelligence problems and the generation of creative art. There are two main reasons for the extreme usefulness of CG for many applications. The first reason is that the computer graphic representation of information may be not only an appropriate but the only reasonable method of handling information. This fact is tersely expressed by the saying: "A picture is worth a thousand words." The second reason is given by the special kind of person–machine interaction that only computer graphics provides.

In the early years of CG, the main emphasis of the work carried out in this field was on the visual aspects, such as the generation and transformation of pictures. These aspects were very intriguing, for there is probably no more direct way of person–machine communication than through the medium of visualization provided by CG. As CG became the primary tool in CAD, the pictorial aspects of CG receded in favor of aspects that arise automatically whenever CG is embedded into larger systems designed for various CAD purposes. These aspects include the question of how a proper problem-oriented language for interactive CG shall be constructed, the question of data base organization and management, and the question of overall system organization seen under the aspects of software portability and device independence and the valid principles of system software construction in general. This book is an attempt to answer some of these questions.

1.3 MODEL OF INTERACTIVE GRAPHICS SYSTEMS

In the following text we present an abstract model of an interactive graphics system, based on the notion of processing systems. A processing system is defined as a pair

$$P=(L,I),$$

where L is a language and I is an interpreter for that language. In our model, an interactive graphics system consists of a stratified hierarchy of such systems, linked

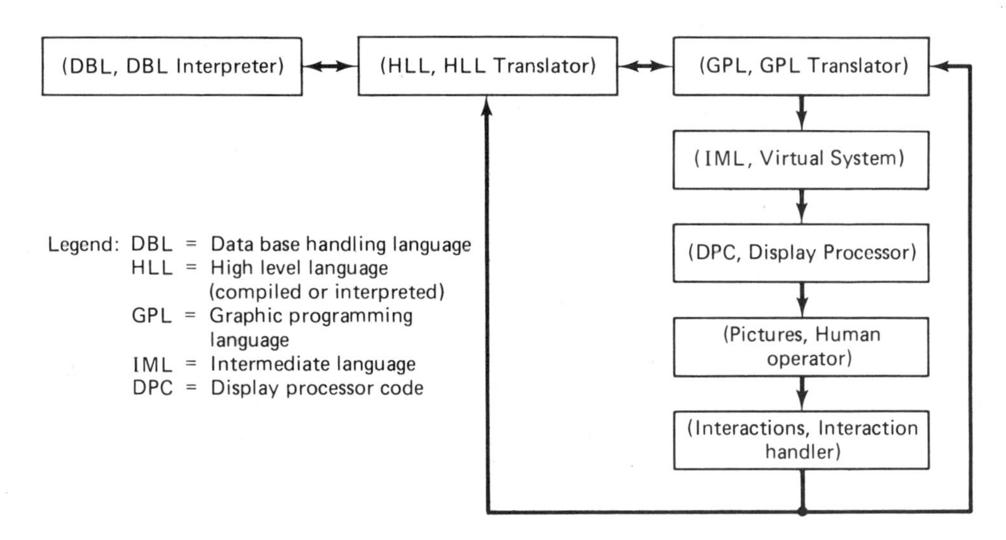

Figure 1.3 Abstract model of an interactive graphics system.

by the fact that, at level i, the information represented in language L_i is translated by the interpreter I_i into a representation in language L_{i+1} of the succeeding processing system (L_{i+1}, I_{i+1}), and so on. Hence, we introduce the notion of the transformation of information from an initial representation to a lower-level representation, a basic concept of any computer system.

On top of such a model for an interactive graphics system as depicted in Figure 1.3, we find the *high-level programming language* (HLL) that is chosen as the *host language* for the system, and the translator for that language. The translator is a compiler if the host language is a compiler language (e.g., FORTRAN, PL/I, PASCAL), or an interpreter if the host language is an interpreter language (e.g., APL). In connection with the selected high-level language, we have a *data base handling language* (DBL) and a *graphic programming language* (GPL). The respective processing systems for DBL and GPL are at the same level as is HLL and its translator. Actually, DBL and GPL are integrated into the central processing system (HLL, HLL translator), either in the form of appropriate subroutine packages or in the form of a language extension. However, for the sake of discussion, it is useful to distinguish these three top-level systems logically. The display system proper is thus represented by the right-hand side of the diagram, which constitutes a stratified hierarchy of processing systems (Fig. 1.3).

In this model, the GPL representation of a *display program* is translated into a representation in an *intermediate language* (IML). The interpreter for the IML program is an *abstract* or *virtual display system*. It should be mentioned at this point that the decision to include the processing system (IML, virtual system) is optional. As a matter of fact, there are many display systems in which the GPL representation of a program is directly translated into *display processor code* (DPC), without any intermediate representation. However, the introduction of an intermediate language is a requirement for a portable and device-independent HLL programming system. However variant the hardware of different display systems may be (see Section 1.7), the "front ends" of these systems (i.e., the virtual display systems) will all look alike, and the idiosyncrasies of the various display processors must only be taken into account in the process of translating an IML representation into the corresponding DPC representation.

The display processor is viewed as a (hardware) interpreter that translates display processor code into pictures. Whereas the specific structure of the software portions of a display system is widely left to the designer's discretion, the two processing systems (DPC, display processor) and (pictures, human operator) are determined basically by the properties of the available hardware or, better yet, by the hardware state-of-the-art. (Of course, the human operator is not determined by the hardware; however, the means offered to him for communicating with the system are so determined.) In the hardware portion of a display system we can distinguish four functional units:

1. Display file memory.
2. Display processor.
3. Display generator.
4. Display console.

Figure 1.4 shows, in block-diagram form, the connections between these four components and the functions they perform. However, not all systems comprise a display file memory, as such a memory is only absolutely necessary in displays with "picture refresh." It is not needed, for example, in storage tube displays, where the CRT screen itself has the capability of storing pictures. For the typical interactive display system, however, the existence of such a memory will be assumed.

The display file memory is typically part of a small-scale computer—in modern systems, a minicomputer. Its task is to store a *display processor program*, i.e., the entirety of instructions that cause the display processor to generate pictures in a certain appearance, in connection with a certain status. Under the term *appearance* we subsume such *attributes* as boldness (beam intensity) or gray level, color, line style, and blink mode. Examples for possible values of appearance parameters are:

- Boldness: BLANK/DIM/MEDIUM/BRIGHT/VERY BRIGHT
- Color: BLANK/GREEN/RED/BLUE/YELLOW/MAGENTA/CYAN/WHITE
- Mode: NOT BLINKING/BLINKING
- Style: SOLID/DOTTED/DASHED/DASH-DOTTED.

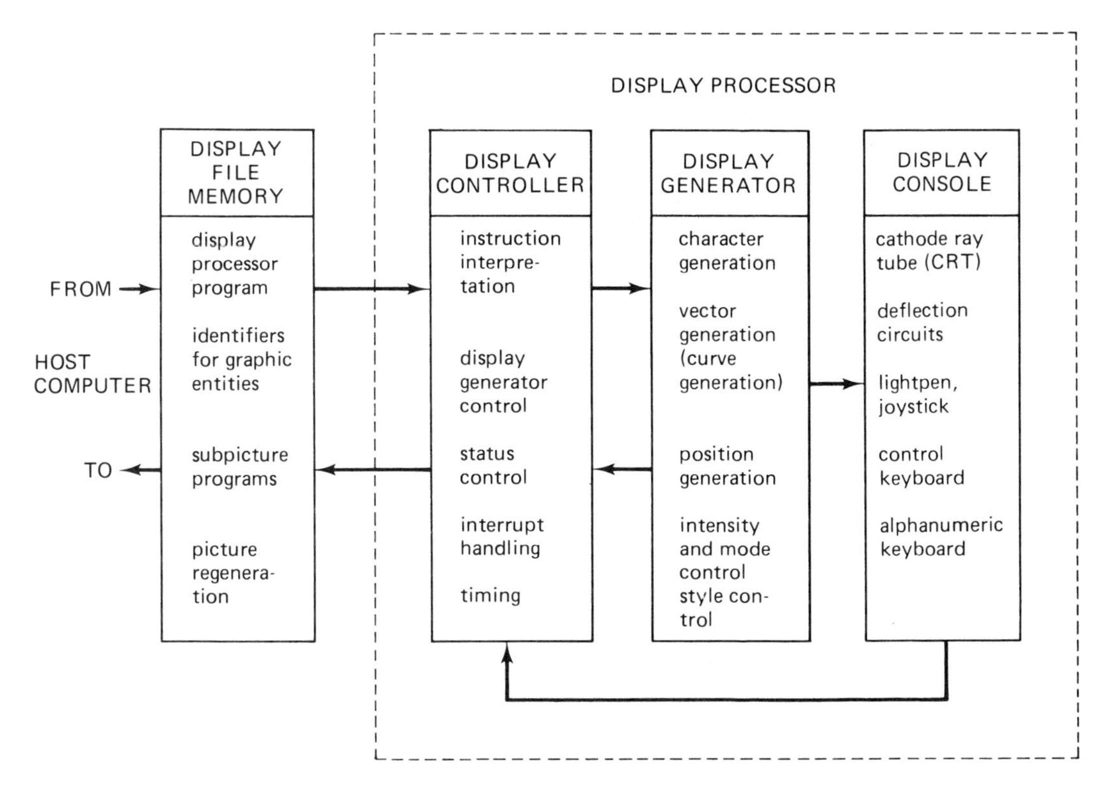

Figure 1.4 Block diagram of the hardware part of a typical "refreshed picture" display system.

In addition to the appearance, a graphic object may be given a certain *status* with respect to certain operations, such as a lightpen pick, a deletion, or a picture transformation. Therefore, the programmer can control through an associated status parameter whether the object may be subjected to the respective operation.

The display controller interprets these instructions and initiates their execution by activating the appropriate units of the display generator. The display generator outputs analog signals to the CRT deflection and intensity-control circuits, causing the beam in the CRT to write certain figures at the desired locations with the desired appearance. A feedback path leads back from the display console to the display controller. This feedback is given by actions which the human observer of the graphic representation on the CRT screen may take, for example, by using a lightpen, a "joystick," a keyboard, or other interactive devices. Thus, the human operator is part of the feedback loop.

From this functional description of a display processor, it becomes obvious that there is an essential difference between a display processor and a general-purpose processor. The only commands which the display processor recognizes—besides JUMP instructions—are commands for the drawing of graphic patterns, and, consequently, the only data types found in a display processor program are integers representing screen coordinates and control bits representing appearance and status data. Unlike a general-purpose processor, the display processor is not equipped to perform arbitrary transformations on data of various types, with the exception that it is capable of performing conversions of relative or incremental coordinate values into absolute coordinates, and vice versa. Thus, it is not meaningful to discuss the instantaneous representation of a program during the various steps of execution, as in the case of the computation of a regular program.

1.4 THE DISPLAY FILE

The term *display file* is often used as a synonym for what we have called a display processor program. The simplest possible structure of such a program is the linear list, but in most systems the display processor program may include subroutine jumps (e.g., for the subpicture technique to be discussed later). A display processor program, which includes the possibility of (nested) subroutine calls, is sometimes called a "structured display file." We are not in favor of such terminology for several reasons, some of which are:

1. Even a linear program has a certain defined structure.

2. We fail to see why we should not call a program a program.

3. In interactive graphics the display processor program represents only part of what we have to store in the file that is built and maintained in the display system. We want to reserve the term "display file" for this file, of which the display processor program usually is only a part.

As was pointed out above, interactive graphics requires, in addition to picture–generation capabilities, that the user be able to "pick" an object displayed on the screen (by pointing a lightpen to it or by positioning a cursor on it) and have this object identified to the application program. The identification is possible only if certain structural relationships and certain identifiers are associated with the display processor code and stored in the display file.

Consequently, we introduce the following definition of a display file:

Definition: *Display File*

A *display file* is the union of:

1. The ordered set of all display processor instructions for the actual display (called a display processor program).

2. A set of identifiers associated with the set of graphic entities in the actual display.

3. A set of pointers, denoting the DPC program segments for the generation of the graphic entities and linking these segments to the entity identifiers.

4. A set of special program segments for "symbols" (subpictures).

In special cases, one or more of these sets, except the first one, may be empty. The first set can only be empty if the whole display is empty, i.e., if there are no pictures on the screen. In the definition above we use such terms as "primitives," "entities," and "symbols," which must yet be defined. Therefore, we add the following definitions.

Definition: *Graphic Primitive*

A *graphic primitive* is any graphic element for the generation of which there exists a special hardware unit in the display processor. Examples of possible primitives are: dots, straight-line segments (vectors), alphanumeric characters, special symbols, special curve segments (e.g., arcs), or even "surface patches." A dot is specified by a *point*, i.e., a tuple of coordinate values. A vector is specified by two points, called *start point* and *terminal point*. An arc is specified by three points. A character or special symbol is defined by a code number.

Graphic objects may now be defined as collections of graphic primitives. From a logical point of view, the question arises as to what to choose as identifiable entities. The most general answer would be to make all graphic primitives individually identifiable. However, this is usually not required and would thus unnecessarily burden the system as well as the programmer. Therefore, it is more efficient to define more complex graphic objects as entities.

Definition: *Graphic Entity*

A graphic entity is a set of primitives, displayed with the same appearance and status and identified by a name.

Hence, we may alternatively define a display file as a list of information items representing graphic entities. The display file is the port to the display processor. Any entity entered into the display file is automatically displayed. Conversely, any interactive modification of a graphic object displayed on the CRT screen modifies certain information items in the display file representation of that entity. The modification of pictures on the screen is sometimes called *picture editing*. Thus, picture editing is rigidly linked to a manipulation of the display file.

Our formal definition of a display file comprises additionally a type of program segment, called *symbol* [1], that is distinguished from the graphic entity according to the following definition.

Definition: *Symbol (Subpicture)*

A symbol is an identifiable display processor program segment, written for the generation of a certain set of graphic primitives. Status and appearance attributes are not specified. A symbol resides in the display file but is not included in the display file scan for picture refresh unless specifically invoked by a graphic entity. The invocation of a symbol thus creates a *symbol instance* that becomes part of the invoking entity and, therefore, inherits its status and appearance parameters.

The reader will have noticed that a symbol can be equated with a subroutine of the DPC program, but not with a subroutine in the GPL program. To avoid confusion, we therefore use the term "symbol" rather than "subpicture" (see [1]).

1.5 THE NECESSITY OF A DUAL REPRESENTATION OF GRAPHIC OBJECTS

We defined computer graphics as the act of representing graphic objects in a computer so that they can be visualized in pictorial form, of providing an interactive mode of operation in order to generate and manipulate such objects, and of associating graphic objects with other data pertinent to the application program. These data, not directly connected with picture generation and editing, may be subjected to numerical or nonnumerical processing in the host computer, and are, therefore, stored in a particular file in that computer. Such a file is often shared by more than one application program, and is thus called a *data base* for the programs. As the data in the data base may be related to certain graphic objects, it is certainly reasonable to store the graphic data (i.e., the data that specify graphic objects) in the data base, as well. Hence, a data base encompasses a number of data records containing the data of graphic objects as well as all other related data.

An application program based on computer graphics will include statements and procedure definitions (or the call of procedures from a library) for the purpose of generating graphic entities. The parameters of this part of the program, which we may call the *display program*, are given in the form of point, line, or surface specifications. Of course, these data must be stored in the data base in a structured form so that the structures in the data base correspond with the structure of the graphic objects to which they belong. The basic difference between a display program and other parts of the application program is that the display program is not executed by the host computer but rather by a special piece of hardware called the display processor. To this end, a display program must first be translated into a display file. The translation takes place in the host computer and is carried out by a special software module for which we use the name *GPL translator* (see Fig. 1.3). Depending on whether the translation is a compilation or an interpretation, we may more specifically use the term "GPL interpreter." †

†Occasionally, the term "display file compiler" is used; however, it is common practice to name a compiler after its source language, not after its object language.

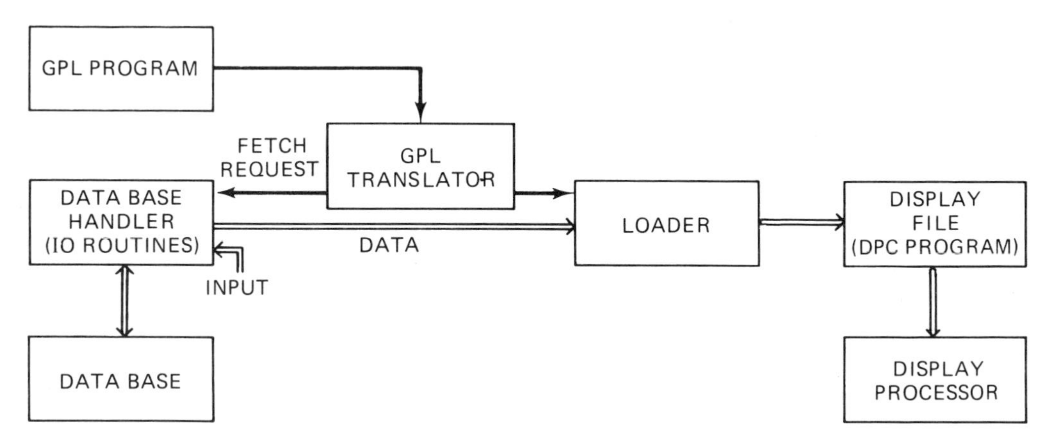

Figure 1.5 Interrelationship between data base and display file in a noninteractive mode of operation (GPL: graphic programming language).

The process of translating a display program into the display file representation is illustrated by Figure 1.5. The display program refers to data in the data base. The program and its data are translated into display processor code, either directly or through an intermediate, device-independent representation. In any case, we have two representations of the data for graphic objects, one in the data base and one in the display file, one for the *model* of an object and one for its *image*. In the display file representation, data are bound to display processor instructions, whereas the data base representation of graphic data contains, besides the data, additional information establishing relationships among the data (see Chapter 2). Graphic data alone do not constitute a model, but must for this purpose be supplemented by an appropriate GPL program. However, this is necessary only for the actually displayed objects.

This dual representation of graphic objects offers a certain advantage. In the course of displaying a sequence of different pictures, any picture that previously existed but had been deleted from the display file at one point can be reproduced whenever it is wanted. This requires, of course, that the result of interactive picture creation or picture editing is entered into the data base. Hence, interactive picture creation and editing implies not only display file manipulation but also permanent updating of the data base. The dual representation of graphic data has the disadvantage that two files must be maintained in memory. However, as memory has become quite inexpensive, this is no longer a decisive cost factor.

1.6 THE PROBLEM OF PICTURE TRANSFORMATIONS

One important question has not yet been raised: What is the domain of the picture-generating functions? On the screen in the display console, we have a certain viewing area which is usually measured in raster units (e.g., 2048×2048 is a

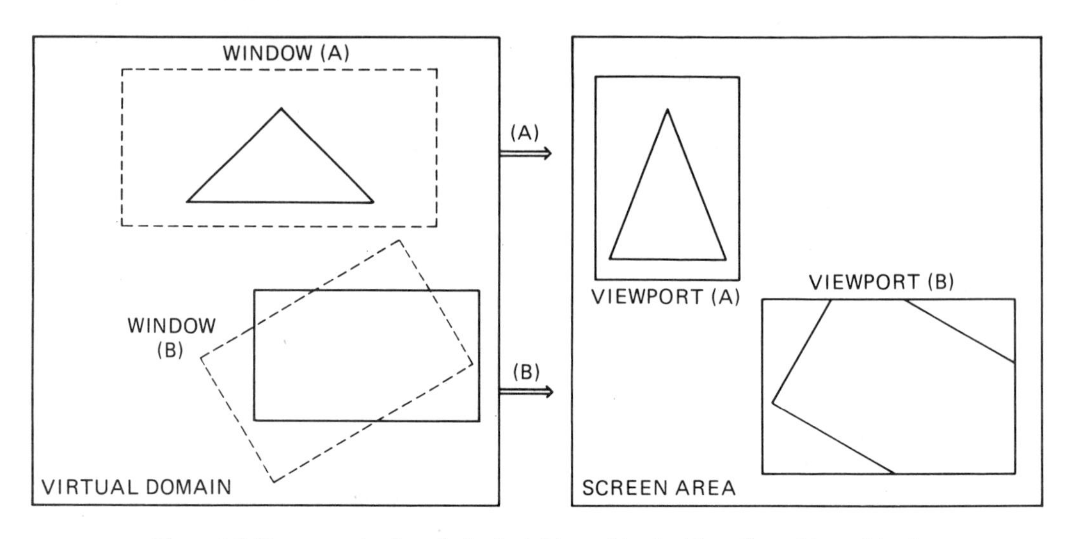

Figure 1.6 Two examples for windowing: (a) combined with scaling; (b) combined with rotation.

typical value). However, this "screen domain" need not necessarily be the domain in which we define pictures when we write the display program. Some systems work in that way, but this is a rather inflexible and unnecessarily restrictive approach. Rather, we would prefer to have the possibility of specifying graphic objects in an arbitrarily chosen domain—preferably the Cartesian plane $\mathbf{R}\times\mathbf{R}$ ($\mathbf{R}$ being the set of real numbers) or a certain subset of it. Subsequently, this area would be mapped onto an arbitrarily chosen part of the screen area. The domain in which the picture-generating functions of the graphic programming language (GPL) are defined will be called the model *world* $\mathbf{R}\times\mathbf{R}$ [1]. Any arbitrary rectangular subdomain of $\mathbf{R}\times\mathbf{R}$ is called a *window*, and the rectangular part of the screen area onto which the window shall be mapped is called a *viewport*.

Let the viewport be bounded by the lines $x=n_1, x=n_2, y=m_1, y=m_2$ (x,y being the screen coordinates and m_1, m_2, n_1, n_2 being given in raster units), and let $W\subset\mathbf{R}\times\mathbf{R}$ denote an arbitrary rectangular window in the world coordinate system. The function w which performs the mapping†

$$w: W \rightarrow [m_1 : m_2] \times [n_1 : n_2]$$

will be called a *windowing function*. Figure 1.6 presents two examples of windowing functions.

Usually, windowing is restricted to the mapping of a rectangular area in the world domain onto a rectangular area in the screen domain. This is done because

†Throughout this text $[a:b]\subset\mathbf{Z}$ will denote the set of integer numbers $\{a,\ldots,b\}$ with $a<\ldots<b$.

such a mapping requires only the *geometric transformations* of translation, rotation and scaling in addition to an operation called *clipping* or *scissoring*. Clipping is an algorithm that determines which parts of a picture lie outside the window boundaries. These parts are subsequently "clipped off" from view.

Windowing, combined with scaling, translation, and rotation, enables the programmer to specify one or more rectangular areas of the world in which all graphic models represented in the data base are defined. On the execution of a graphical program (i.e., when the images of certain models are generated), only the portions of the models that lie inside the windows are mapped into the corresponding images, whereas the portions outside the windows are clipped off. It is obvious that the mapping from windows on viewports implies (intrinsically) a scaling. In so doing, the user may blow up the objects in a selected area, letting the viewport act as a magnifying glass. Certain pictures may be chosen to be subjected to geometric transformations. In this way the programmer may change the relative position of some objects in an image by translating and/or rotating them without changing the remaining image. Moreover, the programmer may have such transformations performed repetitively, for instance, rotating a particular object while the remainder of the image is static. Such an image is said to be *animated.*

If a display processor is equipped with special transformation hardware, such transformations can be performed "on the fly." In the absence of transformation hardware, transformations are usually performed by appropriate routines of the HLL program. In the case of animation, this leads to the repetitive execution of the image-generating procedures, each time with transformed data. However, the data stored in the data base should not be modified by transformations in order to maintain an original representation of the graphic objects. Therefore, one has to distinguish between the original models and their modified versions as obtained by a sequence of transformations. The modified versions may be called *instances* of the original models.

As is shown in Chapter 3, geometric transformations such as scaling, translation, and rotation are straightforward in the sense that they affect only the data of the involved GPL procedures but not the procedures themselves. Hence, these transformations may be performed by fetching points from the data base and subjecting them to the matrix operations outlined in Chapter 3. If (X, Y, Z) are coordinate triples of a three-dimensional world coordinate system (the two-dimensional case is obtained by simply setting $Z=0$), and if (X_T, Y_T) are the corresponding coordinate pairs of a two-dimensional coordinate system obtained after picture transformations, then a picture transformation is in the two-dimensional case a bijective mapping $T_2: X \times Y \rightarrow X_T \times Y_T$. In the three-dimensional case, the mapping must include a projective transformation that reduces the dimensionality of the world space by 1, leading to a mapping $T_3: X \times Y \times Z \rightarrow X_T \times Y_T$ that is not injective. A window in the three-dimensional world space now is a three-dimensional subspace that is usually called a *viewing pyramid* (see Chapter 3).

The combined operation of windowing and picture transformations can be carried out in two different ways, distinguished by the order of execution:

1. Execution of transformations in the world domain followed by a windowing from the world to the screen domain.

2. Windowing in the world domain followed by transformations.

The second approach seems to be more efficient, as in the first case a number of graphical objects may have to be transformed only to be subsequently clipped off. In contrast, if clipping is performed first, all the objects lying entirely outside the window need not be transformed, and a considerable amount of computation may be saved. However, the second scheme requires the inverse mapping of the viewport into the world space. Therefore, a tilted window is obtained if the geometric transformations include rotation, a fact that considerably complicates the clipping procedure. For this reason, the first approach usually is preferred, and it becomes almost mandatory if the geometric transformations include rotation. For this case, Figure 1.5 has to be modified as shown in Figure 1.7. If the second approach is taken, the order of transformation and windowing is interchanged. In the case that geometric transformations are succeeded by windowing, the efficiency of program execution can be enhanced by first performing a preclipping in the world domain, thus culling the objects lying outside a certain "area of interest." Subsequently, the remaining objects are geometrically transformed, and the result is windowed onto the desired viewport in the screen domain.

So far, we assumed that picture transformations are performed on the GPL representation of a display program (i.e., prior to the translation into the DPC representation). In principle, picture transformations might as well be executed on the DPC representation. In fact, this case occurs whenever picture transformations are performed by hardware in the display processor. However, if the geometric transformations must be performed by software, this would be a very awkward procedure, as in the DPC representation, data are bound to the display processor instructions. Therefore, their transformation would require a separation of data and op code prior to a transformation, and a rebinding afterward. The necessary bit manipulations are easily performed by hardware but awkward when performed by software.

As pointed out above, scissoring affects not only the data but also the instructions. Therefore, picture transformations that include windowing may require the generation of new display processor code each time a transformation is performed. Such a complication could be avoided by modifying only the parameter that controls the beam intensity for a graphical entity rather than the DPC program: e.g., entities outside the window would still be drawn but would be rendered invisible by setting their beam intensity to "blank." As a matter of fact, the analog scissoring circuit which is built into most displays, and which scissors an

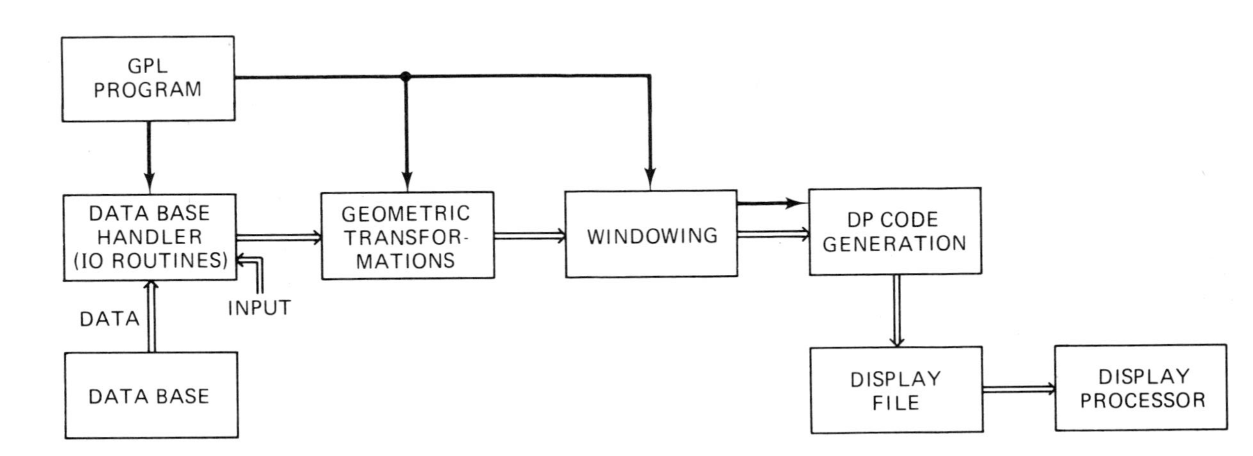

Figure 1.7 Generation of the display file in the presence of picture transformations (noninteractive mode of operation).

image at the screen boundaries, works in exactly that way. However, such an approach has the disadvantage that the picture refresh rate may be drastically reduced if a large number of invisible objects must be drawn.

So far, we have not considered an interactive mode of operation. In interactive computer graphics, the user should be able to command the system to perform certain transformations on the objects displayed on the screen, to delete graphical entities, and to create new ones. Such a system is governed by two basic rules:

1. Anything visible on the screen must be represented in the display file. Thus, picture editing means display file manipulation.

2. Any newly created object must be entered in an appropriate form into the data base.

Of course, we could modify a display strictly by manipulating the display file, yet this would be a poor policy, as the application program would have no knowledge about these modifications or about the creation of new objects. Therefore, it certainly is a better approach to first call certain input routines and put in data and identifiers of new objects and then have these new information items stored in the data base. Subsequently, the corresponding augmentation of the display file may take place, exercised by appropriate procedure calls, thus producing the visible rendering of these objects on the screen. However, in the interactive use of the display screen, it is highly desirable to have the dialog organized in such a way that the result of any user action becomes manifest on the screen as soon as possible. Therefore, it is an even better policy to first call GPL functions for the generation of objects, supply them with data solicited from the user, and enter data into the data base only after completion of the process of interactive object creation.

So far, our discussion is based on the assumption that a display program is compiled into its display processor code (DPC) representation. In order to generate the continuous word stream required for picture regeneration, the DPC program is repetitively scanned at a rate limited by the processing speed of the display processor. A fixed picture refresh rate can be obtained—up to a certain upper bound regardless of the length of the DPC program—by synchronizing each start of a refresh cycle with a *frame time clock*. Since the compilation of a display program as well as the execution of picture transformations is performed before picture regeneration is started, the execution of these operations is uncritical as long as the picture or certain parts of it must not be constantly transformed (i.e., as long as no animated pictures are to be produced).

The question arises whether such a scheme, which necessitates the storing of two representations of the same program, is not somewhat wasteful, or whether picture regeneration could not rather be carried out directly from the high-level display program. In fact, this is possible if the following conditions are fulfilled:

1. The translation of the high-level display program must be performed in a linear scan mode and under real-time conditions, so that the proper word stream is produced as required for picture refresh.

2. A given picture structure must correspond with a respective sequence of (nested) procedure calls such that, on the occurrence of an interrupt, a structural entity can be identified via the procedure currently under execution.

This implies that the high-level language (HLL) code and the display processor code are by and large in a one-to-one correspondence, a condition that results from the necessity of real-time scanning. For example, it would not be possible to have an HLL instruction for circle generation which is translated into a hundred DPC instructions for a straight-line-segment approximation of that circle. Thus, a circle-generation instruction in HLL would only be possible if the display processor includes a hardware circle generator. Similar considerations hold true for all picture transformations. Therefore, the display processor must now be equipped with hardware for translation, rotation, scaling, and clipping. Furthermore, high-level programming cannot be exercised to the point that it really deserves such a name, for the level of programming in HLL must now be matched to the level on which the display processor operates. Costwise, the picture transformation hardware is much more expensive than the memory required for storing a display file. Such a system benefits from greater speed of operation but at the expense of more costly hardware. The fast, real-time execution of picture transformations provides for an immediate response to any command that the user may issue. The block diagram of such a system is basically the same as the one shown in Figure 1.7, but with the borderline between hardware and software shifted farther toward the front end.

1.7 TAXONOMY OF DISPLAY SYSTEMS

Newman and Sproull classify graphics systems into the following categories [101]:

1. Storage tube display systems.
2. Simple refresh display systems.
3. Buffered refresh display systems with transformation hardware.
4. High-performance display systems.

These four types will be discussed briefly.

Storage Tube Display Systems: The major characteristic of storage tube displays is that the CRT screen itself has the capability to store a picture for an arbitrary period of time. Hence, once a picture is generated, it need not be constantly refreshed, and thus the display system need not comprise a display file or buffer memory. As no real-time requirements exist for the generation of a picture, the hardware for the generation of graphic primitives (vectors, characters, etc.) and (if existent) for picture transformations can be slow and, thus, inexpensive. The picture resolution of modern storage tube displays matches that of high-performance displays. There is a price to pay for the decisive cost advantage of storage tube displays, consisting in rather limited interactive capabilities: a picture can only be deleted as a whole.† Therefore, while it is possible to add new subelements to an existing picture, it is not possible to selectively erase parts of it. Except for the simple addition of new objects, picture editing necessitates regeneration of the whole picture, even after the smallest change. As picture generation is relatively slow, this rules out quite a number of interactive computer-aided design applications as well as the generation of animated pictures. Therefore, a storage tube display is primarily a device for the *presentation* of graphical information, whereas its applicability for interactive information *processing* is relatively limited.

Since there is no picture refresh, it is not possible to equip a storage tube display with a lightpen. Fortunately, it is possible to have a special symbol, called a *cursor*, displayed on the screen without being stored (simply by writing the cursor symbol with an intensity that is below the threshold intensity required for the storage effect to take place). Hence, the possibility of controlling the cursor position can be exploited as a substitute for the "pointing" capability of the lightpen (see Chapter 6). The task of identifying a graphic object through a cursor position requires that an appropriate description of the displayed objects, a picture file, is maintained by the host computer.

Simple Refresh Display Systems: In this class we find the line-drawing display with picture refresh (i.e., with a display file) but without transformation hardware. Figure 1.4 delineates such a system. We shall later also find in this class an entirely different type of simple refresh display, the TV raster display, which is capable of generating lines as well as shaded surfaces, in black and white or in color.

Buffered Refresh Display Systems with Transformation Hardware: This type of display system is furnished with special hardware for picture transformations. The data of the transformed picture are stored in a buffer memory, and the display is refreshed from this buffer. The use of a buffer is advisable in cases where the rate at which graphic primitives can be transformed cannot cope with the rate at which these elements can be generated by the display generators. Without a buffer, picture transformations would in this case reduce the refresh rate (or the maximal number of elements that can be displayed in a frame, respectively). In cases where

†Split-screen tubes exist where one-half of the screen can be selectively erased.

the special-purpose hardware of the display processor does not possess the computing capabilities required for the construction of a structured display file in the buffer, lightpen operation must be sacrificed. With faster components becoming available, the speed gap between the transformation and the generation of graphic primitives can be closed and, thus, this display type can be expected to disappear.

High-Performance Display Systems: High-performance systems encompass hardware for picture transformations which is fast enough to allow for the *viewing algorithm approach* [102], i.e., for the generation of a stream of DPC instructions through the repetitive execution of picture-generating procedures. The biggest asset of such a system is that it allows for the generation of animated pictures under real-time conditions.

Rather than classifying computer graphics equipment on the basis of certain technical idiosyncracies such as the kind of display surface, or the existence or nonexistence of a buffer or of transformation hardware, a taxonomy should be based on the functional characteristics of the graphical devices. Such a taxonomy, as presented by SIGGRAPH's Graphics Standards Planning Committee (GSPC) [1], distinguishes the following categories:

1. Plotting devices.
2. Low-interaction-rate displays.
3. High-interaction-rate, high-performance displays.

Storage tube displays, plasma displays, and low-capability refresh displays are all contained in the second category.

1.8 STAND-ALONE VERSUS SATELLITE SYSTEMS

Many interactive display systems presently in use consist of a "dedicated" general-purpose computer in connection with one or more display consoles and other appropriate peripheral devices. These systems may be used either as stand-alone units or as satellites of larger computer systems. In any case, they represent fairly large investments. Most applications in the realm of computer-aided design (CAD), where computer graphics is used, necessitate a vast amount of computation, requiring either a powerful dedicated computer or an expensive high-speed link to a powerful "host computer." The solution of sharing the resources of the host computer with many users in order to increase the economy of their use, a concept that has proven so valuable in educational and commercial time-sharing systems, has not yet been applicable in CAD. The major obstacle is the prohibitively long response time of the conventional time-sharing system. This response time is not a decisive factor in ordinary computing, where the limiting time element is the speed at which the user types in his statements and data. Conversely, the users of CAD systems work with instruments such as, for example, lightpens, digitizers, etc., which allow them to operate at a much faster pace. Most of the design procedures consist of a large number of primitive steps to be performed by the user, while the system must respond to each step immediately to provide the necessary feedback. If the user had to wait seconds for this response to each simple action, she/he would soon lose patience. This is aggravated even further by the fact that the amount of data to be transferred between host computer and terminal is, in the CAD case typically, much higher than in ordinary computing. Hence, with transmission rates as provided by the customary voice-grade telephone connections, the wait time for data transmission may exceed the already annoyingly long time during which the program is waiting in the CPU service queue by as much as an order of magnitude.

This unfavorable situation can be improved in several ways if local computing power is available in the terminal and if the right use is made of it [49]. To a certain extent, the user's actions can be handled and the required feedback can be provided locally. Of course, a minicomputer (or in the future even a microprocessor) in the terminal may, in terms of computing power as well as of memory capacity, not be powerful enough to accommodate and to execute the entire interactive program. This program can be represented by a finite-state automaton (i.e., as a graph). If the whole graph is too complex to be handled in the terminal, it must be decomposed into suitable subgraphs. Hence, the natural breakpoints in an interactive program are the branch instructions or other control statements.

EXERCISES

1. (a) Enumerate the hardware parts of a typical display system. What parts constitute the display processor?
(b) Compare the required capabilities of a typical display processor with those of a central processor of a typical minicomputer, with respect to (1) commands, (2) data types, (3) data transformations.
(c) Enumerate the differences between a graphic entity and a subpicture.

2. (a) What is a display file?
(b) What causes the necessity for a dual representation of graphic objects, and in which form does a dual representation exist?
(c) In which cases can a display file be omitted?

3. (a) Why is it more convenient to perform picture transformations on the GPL representation of a display program, if the display processor is not equipped with special transformation hardware?
(b) Enumerate the geometric transformations.

(c) What is the effect of geometric transformations on the translation of the GPL program into the DPL program? What is the effect of the windowing operation on the translation of the GPL program into the DPL program?
(d) What geometric transformation must always be included in windowing?
(e) Which are the two alternative sequences of actions that can be applied to perform picture transformations, including windowing? Which one is more advantageous?

2

Data Structures, Data Bases, and List Handling

2.1 FORMAL DEFINITION OF DATA STRUCTURES†

2.1.1 A Data Structure Definition

Knuth writes in the introduction to Chapter 2 of Volume 1 of *The Art of Computer Programming* [84]:

> Computer programs usually operate on tables of information. In most cases these tables are not simply amorphous masses of numerical values; they involve important *structural relationships* between the data elements.
>
> In its simplest form, a table might be a linear list of elements, when its relevant structural properties include the answers to such questions as: which element is first in the list? which is last? which elements precede and follow a given one? There is a lot to be said about structure even in this apparently simple case.
>
> In more complicated situations, the table might be a two-dimensional array (i.e., a matrix or grid, having both a row and a column structure), or it might be an n-dimensional array for higher values of n; it might be a tree structure, representing hierarchical or branching relationships; or it might be a complex multilinked structure with a great many interconnections, such as we may find in a human brain.

†This chapter was written jointly with Helmut Berg.

> In order to use a computer properly, it is important to acquire a good understanding of the structural relationships present within the data, and of the techniques for representing and manipulating such structures within a computer.†

Data may consist of numerical values, names, alphanumeric characters, codes, instructions, or, in general, it may be any "representation in a precise, formalized language of some facts or concepts" (Knuth). According to the definition above, the data base or the display file of a graphic representation, for example, are such data structures.

The paragraphs above quoted from Knuth's book illustrate not only the importance of the concept of data structures but they make a point clear that is sometimes confused: the necessity to distinguish between a data structure and its representation within a computer. A certain structure, such as, for instance, a *linear list* or a *tree*, can be represented in a computer file in a variety of ways.

Data can be considered as a collection of data items, i.e., a *set*; and the relationships between the elements of such a set can consequently be expressed by the mathematical concept of *relations* between the objects of sets. Thus, we define a data structure formally as a pair‡

$$(S,\rho),$$

where $S=\{s_1,\ldots,s_n\}$ is a set of data objects and $\rho=\{R_1,\ldots,R_r\}$ is a set of binary relations such that¶ $\bigwedge_{1\leqslant i\leqslant r}: R_i \subseteq S\times S$. Binary relations may exhibit certain properties. We shall see in the following that a particular data structure is exactly determined by the particular properties to which the "generating" relations are constraint. The more constraints are imposed on the relations, the "simpler" is the resulting data structure.

2.1.2 Definitions of Relations and Their Properties

Although we assume that the reader is familiar with the concept of relations, we list in the following all definitions pertinent for our considerations. The definitions are numbered for later reference. The logical connectives AND, OR, and NOT are denoted by the symbols $\vee$, $\wedge$, and $\neg$, respectively. The symbol $\Leftrightarrow$ denotes the valid proposition of implication. "iff" stands for "if and only if"

†The topics of data structures are most excellently and exhaustively treated in Volume 1 of Knuth's magnificent oeuvre, *The Art of Computer Programming*. A supplementary reading of Chapter 2 of the cited book is strongly recommended.

‡Readers not so familiar with the mathematical formalism used in the following are advised to skip the remaining part of Section 2.1.

¶We use the *existence quantifier* $\bigvee_x$ (reading: "There is an x such that...") and the *all quantifier* $\bigwedge_x$ (reading: "For all x...") in the customary manner. $\emptyset$ denotes the empty set.

Definition D1: *Cartesian Product*

Let A and B be sets. The Cartesian product $A \times B$ of the sets A and B is the set of all ordered pairs (a,b) that can be formed with $a \in A$ and $b \in B$.

$$A \times B = \{(a,b) | a \in A \wedge b \in B\}.$$

Definition D2: *Binary Relation on two Sets*

A binary relation R on two sets, A and B, is a subset of $A \times B$ defined by a proposition $p(a,b)$, $a \in A$ and $b \in B$.

$$R = \{(a,b) | p(a,b)\} \subseteq A \times B.$$

The relation R between two particular elements $a \in A$ and $b \in B$ is denoted aRb.

The domain of a relation R is the set of all first components of all pairs $(a,b) \in R$, denoted $\mathfrak{D}(R)$.

$$\mathfrak{D}(R) = \left\{ a \in A \mid \bigvee_{b} : (a,b) \in R \right\} \subseteq A.$$

The range of a relation R is the set of all second components of all pairs $(a,b) \in R$, denoted $\mathfrak{R}(R)$.

$$\mathfrak{R}(R) = \left\{ b \in B \mid \bigvee_{a} : (a,b) \in R \right\} \subseteq B.$$

Definition D3: *Domain and Range Totality of Binary Relations*

A binary relation $R \subseteq A \times B$ is called domain-total iff $\mathfrak{D}(R) = A$; it is called range-total iff $\mathfrak{R}(R) = B$.

Definition D4: *Binary Relation in a Set*

A binary relation R *in a set* S is defined as a subset of the Cartesian product $S \times S$ defined by the proposition $p(s_i, s_j)$, $s_i, s_j \in S$.

$$R = \{(s_i, s_j) | p(s_i, s_j)\} \subseteq S \times S = \{(s_i, s_j) | s_i \in S \wedge s_j \in S\}.$$

Definition D5: *Properties of Binary Relations in a Set*

1. *Reflexivity:* A binary relation R in a set S is called reflexive

$$\text{iff} \bigwedge_{s \in S} : sRs.$$

2. *Symmetry:* A binary relation R in a set S is called symmetric

$$\text{iff} \bigwedge_{s_i \in S} \bigwedge_{s_j \in S} : s_i R s_j \Leftrightarrow s_j R s_i.$$

3. *Transitivity:* A binary relation R in a set S is called transitive iff

$$\bigwedge_{s_i \in S} \bigwedge_{s_j \in S} \bigwedge_{s_k \in S} : s_i R s_j \wedge s_j R s_k \Rightarrow s_i R s_k.$$

4. *Antisymmetry:* A binary relation R in a set S is called antisymmetric

$$\text{iff} \bigwedge_{s_i \in S} \bigwedge_{s_j \in S} : s_i R s_j \wedge s_j R s_i \Rightarrow s_i = s_j.$$

(*Note:* Reflexivity implies domain-totality and range-totality of a binary relation in a set.)

Definition D6: *Ordering Relations*

1. *Partial ordering:* A binary relation in a set S that is reflexive (D5.1), transitive (D5.3), and antisymmetric (D5.4) is called a partial ordering. We will use the symbol $\leqslant$ for a partial ordering.. The notation $s_1 \leqslant s_2$ may read "s_1 precedes or equals s_2" or "s_2 includes s_1." Partial orderings may also be referred to as *reflexive* orderings.

2. *Irreflexive ordering:* A binary relation in a set S that is transitive (D5.3) and irreflexive, i.e., $\bigwedge_{s \in S} : \neg sRs$, is called an irreflexive ordering. We will use the symbol $<$ for an irreflexive ordering. The notation $s_1 < s_2$ may read "s_1 precedes s_2" or "s_2 covers s_1." For each partial ordering a corresponding irreflexive ordering can be found and vice versa [8].

3. *Linear ordering:* A linear ordering of a set S is a partial ordering $\leqslant$ which satisfies the additional condition that for any two objects $s_1, s_2 \in S$, at least one of the two propositions, $s_1 \leqslant s_2$ or $s_2 \leqslant s_1$, is true. A linear ordering of a set S is called well-ordering if for any nonempty subset A of S: $\bigvee_{a \in A} \bigwedge_{b \in A} : a \leqslant b$.

Definition D7: *Equivalence Relation*

A binary relation in a set S that is reflexive (D5.1), symmetric (D5.2), and transitive (D5.3) is called an equivalence relation. An equivalence relation may be denoted by the symbol $\equiv$. Let $s_1, \ldots, s_n$ be the elements of a set S. For all elements $s_i \in S$ we can form the set $[s_i] = \{s_j \in S \mid s_i \equiv s_j\}$. A set $[s_i]$ is called the *equivalence class* of an element $s_i \in S$ and consists of the subset of all elements $s_j \in S$ which are in equivalence relation $\equiv$ to s_i; i.e., $s_i \equiv s_j \Leftrightarrow [s_i] = [s_j]$. Any equivalence relation on a set S partitions (D8) the set S into a set of mutually disjoint and collectively exhaustive equivalence classes.

Definition D8: *Partition of a Set*

A partition $P(S)=\{S_1,\ldots,S_m\}$ of a set S decomposes S into a set of mutually disjoint and collectively exhaustive subsets $S_i \subseteq S$, i.e.,

$$\left(\bigwedge_{i\neq j} : S_i \cap S_j = \varnothing\right) \wedge \left(\bigcup_i S_i = S\right).$$

Thus, any partition of a set S defines an equivalence relation in S and vice versa. The partition of the sets $S_1,\ldots,S_m \in P(S)$ into a set of partitions $P(S_1),\ldots,P(S_m)$, such that $P(S_i)=\{S_{i,1},\ldots,S_{i,k(i)}\}$ is called a *refinement* of $P(S)$. The sets $S_{i,j}$ may again be partitioned, and so on. Each partition is called the *descendant* of its set, and its set is called the *ancestor* of the partition. Any proposition $p(s), s\in S$, defines a subset $\{s|p(s)\}\subseteq S$ which contains exactly the objects $s\in S$ for which $p(s)$ is true. Let $e_p=\{(s_i,s_j)|p(s_i)=p(s_j)\}\subseteq S\times S$ be the equivalence relation which partitions a set S into two equivalence classes $\{s\in S|p(s)\}$ and $\{s\in S|\neg p(s)\}$ (one of them may be empty, i.e., does not exist as an equivalence class). A partition of a set S by a proposition $p(s)$ is called a *simple partition.* A simple partition may be refined by further propositions on S, such that r propositions partition S into 2^r equivalence classes, of which some may be empty. Each class is the intersection of r of the $2r$ subsets of S as defined by the r propositions.

Definition D9: *Nested Set*

A *nonempty* collection $C=\{C_1,\ldots,C_p\}$ of *nonempty* sets C_i is said to be a nested set, if, given any two sets $C_i, C_j \in C$, either $C_i \cap C_j = C_i$ or $C_i \cap C_j = C_j$ or $C_i \cap C_j = \varnothing$.

It follows from definition D8 that a set, its partition, and possible refinements generate a nested set $C=\{C_1,\ldots,C_p\}$, with the additional condition that for each $C_i \in C$ exactly one of the following conditions is true:

1. There exists a partition $P(C_i)\subset C$ of C_i, i.e.,

$$\bigvee_{P(C_i)\subset C} : \left(C_i = \bigcup_{C_k\in P(C_i)} C_k\right) \wedge \left(\bigwedge_{C_k\in P(C_i)} \bigwedge_{C_j\in P(C_i)} : C_k \cap C_j = \varnothing \text{ for } i\neq j\right).$$

2. There exists no subset of C_i in C, i.e., $\bigwedge_{i\neq j}: C_j \not\subset C_i$. Thus any partition generates a collection of nested sets, but not any collection of nested sets constitutes a partition.

2.1.3 Linear Lists

A *linear list* is defined [84] as a set of $n \geqslant 0$ data objects (nodes) $X[1],\ldots,X[n]$ whose structural properties involve only the linear (one-dimensional) relative positions of the nodes. That is, if $n>0$, $X[1]$ is the first node; for $1<k<n$, the kth

node $X[k]$ is preceded by $X[k-1]$ and followed by $X[k+1]$; and $X[n]$ is the last node.

Let $S=\{s_1,\ldots,s_n\}$ be a set of data objects. Let $\leqslant \subseteq S \times S$ be a *linear ordering* of the elements $s_i \in S$ (D6.3). For the sake of convenience we assume that $\leqslant = \{(s_i,s_j)|i \leqslant j\}$ ($\leqslant$ is a well-ordering). Then, the nodes $X[i]$ of a linear list X represent the singletons $\{s_i\}, 1 \leqslant i \leqslant n$. Consequently, the data structure of a linear list is defined by a pair

$$(S,\{\leqslant\}).$$

Hence, a linear list X is an n-tuple $(X[1],\ldots,X[n])$ whose components are identified by an ordinal number specifying their relative position in X.

A simple generalization of a linear list is a two-dimensional or higher-dimensional *array* of data objects. In a rectangular two-dimensional $m \times n$ array we have the linear row lists $(R_i,\{\leqslant_{R,i}\}), i \in [1:m]$, with $R_i=\{R_i[1],\ldots,R_i[n]\}$, or the linear column lists $(C_j,\{\leqslant_{C,j}\}), j \in [1:n]$, with $C_j=\{C_j[1],\ldots,C_j[m]\}$. These linear lists are orthogonally connected such that the linear row lists $(R_i,\{\leqslant_{R,i}\})$ are the n-tuples $R_i=(C_1[i],\ldots,C_n[i])$ and the linear column lists $(C_j,\{\leqslant_{C,j}\})$ are the m-tuples $C_j=(R_1[j],\ldots,R_m[j])$. Thus, the linear orderings $\leqslant_{R,i}$ in the row lists $(R_i,\{\leqslant_{R,i}\})$ imply the same linear ordering on the set of column lists $\{(C_j, \{\leqslant_{C,j}\})|j=1,\ldots,n\}$, and vice versa. The intersection of the linear lists $(R_i, \{\leqslant_{R,i}\})$ and $(C_j,\{\leqslant_{Cj}\})$ determines the element in position (i,j) of the $m \times n$ array. Thus, a two-dimensional $m \times n$ array may be formally defined as a pair

$$\left(\bigcup_{i=1}^{m} R_i,\{\leqslant_{R,1},\ldots,\leqslant_{R,m},\leqslant_{C,1},\ldots,\leqslant_{C,n}\}\right).$$

This definition can easily be extended to n-dimensional arrays ($n \geqslant 3$).

2.1.4 Tree Structures

A *tree* is (recursively) defined [84] as a finite set T of one or more nodes such that:

1. There is one specially designated node called the root of the tree and denoted root(T).

2. The remaining nodes (excluding the root) are partitioned into $m \geqslant 0$ disjoint sets $T_1,\ldots,T_m$, and each of these sets, in turn, is a tree.

The trees $T_1,\ldots,T_m$ are called the *subtrees* of the root. It follows from this definition that every node of a tree is the root of some subtree contained in the whole tree. The number of subtrees of a node is called the *degree* of that node. A node of degree zero is called *terminal node*. A nonterminal node is called a *branch node*.

Let $S=\{s_1,\ldots,s_n\}$ be a set of data objects. Let $C=\{C_1,\ldots,C_p\},p \geqslant n$, be a collection of nested sets generated by S, its partition $P(S)=\{S_1,\ldots,S_m\}$, and possible refinements (D.8), such that $\{s_1\},\ldots,\{s_n\}\in C$. We assume that $\{\{s_1\},\ldots,\{s_n\}\}=\{C_{p-n+1},\ldots,C_p\}$. Then, for each $C_k \in C, 1 \leqslant k \leqslant p-n$, there exists an *equivalence relation* q_k which defines the partition $P(C_k)\subset C$. The nodes of a tree represent the equivalence classes $C_1,\ldots,P_p \in C$. That is, a *tree* is defined by a pair

$$(S,\{q_1,\ldots,q_{p-n}\}),$$

where $\rho=\{q_1,\ldots,q_{p-n}\}$ is the set of equivalence relations which generate the collection of nested sets $C=\{C_1,\ldots,C_p\}$. Note that the definition of a tree as a nonempty set is in compliance with the nested set definition.

Corresponding with Knuth's definition of a tree, root(T) is a representative of the set $S=C_1$ of all data objects. The equivalence relation q_1 partitions S into $m \geqslant 1$ equivalence classes $P(S)=\{S_1,\ldots,S_m\}=\{C_2,\ldots,C_{m+1}\}$, which are represented by the root of the *subtrees* of root(T). The cardinality of the partitions $P(C_h \in C)$, $1 \leqslant h \leqslant p$, is the *degree* of the nodes C_h. That is, the equivalence classes C_k, $1 \leqslant k \leqslant p-n$, are represented by the *branch nodes* of the tree. The equivalence classes $\{s_i\}=C_{p-n+i}$, $1 \leqslant i \leqslant n$, represented by the *terminal nodes* are not partitioned, and hence the degree of the terminal nodes is zero.

Each equivalence class C_j is associated with a *reference element* $s_i \in C_j$; and the equivalence classes $C_{p-n+i}=\{s_i\}$ contain solely the reference element s_i. That is, any equivalence class C_k, $1 \leqslant k \leqslant p-n$, can be represented by a reference element $s_i \in C_k$. The representation of an equivalence class C_k with $\{s_i\}\in P(C_k)$ by its reference element $s_i \in S$ allows for the removal of the terminal node representing $\{s_i\}$ from the tree. Thus, the representation of a partition $P(C_k)$ by the equivalence classes $C_j \in P(C_k)$ in a tree forms not necessarily a set of collectively exhaustive subsets.

The definition of the collection of nested sets $C=\{C_1,\ldots,C_p\}$ does not imply an ordering of the equivalence classes $C_j \in P(C_k)$, $1 \leqslant k \leqslant p-n$, but only indicates the ancestor–descendant relationship among the equivalence classes $C_h \in C$. Trees that are equivalent to a collection of nested sets are called *oriented trees*, since only the relative orientation of the nodes is being considered. An *ordered tree* is obtained by ordering the equivalence classes $C_j \in P(C_k)$ in some ad hoc manner, for example, by embedding the partial ordering $\subseteq$ on the collection of nested sets C into a linear ordering $\leqslant$ [84]. An ordering of all equivalence classes $C_j \in P(C_k)$ implies an ordering $\leqslant$ on the data objects $s_i \in S$. Thus, an *ordered tree* is defined by a pair

$$(S,\{\leqslant, q_1,\ldots,q_{p-n}\}).$$

As an example of an ordered tree, we may consider the representation of a two (or higher)-dimensional array A as a tree structure. Here, the linear row lists $(R_1, \{\leqslant_{R,1}\}),\ldots,(R_m,\{\leqslant_{R,m}\})$ represent the subtrees of a tree with root(A) as a representative for all elements in the array. As discussed in Section 2.1.3, the row lists $(R_i, \{\leqslant_{R,i}\})$ are ordered by the linear orderings $\leqslant_{C,j}$ of the column lists $(C_j, \{\leqslant_{C,j}\})$. The terminal nodes $A\ [i;j]$ are ordered by the linear orderings $\leqslant_{R,i}$ of the row lists $(R_i, \{\leqslant_{R,i}\})$. A similar tree structure may be constructed by representing the column lists $(C_j, \{\leqslant_{C,j}\})$ by the roots of the subtrees. In the case of a higher-dimensional array, the corresponding tree structure has as many levels in addition to the root as is the dimensionality (rank) of the array.

The exclusion of the set S of all data objects [represented by root(T) of a tree T] from the collection of nested sets $C = \{C_1, \ldots, C_p\}$ results in a structure called a *forest*. That is, a forest is a set of zero or more disjoint trees [84]. Again, we distinguish between *oriented forests* defined

$$(S, \{q_2, \ldots, q_{p-n}\}),$$

and *ordered forests* defined

$$(S, \{\leqslant, q_2, \ldots, q_{p-n}\}).$$

The set of equivalence classes $P(S) = \{S_1, \ldots, S_m\}$ is represented by the roots of the trees in the forest. Thus, the concept of forest equals that of equivalence classes.

Another concept of a tree structure is the concept of *binary trees*. A binary tree B is a finite set of nodes which is either empty or consists of a root and two disjoint binary trees called the *left* and the *right subtrees* of the root [84]. The conceptual differences between a tree and a binary tree are:

1. A tree is never empty; and each node of a tree may have an arbitrary number of descendants.

2. A binary tree can be empty; and each of its nodes can have not more than two descendants. We distinguish between the "left" and "right" descendants.

Let $S = \{s_1, \ldots, s_n\}$ be a set of data objects. A *binary tree* is defined by a pair

$$(S, \{e_1, \ldots, e_r\}),$$

where $\rho = \{e_1, \ldots, e_r\}$ is a set of *equivalence relations*, $e_i = \{(s_j, s_k) | p_i(s_j) = p_i(s_k)\} \subseteq S \times S$, defined by a set $\pi = \{p_1, \ldots, p_r\}$ of *propositions* $p(s)$, $s \in S$ (D8). Each node on the kth level of a binary tree is defined by a set $\pi' \subseteq \pi$ of k propositions $p_j \in \pi$, such that the set $S' \subseteq S$ represented by such a node is defined as an intersection

$$S' = \bigcap_{p_j \in \pi'} S_j', \quad \text{where either} \quad S_j' = \{s \in S | p_j(s)\} \quad \text{or} \quad S_j' = \{s \in S | \neg p_j(s)\}.$$

As such intersections can be *empty*, π does not generate a collection of nested sets (D9).

The proposition $p_1(s)$, $s \in S$, partitions S into two equivalence classes, $S_1 = \{s \in S | p_1(s)\}$ and $\bar{S}_1 = \{s \in S | \neg p_1(s)\}$ (one of which may be empty), which are defined by the equivalence relation $e_1 \in \rho$. Contrasting to trees, S is not represented in a binary tree B, but root(B) represents S_1. The two binary subtrees of root(B) are defined:

1. The root of the left subtree represents $S_2 = S_1 \cap \{s \in S | p_2(s)\}$, i.e., $S_2 \subseteq S_1$. The set represented by the root of the left subtree is a subset of the set represented by root(B).

2. The root of the right subtree represents $S_3 = \bar{S}_1 \cap \{s \in S | p_3(s)\}$, i.e., $S_1 \cap S_3 = \varnothing$. The intersection of the sets represented by root(B) and by the root of its right subtree is empty.

The recursive application of this definition generates a binary tree, and there exists exactly one of the following alternatives:

1. $(S_2 \neq \varnothing) \wedge (S_3 \neq \varnothing)$ defines a binary tree with nonempty right and left subtrees.

2. $S_1 = S$ defines a binary tree with an empty right subtree.

3. $S_1 = \varnothing$ defines a binary tree with an empty left subtree.

4. $S = \varnothing$ defines an empty binary tree.

Criteria for terminal nodes are:

1. $S \supseteq S_i = \{s_k\} \wedge \bar{S}_i = \varnothing$ defines a terminal node represented by a left descendant.

2. $S \subseteq S_j = \{s_k\}$ defines a terminal nodes represented by a right descendant.

There exists a natural correspondence between forests and binary trees. Let $P(S) = \{S_1, \ldots, S_m\}$ be the collection of equivalence classes represented by a forest. The same equivalence classes $S_i \subseteq S$ can be generated by an m-tuple of propositions $(p_1, \ldots, p_m)$ such that

$$\left(S_1 = \{s \in S | p_1(s)\}\right) \wedge \left(S_i = \bar{S}_{i-1} \cap \{s \in S | p_i(s)\}\right), \qquad \text{for} \quad i = 2, \ldots, m.$$

This definition implies the following construction of a binary tree corresponding to a given forest (see [84]). Let $B(P(S))$ denote the binary tree which corresponds to a forest whose roots represent $P(S) = \{S_1, \ldots, S_m\}$, then

1. If $P(S) = \varnothing$, $B(P(S))$ is empty.

2. If $P(S \neq \varnothing$, root($B(P(S))$) represents S_1; the left subtree of $B(P(S))$ is $B(P(S_1))$; and the right subtree of $B(P(S))$ is $B(\{S_2, \ldots, S_m\})$.

Let $B(T)$ denote the binary tree that corresponds to a tree T. The definitions

1. $\text{Root}(B(T)) = \text{root}(T)$,

2. The left subtree of $B(T)$ is $B(P(S))$,

3. The right subtree of $B(T) = \varnothing$, imply a correspondence between trees T and binary trees $B(T)$ which have no right subtree.

2.1.5 Generalized List Structures

List structures are a very general type of structure that combine the general characteristics of trees, forests, and binary trees, on one hand, and of linear lists, on the other hand. Knuth [84] uses the capitalized term "List" in order to distinguish it as a specific structure rather than the generic "list." A *List* L is recursively defined as a finite sequence of zero or more atoms or Lists. Here, *atom* is an undefined concept referring to elements from any universe of objects that might be desired, as long as it is possible to distinguish an atom from a List.

Let $S = \{s_1, \ldots, s_n\}$ be a set of data objects. Let $\rho = \{R_1, \ldots, R_r\}$ be a set of relations $R_i \subseteq S \times S$. For each relation R_i we define the set $A_i \subseteq S$ of all elements $a_j \in S$ that may occur as first component in the pairs of R_i. The a_j shall be called the *reference elements* of R_i. Let $B_i \subseteq S$ denote the set of all elements b_k which may occur as second component in the pairs of R_i. The subsets $A_i \subseteq S$ and $B_i \subseteq S$ are a priori defined for each $R_i \in \rho$ when a List is set up for a certain application. Let the relations R_i be *range-total* on $A_i \times B_i$; i.e., $\mathcal{D}_i \subseteq A_i$ and $\mathcal{R}_i = B_i$. The set ρ of relations $R_i = \{(a_j, b_k) | p_i(a_j, b_k)\} \subseteq \mathcal{D}_i \times \mathcal{R}_i \subseteq A_i \times B_i \subseteq S \times S$ is defined by a set $\pi = \{p_1, \ldots, p_r\}$ of propositions. Then, a relation R_i generates for each reference element $a_j \in A_i$ a subset of $\mathcal{R}_i = B_i$ that will be denoted $\mathcal{R}_i / a_j$ (read: "the subset of $\mathcal{R}_i$ with respect to a_j"), such that

$$\mathcal{R}_i / a_j = \{ b_k \in \mathcal{R}_i | p_i(a_j, b_k) \}.$$

$\mathcal{R}_i / a_j = \varnothing$ if $a_j \notin \mathcal{D}_i$ and $\bigcup_{a_j \in \mathcal{D}_i} \mathcal{R}_i / a_j = \mathcal{R}_i$. The relations $R_i \in \rho$ define a set $N = \{\mathcal{R}_i / a_j | 1 \leqslant i \leqslant r \wedge a_j \in A_i\}$ such that $\{s_1\}, \ldots, \{s_n\} \in N$. The nodes of a List represent the sets $\mathcal{R}_i / a_j \in N$. Thus, a List is defined by a pair

$$(S, \{R_1, \ldots, R_r\}).$$

A List L represents the set S of all data objects. The relation $R_1 \subseteq A_1 \times S$ defines a subcollection $\{\mathcal{R}_1 / a_j | a_j \in A_1\} \subseteq N$ of subsets of S which are represented by the nodes at the first level of the List L. The recursive application of this definition to the sets $\mathcal{R}_1 / a_j \in N$ generates all sets $\mathcal{R}_i / a_j$ represented by the nodes of the List L, and we have exactly one of the following alternatives for each $\mathcal{R}_i/a_j \in N$:

1. If there exists *no* relation R_k with $R_k \subseteq A_k \times \mathcal{R}_i/a_j$ for a particular $\mathcal{R}_i/a_j \in N$ and if $C(\mathcal{R}_i/a_j)=1$ [$C(\mathcal{R}_i/a_j)$ denotes the cardinality of $\mathcal{R}_i/a_j$], then $\mathcal{R}_i/a_j$ is represented by an *atom*, i.e., $\mathcal{R}_i/a_j \in \{\{s_1\},\ldots,\{s_n\}\} \subseteq N$.

2. If there exists a relation R_k with $R_k \subseteq A_k \times \mathcal{R}_i/a_j$ for a particular $\mathcal{R}_i/a_j \in N$, then $\mathcal{R}_i/a_j$ is represented by a *(sub)-List*.

Note that there must exist a relation $R_k \subseteq A_k \times \mathcal{R}_i/a_j$ for each $\mathcal{R}_i/a_j$ with $C(\mathcal{R}_i/a_j) \neq 1$ as implied by the above conditions. All sets $\mathcal{R}_i/a_j = \varnothing$ are represented by an *empty List*, but there exists no empty atom. Lists containing only a single data object $s_j \in S$ may exist if there exists a relation $R_k \subseteq \{s_j\} \times \{s_j\} = A_k \times \mathcal{R}_i/a_j$.

The above definition of a List indicates a hierarchy among the sets $\mathcal{R}_i/a_j \in N$ that was also found in tree structures (see Section 2.1.4). However, List structures generalize tree structures, exhibiting the following idiosyncrasies:

1. Lists may *overlap*; i.e., sub-Lists need not be disjoint: $\mathcal{R}_i/a_j \cap \mathcal{R}_k/a_m \neq \varnothing$, for $1 \leqslant i \leqslant r$, $1 \leqslant k \leqslant r$, $a_j \in A_i$, $a_m \in A_k$.

2. Lists may be *recursive*; i.e., they may contain themselves: $\mathcal{R}_k/a_m = \mathcal{R}_i/a_j$, for $R_k \subseteq A_k \times \mathcal{R}_i/a_j$ and $R_i \subseteq A_i \times \mathcal{R}_k/a_m$.

Each node in a sub-List representing a set $\mathcal{R}_i/a_j \in N$ is identified by a common relation $R_k \subseteq A_k \times \mathcal{R}_i/a_j$ and the unique reference element $a_m \in A_k$. Note that, in contrast to the reference elements in tree structures, we may have $a_m \notin \mathcal{R}_k/a_m$, as range-total relations need not be reflexive. The definition of some ad hoc ordering $\leqslant$ on the data objects $s_i \in S$ implies an ordering of the nodes representing the sets $\mathcal{R}_i/a_j \in N$ in all sub-Lists of a List L. In this case, L is defined by a pair

$$(S, \{\leqslant, R_1, \ldots, R_r\}).$$

Such a List may be considered as a generalization of a linear list $(S, \leqslant)$ with the proviso that each element $S[j]$ may represent atoms or Lists $\mathcal{R}_i/a_j$, $a_j \in A_1$.

2.1.6 Associative Structures

So far, we have considered data structures (S, ρ) which are generated by binary relations $R_i \in \rho$ with certain constraints, such as being an ordering relation (linear lists), an equivalence relation (tree structures), or a range-total relation on some predefined Cartesian products $A_i \times B_i$; $A_i, B_i \in S$; (List structure). Additionally, the definition of these relations implies a hierarchy among the subsets of data objects $S_i \subseteq S = \{s_1, \ldots, s_n\}$ generated by the relations R_i, with the property that all singletons $\{s_i\}$, $s_i \in S$, are explicitly represented by the nodes of the corresponding data structure. This offers the possibility of introducing an ordering on the data objects $s_i \in S$ which, together with the defined hierarchy, implies an ordering on all nodes in the data structure.

In the following, we will generalize our considerations, introducing data structures (S,ρ) which are generated by *binary relations without any restrictive properties*. That is, we define such a data structure by a pair

$$(S,\{R_1,\ldots,R_r\}),$$

where ρ is a set of relations $R_i \subseteq S \times S$ such that

$$S \supseteq A = \bigcup_{1 \leqslant i \leqslant r} \mathcal{D}_i \quad \text{and} \quad S \supseteq B = \bigcup_{1 \leqslant i \leqslant r} \mathcal{R}_i \quad \text{and} \quad S = A \cup B.$$

We call the elements $a_j \in A \subseteq S$ *domain elements* and the elements $b_k \in B \subseteq S$ *range elements*. The set ρ of relations $R_i = \{(a_j,b_k) | p_i(a_j,b_k)\} \subseteq \mathcal{D}_i \times \mathcal{R}_i \subseteq A \times B \subseteq S \times S$ is defined by a set $\pi = \{p_1,\ldots,p_r\}$ of propositions. Thus, the data structure under consideration may be specified by the *triad* (A,π,B) [22].

As the data structure definition above does not distinguish the domain elements $a_j \in A$ as reference elements, any combination of elements $a_j \in A$, $b_k \in B$, and $p_i \in \pi$ may be defined as a *reference element* for the subsets of data objects $S_i \subseteq S$ represented by the nodes of the data structure. Hence, the subsets $S_i \subseteq S$ represented by the data structure under discussion are defined by the *eight basic triads* listed below.

1. A triad of the form $(\{a_j\},\{p_i\},\{b_k\})$ defines the relationship $a_j R_i b_k$ between two particular data objects $a_j \in A$ and $b_k \in B$. (Q0)

2. The subset of a relation $R_i \in \pi$ which consists of all pairs with the same element $a_j \in \mathcal{D}_i$ as the first component is specified by a triad $(\{a_j\},\{p_i\},B)$. The subset of a relation R_i with respect to an element $a_j \in \mathcal{D}_i$ is

$$R_i / a_j = \{(a_j,b) | p_i(a_j,b) \wedge b \in \mathcal{R}_i\}.$$

In practical applications, it would be redundant to represent the pairs in R_i/a_j, as the first component (a_j) in such a set of pairs is a priori known. Therefore, it is sufficient to represent, instead, the subset of $\mathcal{R}_i$:

$$\mathcal{R}_i / a_j = \{b \in \mathcal{R}_i | p_i(a_j,b)\}. \tag{Q1}$$

3. Similarly, the triad $(A,\{p_i\},\{b_k\})$ defines the subset of $\mathcal{D}_i$:

$$\mathcal{D}_i / b_k = \{a \in \mathcal{D}_i | p_i(a,b_k)\}. \tag{Q2}$$

4. Any triad $(\{a_j\},\pi,\{b_k\})$ defines the subset of the set π:

$$\pi/a_j,b_k=\{p_i\in\pi|p_i(a_j,b_k)\}. \quad \text{(Q3)}$$

5. The specification of an element $a_j\in A$ in a triad $(\{a_j\},\pi,B)$ defines a subset R/a_j of the Cartesian product $\pi\times B$:

$$R/a_j=\{(p_i,b)|p_i(a_j,b)\wedge p\in\pi\wedge b\in\mathcal{R}_i\}.$$

Actually, it is more economical to represent only the subset of π:

$$\pi/a_j=\{p_i\in\pi|p_i(a_j,b)\wedge a_j\in\mathcal{D}_i\} \quad \text{(Q4)}$$

and to associate with each proposition $p_i\in\pi/a_j$ the subset $\mathcal{R}_i/a_j$ (see Q1). The union of the sets $\mathcal{R}_i/a_j$ forms the subset of B:

$$B/a_j=\bigcup_{p_i\in\pi/a_j}\mathcal{R}_i/a_j.$$

6. Analogously, the specifications given by a triad $(A,\pi,\{b_k\})$ result in the subset of π:

$$\pi/b_k=\{p_i\in\pi|p_i(a,b_k)\wedge b_k\in\mathcal{R}_i\}, \quad \text{(Q5)}$$

from which the subset of A,

$$A/b_k=\bigcup_{p_i\in\pi/b_k}\mathcal{D}_i/b_k$$

can be derived.

7. The specification of a proposition $p_i\in\pi$ in a triad $(A,\{p_i\},B)$ defines the relation R_i as a subset of the Cartesian product $A\times B$. From the definitions in cases 2 and 3, we have

$$R_i=\bigcup_{a_j\in\mathcal{D}_i}R_i/a_j=\bigcup_{b_k\in\mathcal{R}_i}R_i/b_k. \quad \text{(Q6)}$$

8. For the sake of completeness, the definition of the entire data structure by the triad (A,π,B) in which no component is explicitly specified may be repeated.

The triads defined in cases 1–7 correspond to the seven basic forms of associative queries possible in systems of binary relations, as defined in [37]. Therefore, we call the data structure $(S,\{R_1,\ldots,R_r\})$, with $R_i\subseteq A_i\times B_i$, $1\leqslant i\leqslant r$, an *associative data structure*. Summarizing this discussion, we list for the reader's convenience in the following table the seven associative queries, Q0,...,Q6, and the subsets defined by them

Triad	*Specified subset*	*Query*
$(\{a_j\},\{p_i\},\{b_k\})$	$a_j R_i b_k$	Q0
$(\{a_j\},\{p_i\},B)$	$\mathcal{R}_i/a_j=\{b\in\mathcal{R}_i \mid p_i(a_j,b)\}$	Q1
$(A,\{p_i\},\{b_k\})$	$\mathcal{D}_i/b_k=\{a\in\mathcal{D}_i \mid p_i(a,b_k)\}$	Q2
$(\{a_j\},\pi,\{b_k\})$	$\pi/a_j,b_k=\{p_i\in\pi \mid p_i(a_j,b_k)\}$	Q3
$(\{a_j\},\pi,B)$	$\pi/a_j=\{p_i\in\pi \mid p_i(a_j,b) \;\wedge\; a_j\in\mathcal{D}_i\}$	Q4
$(A,\pi,\{b_k\})$	$\pi/b_k=\{p_i\in\pi \mid p_i(a,b_k) \;\wedge\; b_k\in\mathcal{R}_i\}$	Q5
$(A,\{p_i\},B)$	$R_i=\bigcup_{a_j\in\mathcal{D}_i} R_i/a_j=\bigcup_{b_k\in\mathcal{R}_i} R_i/b_k$	Q6

2.2 REPRESENTATION OF DATA STRUCTURES IN A COMPUTER

2.2.1 Sequential or Linked Memory Allocation

Any representation of a data structure in the computer memory must encompass the data to be stored as well as the implicitly given relationships which constitute the structuring of the data. In addition to data processing, i.e., value-transforming operations, structure-transforming operations must exist as well. Few existing high-level programming languages provide such operations as primitive operators of the language (e.g., APL). Hence, in most cases the user not only needs to create the data structures required by his applications but also the necessary value-transforming and structure-transforming operators. This is accomplished by writing appropriate procedures.

Many programming systems have been designed to facilitate working with particular structures. Most of them are tailored for specific applications and, therefore, impose certain constraints on the programmer. A major goal in the development of data base management systems is the achievement of data independence in order to mitigate the constraints imposed on a programming system. Yet, it often is more efficient to tailor the data format and the processing algorithms to a particular application in one's own program.

The form of a computer representation of data structures depends on the intended use of the data; and there is no distinguished representation in which all operations on a structure are equally efficient. Basically, there is the distinction between *location-addressed* and *content-addressed* memory representations of data structures. In the first case, data are addressed by providing logical or physical addresses determining the location of data storage in physical memory. In a content-addressed or associative memory, data are accessed through a key, i.e., a

(a) Sequential Allocation

Address	Contents
L+M	Item 1
L+2M	Item 2
L+3M	Item 3
L+4M	Item 4
L+5M	Item 5

(b) Linked Allocation

Address	Contents	
A	Item 1	B
B	Item 2	C
C	Item 3	D
D	Item 4	E
E	Item 5	Λ

Λ = 'null link'

Figure 2.1 Example for the memory representation of five data items: (a) in sequential allocation; (b) in linked allocation.

known part of their value. However, even an associative memory representation internally applies an implicit ordering of memory locations. Hence, the one-dimensional linear list, the *data vector*, certainly is the simplest possible form of internal data storage. We may call this the "physical" data structure; and the problem of representing (logical) data structures in a computer is to find an efficient scheme of mapping the desired logical structure to the physical structure. We call such a mapping the *addressing function*. Two basic methods may be employed in the implementation of an addressing function: the *sequential allocation* and the *linked allocation* (Fig. 2.1).

In the sequential allocation scheme, consecutive memory locations are assigned to the records in the data vector. The data vector is logically separated from the stored structure specification. In the simplest possible case, the *linear list*, a structure specification may consist in a special record containing the dimension d of the data vector V, the number n of memory words occupied by each record, and the base address β indicating the beginning of V in memory. Under the dimension of a data vector V, we understand its cardinality $C(V)$, i.e., its number of elements.

The address under which a certain record can be accessed can be calculated on the basis of an addressing function. The addressing function maps a logical *index*, designating a particular record in the structure, to a physical memory address. In the example of a linear list, the addressing function consists in a simple displacement and scaling operation; i.e., we have in this case

$$\alpha(i) = \beta + (i-1)\cdot m.$$

The term "linear list" stands for "linearly ordered set." A set may be ordered one-dimensionally or multidimensionally. An r-dimensionally ordered set, $r \in \mathbf{N}$ ($\mathbf{N}$ denoting the set of positive integers), can be defined by the mapping

$$\sigma: \mathbf{N}^r \rightarrow \mathbf{N},$$

which we call the *structuring function*. The positions in the r-tuples $(n_1, \ldots, n_r) \in \mathbf{N}r$ are called the *coordinates* of the structure, and r is called the *rank*. An r-tuple $(n_1, \ldots, n_r) \in \mathbf{N}^r$ is called an *index r-tuple* and contains an index value for each coordinate. The structure of an ordered set is determined by specifying the admissible range $[L_i : U_i]$ for each index value n_i. The addressing function is in this case a mapping (M denotes the set of memory addresses)

$$\alpha: \mathbf{N}^r \rightarrow M,$$

mapping an index-r tuple to the memory address where the designated element of the r-dimensional ordered set is stored. As will be shown, multidimensional ordered sets are an underlying concept for arrays as well as for trees.

The specification of an index r-tuple for an r-dimensional ordered set A shall be called *indexing* and denoted $A[n_1; \ldots; n_r]$. $[n_1; \ldots; n_r]$ is the *index list*; and the designated record in the data vector V is $V[\sigma(n_1; \ldots; n_r)]$. The value of a multidimensional ordered set is given by a pair $(\langle\text{structure}\rangle, \langle\text{data}\rangle)$, indicating that the information about an ordered set is more than just the data.

In order to retain the set-element relationship in a memory representation of a multidimensionally ordered set, three information items must be stored: (1) the data vector; (2) the structure, i.e., the parameters of the structuring function (e.g., the admissible range of each index value n_1); and (3) a specification of the absolute and relative positions of data records in memory (e.g., the address of the first record, $V[1]$, and the number m of memory words occupied by each record).

In the linked allocation scheme, a structure is built by specifying element–successor relationships or element–predecessor relationships through pointers. Pointers are addresses that are stored in the data records. Hence, contrasting to a sequential allocation where the addressing function allows for the *calculation* of a successor address, the value of an addressing function in a linked allocation is obtained by the look-up of *stored* pointers. This technique requires more memory space than the sequential allocation, yet it offers the often decisive advantage that the order of records may arbitrarily be changed, and that a structure may be expanded or reduced without having to relocate the stored information items (Fig. 2.2).

A linked representation of a linear list is called a *linked list*. SUC($X[i]$) denotes the pointer in the data record $X[i]$ of a linked list X, i.e., the address of the successor node $X[i+1]$. Since $X[C(X)]$ has no successor node, we define

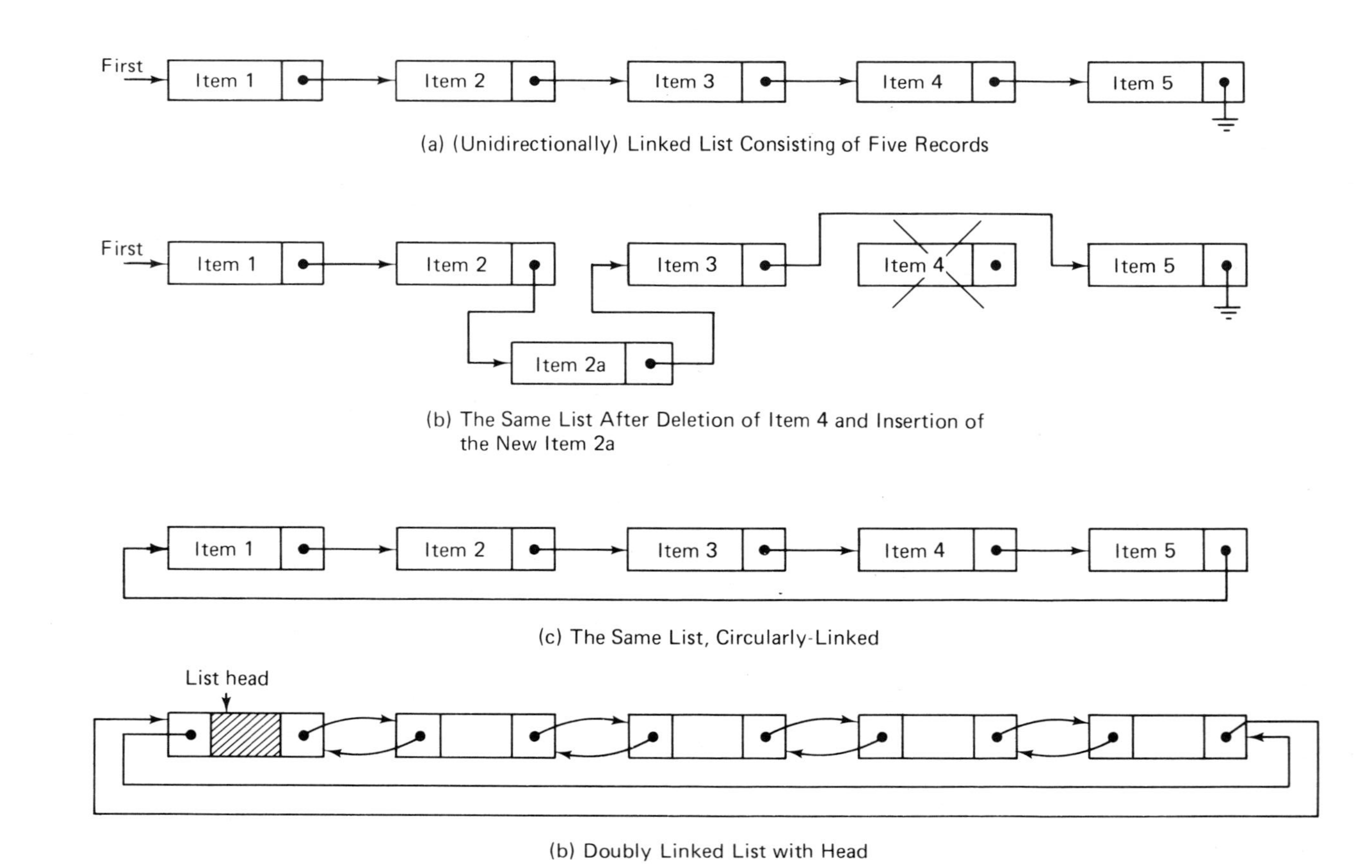

Figure 2.2 Examples of linked linear lists.

$\mathrm{SUC}(X[C(X)]) = \Lambda$, where $C(X)$ is the number of nodes and Λ is the null link indicating the end of the list. The base address is given as a pointer variable TOP(X), which contains the address of the first node as its value. The addressing function for linked lists is defined

$$\alpha(1) = \mathrm{TOP}(X) \quad \text{and} \quad \alpha(i) = \mathrm{SUC}(X[i-1]), \qquad 1 < i \leqslant C(X).$$

Obviously, the addressing function α is independent of the linear ordering of memory locations. Any arbitrary change in the ordering of records and reductions and expansions of the data vector can be made without relocating the data records within the data vector; and only the values in the link fields must be changed. Thus, linked lists are the adequate form for linear structures that may grow and shrink dynamically. As an access to node $X[i]$ is obtained iteratively through TOP(X) and the link field in the nodes $X[k]$ ($k < i$), random access to a particular node may be significantly slower than in sequential allocation.

Two modifications of the described linking mechanism may mitigate this disadvantage:

1. Make TOP(X) a special recognizable node in the list, called the list head. The list head indicates the beginning of the list [$\alpha(1) = \mathrm{SUC(HEAD)}$] and is usually stored in a fixed memory location β.

2. Let the last node link back to the list head, i.e., $\mathrm{SUC}(X[C(X)]) = \beta$. The resulting memory representation of a linear list is called a *circular list*. A circularly linked list always contains the head and thus is never empty. The symmetric linking scheme makes it possible to access all records of a list, starting the iterations through the link fields at any record.

Linear structures may also be built by specifying the predecessor–successor relationship for each node. In this case, two pointers are included in each node, pointing to the nodes on either side of that node. The pointers to the successor and predecessor nodes of $X[i]$ are denoted $\mathrm{SUC}(S[i])$ and $\mathrm{PRE}(X[i])$, respectively. This representation of a linear list is called a *doubly linked list* and is defined by the following symmetric definitions of the addressing function:

$$\alpha(1) = \mathrm{SUC(HEAD)}, \qquad \alpha(i) = \mathrm{SUC}(X[i-1]), \qquad \beta = \mathrm{SUC}(X[C(X)]);$$

$$X[C(X)] = \mathrm{PRE(HEAD)}, \qquad \alpha(i) = \mathrm{PRE}(X[i+1]), \qquad \beta = \mathrm{PRE}(X[1]).$$

Compared with the linking scheme above, the ability to go back and forth at will highly increases the efficiency of accesses, deletions, and insertions of records. The various linking schemes are illustrated in Figure 2.2, using the customary symbols introduced by Knuth [84] for the nodes of a list. The flexibility gained through the links greatly facilitates the representation of structures more intricate than linear lists.

2.2.2 Representation of Arrays

Practically, the most frequently encountered ordered set representation is the homogeneous (all components are of the same data type), rectangular array, resulting from the fact that homogeneous, rectangular arrays occur as a primitive data structure in most high-level programming languages. An r-dimensional array may be considered as an ordered set of r orthogonal linear lists (see Section 2.1.3). If the array is rectangular, all linear lists in the ith coordinate have the same dimension d_i, $i \in [1:r]$. Thus, the admissible range for each of the index values n_i is defined by the *dimension vector* $D=(d_1,\ldots,d_r)$.

Assuming that the elements of an array are stored in the data vector V in *row major order*, we have the structuring function for a sequential allocation of an r-dimensional array (if the "count origin" is 1), as follows:

1-origin:

$$\sigma_1(n_1,\ldots,n_r)=\sum_{i=1}^{r-1}(n_i-1)\prod_{j=i+1}^{r}d_j+n_r=i(n_1,\ldots,n_r),$$

where $i(n_1,\ldots,n_r)$ indicates the location of the element $A[n_1;\ldots;n_r]$ in the data vector. Hence, besides the base address β and the parameter m, two items must be stored in a sequentially allocated representation of rectangular arrays: the data vector and the dimension vector. The addressing function for sequentially allocated rectangular arrays is the linear function

$$\alpha_1(n_1,\ldots,n_r)=\beta+(\sigma_1(n_1,\ldots,n_r)-1)\cdot m.$$

Note that the addressing function is nonlinear if the array is not rectangular. In the case of nonrectangular arrays, the sequential allocation scheme may still be efficient as long as the array is regularly shaped (e.g., a triangular matrix). However, in computer graphics we have hardly to deal with such arrays and thus will not dwell on that point. In the case of irregularly shaped arrays, the linked allocation scheme is more adequate. In this case, each node will contain as many pointers as is the number of orthogonal (and one-dimensional) lists of which the nodes are a part. Moreover, it shall be emphasized that the pointers and the data need not be united in one record, but the structure may be represented strictly by a pointer list, whereas the data are stored in an associated data list. Such an approach increases the flexibility of a structure and allows for data records of variable size. The price for it consists in the additional pointer per node that points to the data record.

2.2.3 Representation of Trees

Ordered trees, as well as arrays, are forms of multidimensionally ordered sets. This suggests an efficient approach toward the description of a sequential allocation of the nodes of a tree: first, to specify the ordered set representation of the tree and second, to use the already established mapping from the resulting multidimensionally ordered set to the data vector. Therefore, the r subtree levels of an ordered tree are equated with the r coordinates of a multidimensionally ordered set; i.e., "subtree level" and "coordinates" become synonymous. Let the nodes in the ith coordinate of a tree T be denoted $T[n_1;\ldots;n_i]$, and let its root be denoted $T[\ ]$. The degree of a node $T[n_1;\ldots;n_i]$ will be called the *dimension* $d[n_1;\ldots;n_i]$ of the node. In contrast to rectangular arrays, the dimensions $d[n_1;\ldots;n_i]$ of the nodes in the ith coordinate of a tree are, in general, not identical. Thus, a dimension $d[n_1;\ldots;n_i]$ must be associated with each node $T[n_1;\ldots;n_i]$ in a tree. By definition, the admissible range of the index value n_i of the subtrees with roots in the ith coordinate is bounded by the dimension of the root $T[n_1;\ldots;n_{i-1};n_i]$, i.e., $n_i \in [1:d[n_1;\ldots;n_{i-1}]]$. Hence, a generalization of the structuring function of sequentially allocated arrays for general tree structures requires the storage of a tree structure of dimensions $d[n_1;\ldots;n_i]$. Therefore, purely sequential allocation schemes are only used when the respective trees exhibit strong regularities.

The most regular tree structure is the complete *binary tree*. All nodes $T[n_1;\ldots;n_i]$, $1 \leqslant i \leqslant r-1$, and the root $T[\ ]$ of a complete binary tree with r subtree levels have exactly two subtrees. Such a binary tree with r coordinates contains $2^{r+1}-1$ nodes; the ith coordinate contains 2^i nodes. Several sequential allocation schemes can be defined for binary trees, which lend themselves to specific traversal algorithms [84].

A general sequential allocation scheme for binary trees which corresponds to the structuring function for r-dimensional arrays is defined by the following structuring function:

$$\sigma(n_1,\ldots,n_i) = 2^i + \sum_{j=1}^{i} n_j \cdot 2^{i-j}.$$

$\sigma(n_1,\ldots,n_i)$ defines the index of the node $T[n_1;\ldots;n_i]$ in the data vector V. By definition, the index of the root node $T[\ ]$ is 1, i.e., $V[1] = T[\ ]$. The index lists $[n_1;\ldots;n_i]$, $n_i \in [0:1]$, $1 \leqslant i \leqslant r$, may be defined such that $T[n_1;\ldots;n_i;0]$ and $T[n_1;\ldots;n_i;1]$ denote the roots of the left and right subtree of the root $T[n_1;\ldots;n_i]$, respectively. With this convention, an index list $[n_1;\ldots;n_i]$ corresponds to the binary representation of the index identifying the relative position of the node $T[n_1;\ldots;n_i]$ in the ith coordinate of the binary tree T. The addressing function above of the sequential allocation scheme for binary trees equals that of rectangular arrays. Since all dimensions $d[n_1;\ldots;n_i]$ are identical, only the base address β and the parameter m, which specifies the number of memory words occupied by each record, must be stored besides the data vector V.

However, even in the case of regularly shaped trees, sequential allocation is not appropriate if the size or shape of the tree changes dynamically during program execution. The disadvantage of a linked allocation is, on the other hand, that each node of the tree may have a different number of children and, hence, a number of pointers varying from record to record. In order to obtain a general linked allocation scheme for trees with a standardized number of link fields, tree structures often are represented in terms of an equivalent binary tree (see Section 2.1.4). In the simplest linked allocation of a binary tree, we have two pointers within each node, pointing to the left and right subtree of that node. Let two such pointers, pointing to the roots of the subtrees of node $T[n_1;\ldots;n_i]$, be denoted by LEFT($T[n_1;\ldots;n_i]$) and RIGHT($T[n_1;\ldots;n_i]$), respectively. If a subtree is empty, the value of the corresponding pointer is Λ. Hence, the addressing function of a linked binary tree is defined as

0-origin:

$$\alpha(n_1,\ldots,n_{i-1},0)=\text{LEFT}(T[n_1;\ldots;n_{i-1}]),$$

$$\alpha(n_1,\ldots,n_{i-1},1)=\text{RIGHT}(T[n_1;\ldots;n_{i-1}]).$$

The value α_T of a pointer variable TOP(T) is the address of the root node $T[\ ]$. Obviously, this linked representation of a binary tree defines a parent–child relationship between subtrees $T[n_1;\ldots;n_i]$ and the mutually disjoint subtrees $T[n_1;\ldots;n_i;0]$ and $T[n_1;\ldots;n_i;1]$. In order to provide the possibility of traversing a binary tree such that each node is visited exactly once, element–successor relationships (preorder, inorder, postorder, etc.) may be defined [84]. This is not the place to discuss the complex problems of tree traversals and tree searches. An example for the linked allocation representation of a simple binary tree is depicted in Figure 2.3.

The given allocation scheme is wasteful of memory space since the number of null links may outnumber the pointers [84]. Therefore, the null links Λ may be replaced by the following explicitly tagged *threads*:

$$\text{LEFT}(T[n_1;\ldots;n_{i-1};0])=\alpha(n_1,\ldots,n_{i-2}); \qquad \text{LEFT}(T[n_1;\ldots;n_{i-1};1])=\alpha(n_1,\ldots,n_{i-1});$$

$$\text{RIGHT}(T[n_1;\ldots;n_{i-1};0])=\alpha(n_1,\ldots,n_{i-1}); \qquad \text{RIGHT}(T[n_1;\ldots;n_{i-1};1])=\alpha(n_1,\ldots,n_{i-2}).$$

From this, a *threaded list* is obtained by two additional modifications:

1. TOP(T) is converted into a list head (stored in memory location β) with the pointers LEFT(HEAD)$=\alpha_T$ and RIGHT(HEAD)$=\beta$.

2. Two special threads are introduced for terminal nodes, LEFT($T[0;\ldots;0]$) $=\beta$ and RIGHT($T[1;\ldots;1]$)$=\beta$.

With respect to traversal, a threaded tree representation is superior to an unthreaded one requiring the same memory space. The threaded list representation of the tree of Figure 2.3 is depicted in Figure 2.4.

Other representations of binary trees may be obtained by using a combination of the linked mechanism discussed for linear lists, as provided through the

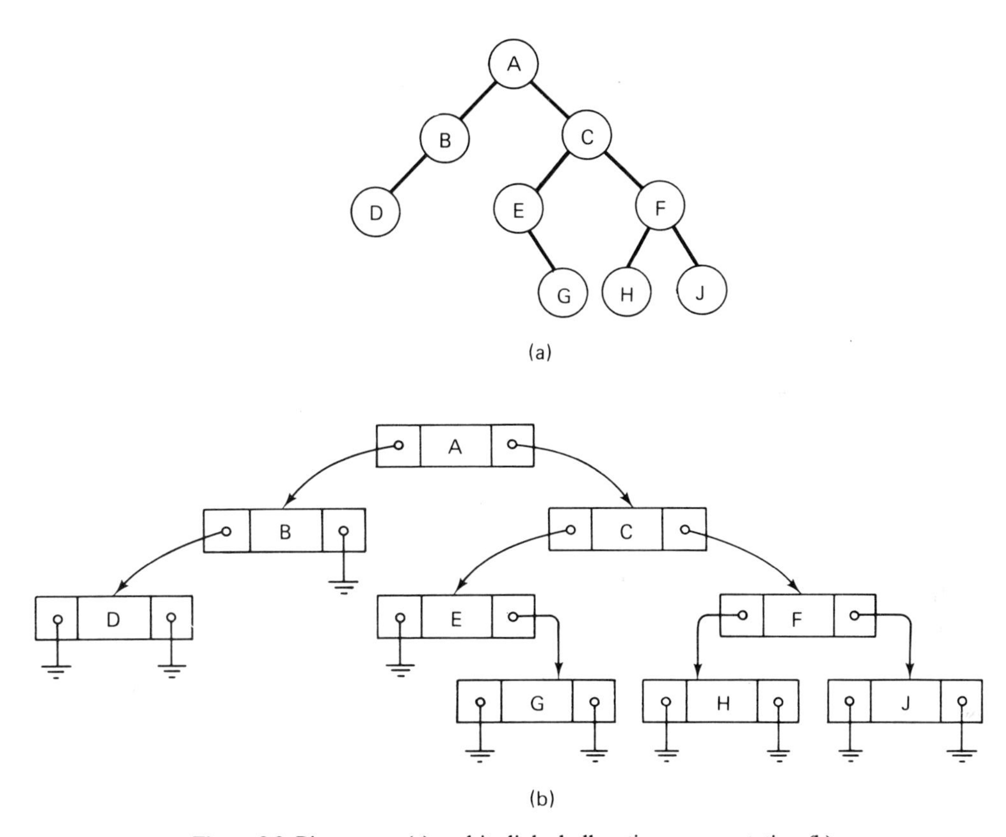

Figure 2.3 Binary tree (a) and its linked allocation representation (b).

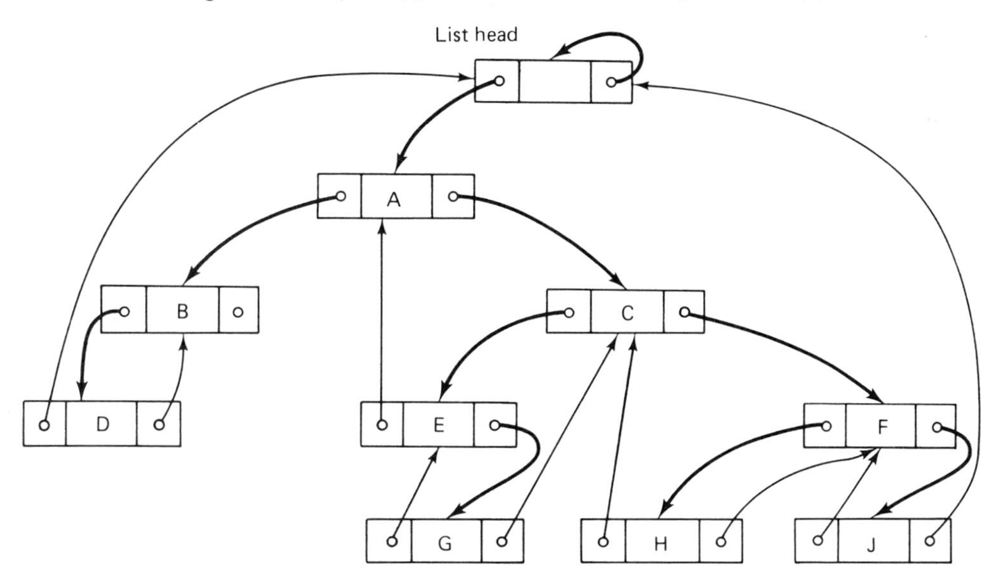

Figure 2.4 Modification of the linked list [Fig. 2.3(b)] into a threaded list.

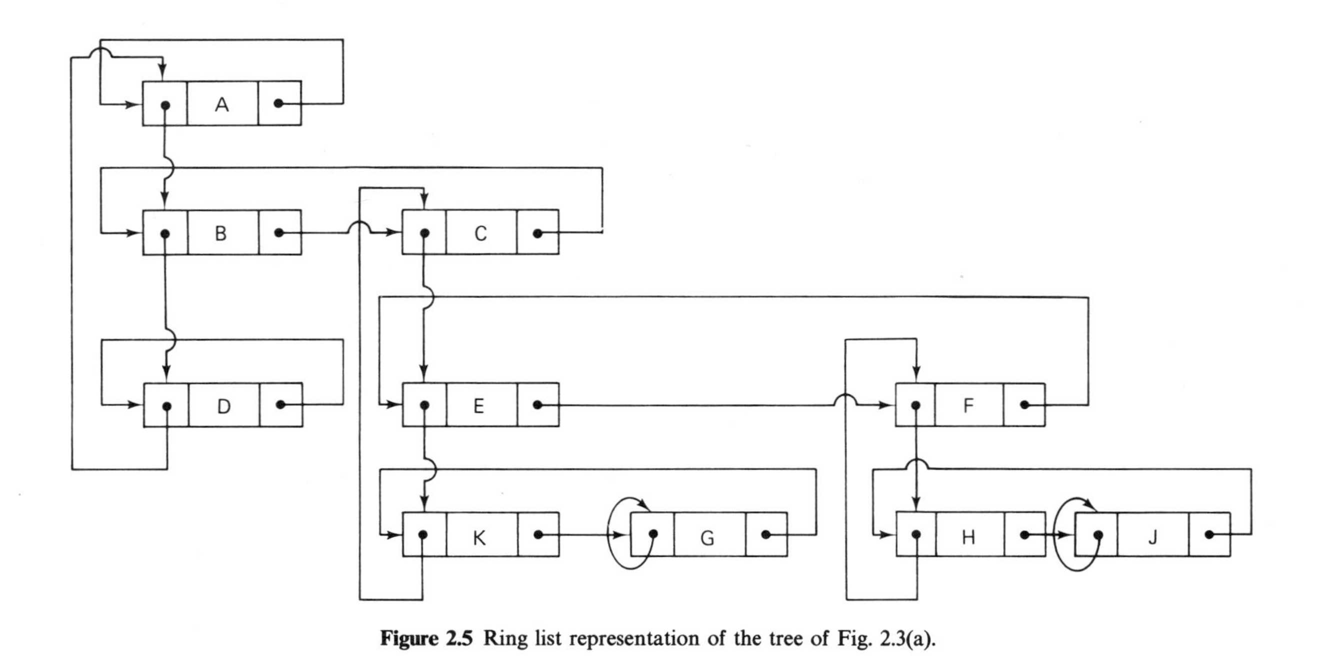

Figure 2.5 Ring list representation of the tree of Fig. 2.3(a).

pointers LEFT and RIGHT. A widely used binary tree representation is the *ring list*, an orthogonally connected linear list, which may also be applied to the representation of general trees and arrays. A ring list representation of a binary tree may be defined, for example, by the following simple addressing functions (for 0-origin):

1. $\text{RIGHT}(T[n_1;\ldots;n_i])=\alpha(n_1,\ldots,n_i+1)$ for $n_i+1<d[n_1;\ldots;n_{i-1}]$.

2. $\text{RIGHT}(T[n_1;\ldots;n_i])=\alpha(n_1,\ldots,n_{i-1},0)$ for $n_i+1=d[n_1;\ldots;n_{i-1}]$.

3. $\text{LEFT}(T[n_1;\ldots;n_i])=\alpha(n_1,\ldots,n_i,0)$ for $d[n_1;\ldots;n_i]>0$.

4. $\text{LEFT}(T[n_1;\ldots;n_j;\ldots;n_i])=\alpha(n_1,\ldots,n_j)$ for $d[n_l;\ldots;n_i]=0$, with j defined by the condition $\bigvee_{T[n_1;\ldots;n_j-1]}: \text{RIGHT}(T[n_1;\ldots;n_j-1])=\alpha(n_1,\ldots,n_j)$.

Obviously, all nodes $T[n_1;\ldots;n_{i-1};0]$ and all nodes $T[n_1;\ldots;n_j]$ of case (4) must be identifiable as head nodes. Therefore, RIGHT pointers are used to connect the roots of the subtrees with other roots in each coordinate by horizontal rings. All leftmost descendants of the roots of subtrees are connected into a vertical ring through the LEFT pointers. The root node $T[\ ]$ (stored in α_T) is defined $\text{LEFT}(T[\ ])=\alpha(0)$, $\text{RIGHT}(T[\ ])=\alpha_T$, and $\text{LEFT}(T[0;\ldots;0])=\alpha_T$, where $T[0;\ldots;0]$ is the leftmost terminal node. With respect to traversal, ring lists are superior to unthreaded lists; however, they are less efficient for random access.

As an example, Figure 2.5 depicts a ring list representation for the binary tree of Figure 2.3(a). In this tree, node E has no left child. In order to remain consistent with the simple construction rule set above, a "dummy" node K must be inserted to provide access to node G. This is the price for the simplicity of construction. For construction rules that avoid dummy nodes, see, for example, Knuth [84].

2.2.4 Representation of Generalized Lists (Hierarchical Structures)

Based on the similarity of the implicit hierarchical relationships of Lists and trees, the nodes of a List L may be denoted by using the index notation for tree structures. Here, the coordinates are the levels of List references. The admissible range of the index value n_i are determined by the List dimensions $d[n_1;\ldots;n_{i-1}]$ of the nodes $L[n_1;\ldots;n_{i-1}]$, i.e., $n_i\in[1:d[n_1;\ldots;n_{i-1}]]$. Memory representations of Lists are usually variations of the schemes applied for the representation of binary trees. However, the following idiosyncrasies of List structures must be taken into account.

1. Since Lists are generalizations of linear lists, the *element–successor relationship* or the *predecessor–element relationship* must be specified by respective pointers $\text{SUC}(L[n_1;\ldots;n_i])$ and $\text{PRE}(L[n_1;\ldots;n_i])$ in each node.

2. Since the nodes $L[n_1;\ldots;n_i]$ either represent a data record or represent another List, we must explicitly distinguish between *atomic* and *nonatomic* nodes, respectively.

3. Since Lists may be empty, each List represented by a node $L[n_1;\ldots;n_i]$ should contain a distinguishable *List head* HEAD$[n_1;\ldots;n_i]$ which can be referenced by nonatomic nodes.

4. The *child–parent relationship* must be specified by the respective pointers CHILD and/or PARENT in all nonatomic nodes and List heads.

5. In contrast to tree structures, the possible *overlapping* of Lists represented by nodes $L[n_1;\ldots;n_i]$ and $L[m_1;\ldots;m_j]$ may prohibit the definition of bijective structuring functions. Nonempty intersections of Lists are treated as single entities in the representation. That is, they are stored only once, and in a List L with r coordinates, the pointers CHILD$(L[n_1;\ldots;n_i])=$CHILD$(L[m_1;\ldots;m_j])$ may both point to HEAD$[k_1;\ldots;k_h]$, $1 \leqslant h \leqslant r$. However, the entity character of the two Lists represented by the nodes $L[n_1;\ldots;n_i]$ and $L[m_1;\ldots;m_j]$ must be preserved since they might be separated by manipulations. In general, we encounter the problem that several types of relationships may be present within a single List. Therefore, nodes are required which have an arbitrary number of link fields. Such List representations are said to be *multilinked.* Not all pointers in the records of multilinked structures are essential, but they are helpful with respect to the efficiency of manipulations.

Any combination of the three basic techniques for representing linked linear lists (straight, circular, and double linkage) can be used throughout the different link fields. Hence, the definition of the appropriate addressing functions and the common operations on Lists can be derived from the preceding discussion of linear lists and tree structures. Since it is necessary to distinguish between List heads, nonatomic, and atomic nodes, each node contains a *specification field.* In addition to the specification of the node type, List heads may contain a List identifier and information which aids garbage collection and traversal algorithms. The insertion of a List identifier into nonatomic nodes corresponds to the representation of information in nonterminal nodes of trees. Variable-size data blocks in atomic nodes may be handled by specifying the block length and the data type or by separating entirely the pointer list specifying the structure and the associated data list. Because of the universality of Lists, a large number of List processing systems have been designed. These systems are based on the concept of List representation given above. However, they often are less efficient than less general systems written for a specific application (only efficiently usable in a small range of applications).

In computer graphics, generalized Lists implemented in the form of (hierarchical) ring lists have proven to be quite popular. As a simple example, consider Fig. 2.6. Here, a picture is given consisting of two items, S and T, which are *connected.* That is, the set of the primitives of S and the set of the primitives of

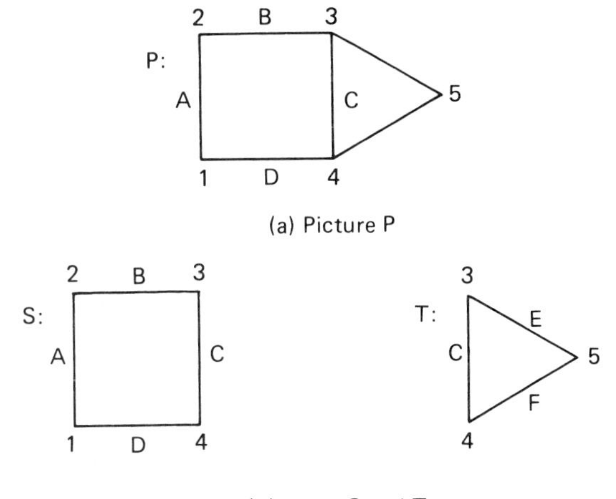

(a) Picture P

(b) Items S and T

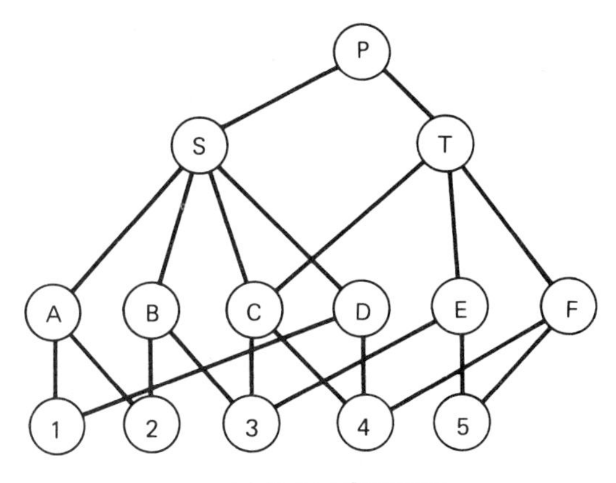

(c) Picture Structure

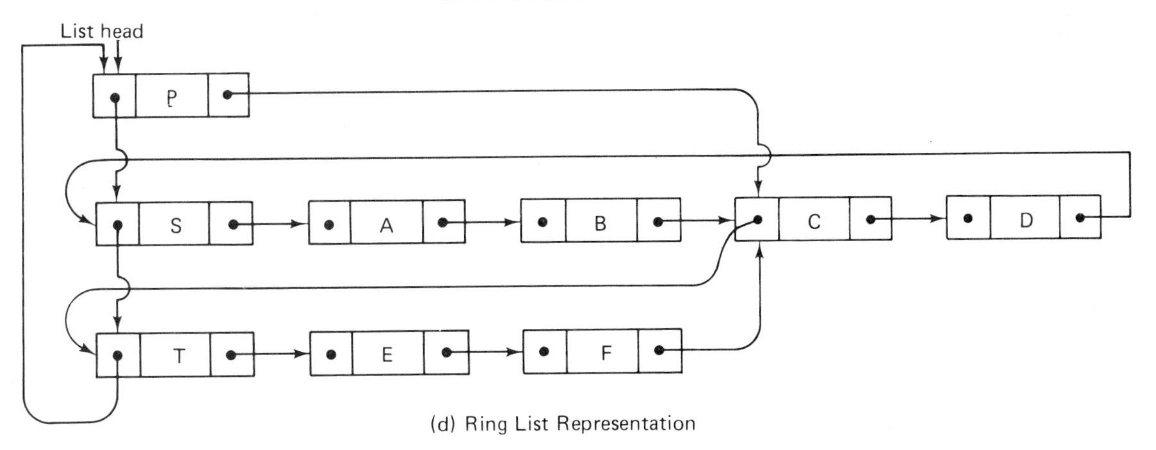

(d) Ring List Representation

Figure 2.6 A picture, consisting of two connected items, and a possible ring list representation using two-pointer nodes.

T have a nonempty intersection (in the example, the intersection consists in line segment C, Fig. 2.6). Because of these interconnections, the resulting graph is not a tree (if we cut the root of the treelike structure, we do not obtain forests). However, the structure certainly is hierarchical.

For its display, the picture P will be treated as an entity, as we do not want to generate line C twice. In the underlying data structure, however, the entity character of the two connected items shall be preserved, as we may want to manipulate them individually (e.g., separate them). A possible ring list representation of P is diagrammed as part of Figure 2.6. Here, use is made of the two-pointer nodes, with the proviso that RIGHT points to the adjacent primitive, whereas LEFT is used to establish the hierarchy. The (unused) right-hand pointer in the head node P may, for instance, be used to indicate which line actually is shared by both items.

2.2.5 Representation of Associative Structures

Unlike other data structures, associative structures exhibit no ordering of the data records based on logical relations. Of course, there exists an inherent ordering in associative structures as well, but this ordering is usually given by the order in which records have been entered into the structure and thus is quite arbitrary. Associative structures represent collections of subsets of a set of data records $S = A \cup B$ and a set of propositions π (see Section 2.1.6). Hence, the set–element relationships for the set S is specified as a *proposition–element* relationship. In other words, as there exist no relations forming orderings or classes through which elements can be identified, the elements must identify themselves through their value or a certain part of it called the *key*. Therefore, an associative data structure representation requires a content-addressed memory organization; i.e., the addressing function is here a mapping

$$\alpha: K \rightarrow M,$$

where K is the set of keys and M is the set of memory addresses.

According to the general definition of an associative structure by a "triad" (A, π, B) (see Section 2.1.6), any data object, when entered into the structure, must be augmented to the complete association as given by a triad $(\{a\}, \{p\}, \{b\})$, $a \in A$, $p \in \pi$, $b \in B$. In a subsequent retrieval, any combination of two components of this triad can be used as a key. Therefore, the triads specifying data records must be represented such that all primitive associative retrieval requests (queries Q1–Q6 in Section 2.1.6) can be honored. In general, we may have the situation $A \cap B \neq \emptyset$, $\mathfrak{D}_i \cap \mathfrak{D}_j \neq \emptyset$, $\mathfrak{R}_i \cap \mathfrak{R}_j \neq \emptyset$, and hence a sensible memory representation will use separate lists for the structure (i.e., pointer lists) and the data. The components of a triad $(\{a\}, \{p\}, \{b\})$ are internally represented by identifiers for the data objects and the propositions.

The associative addressing function $\alpha: K \rightarrow M$ can be established by a table. This approach is called the *dictionary look-up method*. The dictionary entries are pairs (key, value), where the second component is usually a pointer to the separately stored set of data records. In the early List-processing languages (LISP, IPL-V) [66], this method is used to provide for the ability of answering the associative query Q1, $(\{a\},\{p_i\},B)$. For this purpose, a dictionary entry $(\text{key},\text{value})=(a,\mathfrak{R}_i/a=\{b\in\mathfrak{R}_i|p_i(a,b)\})$ is associated with each proposition $p_i\in\pi$. This method exhibits the disadvantage that for each access to a set $\mathfrak{R}_i/a$ (which is empty if $a\notin\mathfrak{D}_i$) the set of keys $\mathfrak{D}_i$ must be searched for the specified domain element $a\in A$. A second disadvantage of this method is the *one-way* link between keys and values; i.e., separate dictionaries are needed in order to answer query Q2, $(A,\{p_i\},\{b\})$, and query Q3 $(\{a\},\pi,\{b\})$. These disadvantages have been mitigated in systems such as CORAL [136], ASP [65], and APL [27], which provide *two-way* links between the keys and the associated sets of data objects, as well as relative addressing of the dictionary entries. However, these systems are still burdened with the inefficiency of dictionary look-up.

As an example, we consider CORAL. The underlying notion of CORAL is to facilitate the associative retrieval of data by providing bidirectional links between the keys and the classes of data identified by the key without the expense of a doubly linked hierarchical list. This is accomplished by representing data entities by unidirectionally linked rings, whereas these rings are linked bidirectionally. To this end, two "building blocks" are defined, the forward ring and the backward ring, as diagrammed in Figure 2.7.

The CORAL structure is formed by consecutively chaining forward rings and backward rings. A ring head contains a pointer, a ring identifier, and certain other

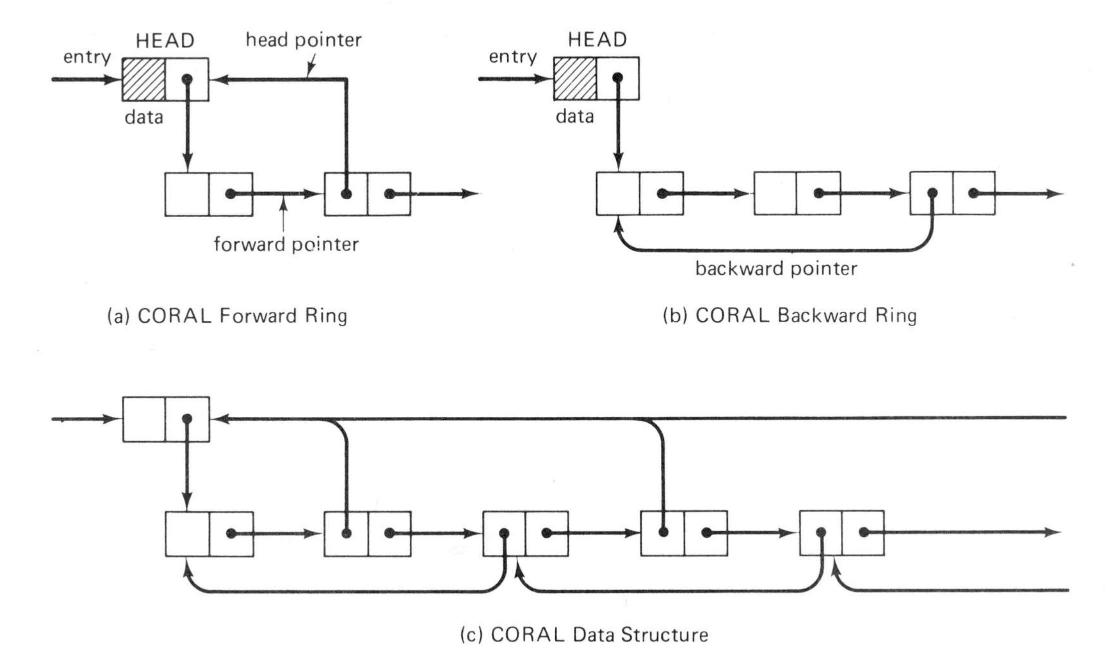

Figure 2.7 Illustration of CORAL, an associative structure with dictionary look-up.

information (e.g., block type and length). Ring elements may be connected with an arbitrary number of data cells. An example of a CORAL structure, representing the picture of Figure 2.6, is depicted in Figure 2.8.

A faster associative addressing scheme, which avoids the storage of a dictionary, is provided by the *hash coding* technique. Hash coding is the application of a numerical operation which converts a key $k \in K$ of a data record into its associated memory address $\mu \in M$, i.e., it is a function $h: K \rightarrow M$. Contrasting to the unique association of addresses and keys in the dictionary method, the injectivity of a hash function cannot be ensured. There is always a chance of *collisions* $h(k_i) = h(k_j)$ (two different keys lead to the same address). The hash function for query Q1 is defined $h_1: A \times \pi \rightarrow M$; i.e., the sets $\mathfrak{R}_i/a$ are accessed by the addresses $h_1(a,p_i)$. The possible collision cases for h_1 are: $h_1(a_j,p_i) = h_1(a_m,p_i)$ or $h_1(a_j,p_i) = h_1(a_j,p_k)$. Thus, the hash function h_1 may define equivalence classes of the sets $\mathfrak{R}_i/a$, called *collision classes*. As the elements of the collision classes cannot be distinguished by the hash function, they must be addressed via dictionaries (*collision lists*). Such a collision list is accessed by the common addresses $h_1(a,p_i)$. However, the number and (most important for the speed of the search process) the cardinality of the collision classes are considerably smaller than those of the dictionaries in the dictionary method. The LEAP system [37] probably was the first associative programming system based on the use of hash coding.

Both associative addressing schemes suggest the memory representation of the triads $(\{a\},\{p_i\},\{b\})$ in the form of triads $(\{a\},\{p_i\},\mathfrak{R}_i/a)$. Since inside a computer even a general set must be ordered in some way, the elements $b \in \mathfrak{R}_i/a$ may be stored in a linear list in the order in which the triads $(\{a\},\{p_i\},\{b\})$ are created. The answer to query Q6, $(A,\{p_i\},B)$, is then given as the union

$$R_i = \bigcup_{a \in \mathfrak{D}_i} R_i/a = \bigcup_{a \in \mathfrak{D}_i} \{(a,b) | p_i(a,b) \wedge b \in \mathfrak{R}_i/a\}.$$

Instead of storing pairs (a,b), the appropriate triads $(\{a\},\{p_i\},\{b\})$ are conveniently obtained by chaining the sets $\mathfrak{R}_i/a$, for all $a \in \mathfrak{D}_i$, thus associating with each $b \in \mathfrak{R}_i/a$ the key (a,p_i) of $\mathfrak{R}_i/a$. For an equally efficient answering of the queries Q2, $(A,\{p_i\},\{b\})$, and Q3, $(\{a\},\pi,\{b\})$, analogous list structures must be built. When the dictionary method is used, entries

$$(\text{key}, \text{value}) = \left(p_i, \mathfrak{D}_i/b = \{a \in \mathfrak{D}_i | p_i(a,b)\}\right) \text{ associated with } b \in B,$$

$$(\text{key}, \text{value}) = \left(b, \pi/a, b = \{p_i \in \pi | p_i(a,b)\}\right) \text{ associated with } a \in A$$

are needed in order to answer queries Q2 and Q3, respectively. Similarly, hash functions $h_2: \pi \times B \rightarrow M$ for Q2, and $h_3: A \times B \rightarrow M$ for Q3 may be defined when the associative addressing is based on hash coding. Chaining all propositions p_i with $b \in \mathfrak{R}_i$ provides the answer

$$\pi/b = \{p_i \in \pi | p_i(a,b) \wedge b \in \mathfrak{R}_i\}$$

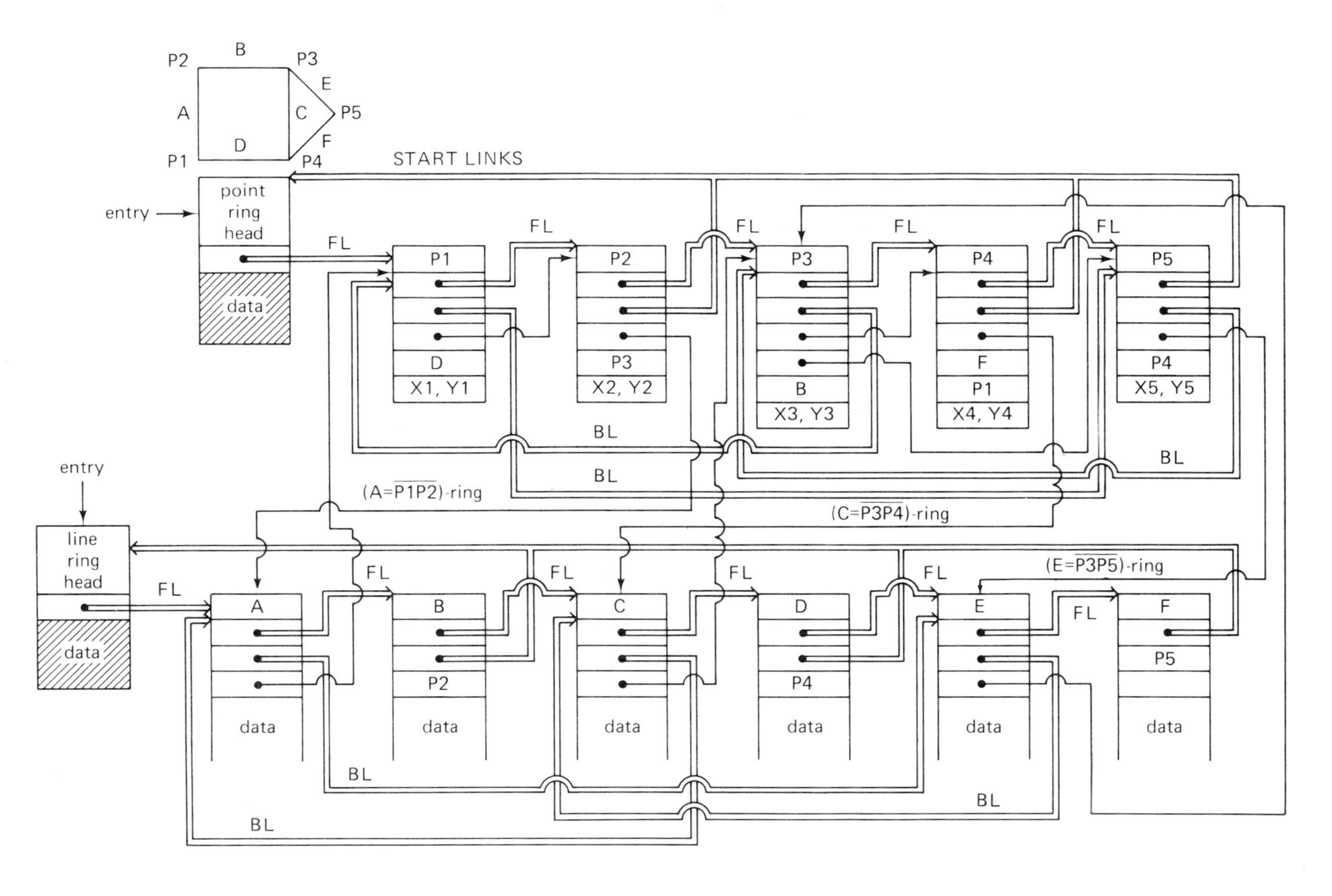

Figure 2.8 Representation of Fig. 2.8 in the CORAL data structure FL = forward link; BL = backward link. The upper ring encompasses all points and the lower ring encompasses all lines (line rings are drawn only for A, C, E, not for B, D, F).

to query Q5, with $A/b=\bigcup_{p_i\in\pi/b}\mathfrak{D}_i/b$. Chaining the sets $\pi/a,b$ for all $b\in\mathfrak{R}_i/a$, $p_i\in\pi/a,b$, provides the answer

$$\pi/a=\{p_i\in\pi|p_i(a,b)\wedge a\in\mathfrak{D}_i\}$$

to query Q4, with

$$B/a=\bigcup_{p_i\in\pi/a}\mathfrak{R}_i/a.$$

A list head may be included in each linear list representing a set of either type, $\mathfrak{R}_i/a$, $\mathfrak{D}_i/b$, or $\pi/a,b$. In the dictionary method, the dictionary entries automatically constitute the list heads. When hash coding is used, storing the key and possibly a collision list pointer in these list heads allows for the incorporation of the collision lists into the inherent data structure. The sets R_i, π/b, and π/a may be formed by appropriately chaining the list heads of the sets $\mathfrak{R}_i/a$, $\mathfrak{D}_i/b$, and $\pi/a,b$, respectively. In order to answer queries Q4, Q5, and Q6 in a direct way, these chained lists should also be accessible through list heads. Since a data record may belong to different sets, each of the separately stored data records may be linked to several list heads. Any combination of the techniques for representing linear lists can be used throughout the various link fields in the records of an associative data structure.

Each triad $(\{a\},\{p_i\},\{b\})$ is represented in each of the three list structures for answering queries Q1, Q2, and Q3. Thus, the associative query Q0, which tests the existence of a triad $(\{a\},\{p_i\},\{b\})$ in the structure, can be answered by a search in any of the sets $\mathfrak{R}_i/a$, $\mathfrak{D}_i/b$, or $\pi/a,b$. This search and an update in all three associated sets must also be performed for the creation of a triad, since an associative data structure, specified by the triad (A,π,B) as a general set of triads $(\{a\},\{p_i\},\{b\})$, should not contain multiple occurrences of any triads.

The implementation of an associative structure that is capable of answering queries Q1 and Q6 shall be demonstrated by a simple example. Let us assume that the following relations of the type "SON OF" be given

JOHN is SON OF TOM
BILL is SON OF TOM
JACK is SON OF TOM
PETE is SON OF TIM
TIM is SON OF SAM
TOM is SON OF DON.

In the first three cases, the identifier TOM is hashed (representing the element a_j), resulting in an index to the list in the SON OF ring where the subset $\mathfrak{R}_i/a_j=\{\text{JOHN, BILL, JACK}\}$ is stored. Likewise, in the other three cases, TIM, SAM, and DON, respectively, is hashed. Here, the associated subsets contain in each case only one element (PETE, TIM, and TOM, respectively). A collision is handled by linking the node representing the colliding case to the node with which this case collides, thus forming a linked list as representative of a collision class. Therefore, we set up for each triad $(\{a_j\}, \{p_i\}, \{b_k\})$ a node with four entries:

Identifier for a_j	Collision list head	Head of list for $\{b \in B \mid p_i(a_j, b)\}$	Proposition ring pointer

The proposition ring pointer points to another node with the same identifier p_i, thus linking all nodes with a given proposition identifier into a ring. Access to the head of that ring list may, in return, be provided by hashing the identifier of p_i. Hence, instead of a hash function $h_l: A \times \pi \rightarrow M$, two hashing functions, $h_1': A \rightarrow M$ and $h_1'': \pi \rightarrow M$, are executed. Such an "orthogonalization" simplifies the handling of collisions. The resulting structure is depicted in Figure 2.9.

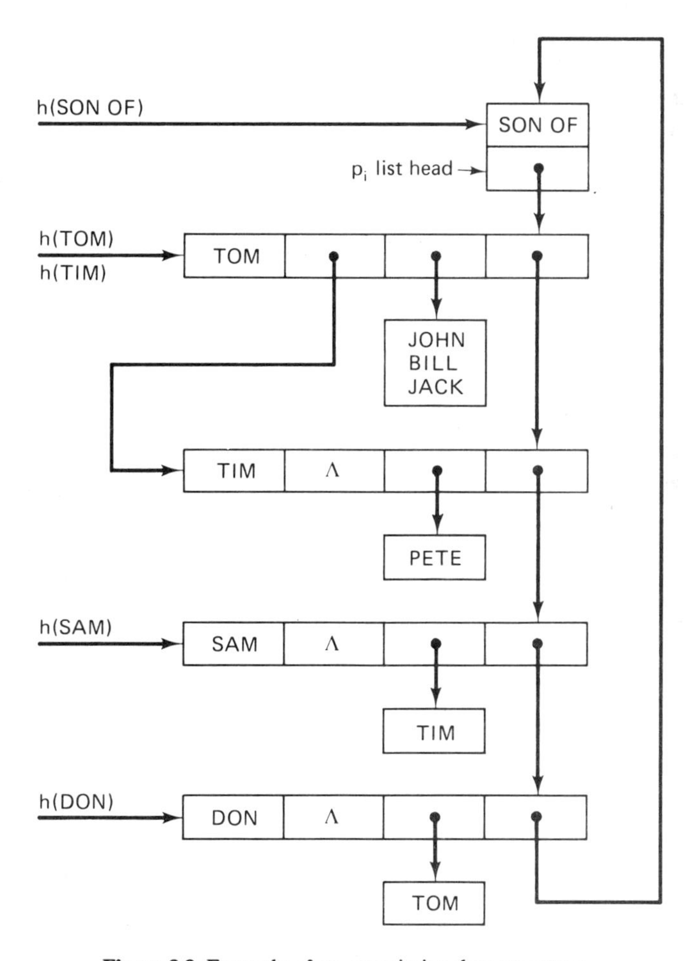

Figure 2.9 Example of an associative data structure.

2.3 DATA MODELS AND DATA BASE ORGANIZATION

2.3.1 Objectives of Data Base Management Systems

So far, we discussed logical structures, resulting from the definition of relations among the elements of data sets, and their conversion to physical structures of memory representations as defined by addressing functions. However, it would be too tedious and time-consuming for the application programmer [50] if he had to program a particular data base management system every time the application program requires a data base. Therefore, generalized data base management systems (DBMS) are often supplied as part of the "utility software" as, for example, various compilers may be supplied. Such systems are independent "operators" of the operating system; that is, data are not "owned" by a program but stored and accessed—and made available to several programs—by the generalized DBMS. This *data integration* is accomplished by defining a data format, storing it as a data definition, and allowing general-purpose DBMS software to access it. Generally, the data definition consists of a statement of the names and properties (e.g., data types) of data elements and their relationships to other elements in the data base. Data are allowed to be shared by a community of users, raising the problem of ensuring data *integrity* over time and providing *security* against unauthorized access. Since a user is only concerned with the logical structure as described in the data definition, a *centralized data administration* may select different physical structures for different parts of the data base, thereby optimizing storage use and retrieval efficiency [127]. To these objectives, *data independence* may be added [45, 9].

Generally, data independence of a DBMS is the independence of program requests to the DBMS from physical or logical data transformations. The need for *physical data independence* stems from the fact that there is no "best" physical data organization. Hence, in order to adapt itself to actual requirements, the system must be enabled to change the physical structure and the access modes of a data base without affecting program requests. *Logical data independence* is mandatory for a system that allows for the addition of new relations to the logical structure of a data base. Logical data independence is considered to be achieved if logical changes in a data base will not significantly affect the programs that access it. The capability of a DBMS to support various logical structures may be interpreted as a first step toward logical data independence. In general, data independence is not achievable without imposing severe restrictions on the classes of "legitimate data" and "legitimate relations" specified in the data definitions.

Definitions of legitimate types of data and relations are called *conceptual data models*. An early attempt in the development of conceptual data models based on set theory is the "information algebra" of CODASYL [16]. The necessity for a standardization of data definitions on the basis of conceptual data models is also discussed [96]. Such a standardized data definition for a specific data base is called a *conceptual schema*; and the storage of the conceptual schema by the DBMS provides the basis for sharing a data base among a community of users with potentially different conceptual views. The DBMS accesses the memory representation of the shared data base through standardized addressing functions corresponding to different conceptual views. Thus, the conceptual view of each user is only represented by the data definition. In order to be generally applicable, a conceptual data model must provide efficient representations of the two constituent components of data structures: *general sets* and *relations* on the objects of these sets.

2.3.2 Conceptual Data Models

One of the first representation-independent, conceptual data models is Child's *Set Theoretical Data Structure* (STDS) [14, 15]. In this model a "reconstituted definition of relations" allows the uniform application of general set operations to general sets as well as to ordered sets. According to the stated objectives, the storage representation of the STDS consists of a data definition (logical structure) and the actual data storage (physical structure). The data definition essentially consists of a collection of set names which are linked to set representations. *Generator sets* have a storage representation in the form of data names, whereas *composite sets* (unions of generator sets) are represented by a set of generator set names. A data name locates a data (single item of a collection of items) in the data storage. Thus, a "physical" set may belong to several "logical" sets and the data storage may be viewed under different data definitions. Since no duplication of storage is necessary and no pointers exist between sets, an STDS is a minimal storage representation for arbitrarily related data. Since there is no linkage between sets, the set operations act as the only structural ties between sets. This unique feature of the STDS is usually replaced by an explicit statement of the relations in the major conceptual data models.

An important class of data models encompasses the *graph-oriented data models*. These models are based on the notion of introducing named *binary relations* between the objects in the data base. Early implementations of standardized binary association models are the LEAP system [37] and the TRAM system [6]. However, the major graph-oriented conceptual data model is the *network model* of the CODASYL Data Base Task Group proposal [17].

Since the relationships between the data items in a record are assumed to be defined by the record type, information can be represented in a data base by the contents of records and relationships between record types. The only binary relationships offered in the network model are binary associations which establish a logical containment among records of heterogeneous types. That is, for each association, there is a distinction between the *owner* record type and the *member* record types. The binary associations must not violate the *rules of unique ownership*; i.e., (1) they can only represent *one-to-many relationships* between one owner record and the member records; (2) no member record can be shared by different owner records for the same association; (3) no record type may participate as both owner and member in the same binary association.

Transformation of complex relationships into the restricted one-to-many relationships of the network models are discussed in [138]. Generally, a record type may serve as a member in one or more binary associations and as owner in one or more other binary associations. The possibility of having the same member record type participating in two different associations with the same owner record type requires an explicit naming of the associations. This naming mechanism also allows the definition of recursive structures through the use of intermediate record types.

The graphical representation of binary associations as named, directed arcs between record types (nodes) results in a network, called a *data structure diagram*. This explicit notation of binary associations as connections between record types is usually preserved in memory implementations of this model. Dictionaries lead to owner record types which link different levels of contained member record types through chains of pointers. However, this chaining mechanism restricts the *physical data independence*, since the user may need to be aware of physical record placement strategies. A measure of *logical data independence* is provided through a schema–subschema mechanism which allows a program to interact only with a subschema describing the relevant part of the data base. Further standardization of binary association models which aid in improving data independence have been developed in the Data Independent Accessing Model (DIAM) [125].

A third model, the *relational data model*, uses named n-ary relations (of assorted degree) as a tool for a generalized data base management. The first rigorous definition of n-ary relations in the data base context dates to a paper by Codd [8], which emphasized their advantage with respect to data independence and symmetry of access. The relational model does not explicitly distinguish between the contents of records and relationships between records. All information is represented by data values in the components of n-tuples which constitute n-ary relations. An n-ary relation is a finite subset $R \subseteq \mathfrak{D}_1 \times \cdots \times \mathfrak{D}_n$ of the Cartesian product of the domains $\mathfrak{D}_k$; and it is viewed as a named set of n-tuples. According to the intuitive notion of information, some of the domains may have relations as elements. However, the relational model restricts the representation of n-ary relations to the *first normal form*. A relation is in first normal form if each component of each n-tuple is nondecomposable, i.e., if the component is not a list or a relation. Thus, normalization results in standardized relations which can be represented as a matrix whose rows are n-tuples [18]. An informal survey of objectives for higher normal forms which aim to provide more favorable update properties is given in [13].

Since relations are only made up of items, they have no containment; i.e., there exist no links between relations. Links are established through the use of the data items as keys. Domains (or domain combinations) whose values uniquely identify each n-tuple of a given relation are called *primary keys*. Thus, many-to-many relationships can be represented by explicit relations between primary keys. A one-to-many relationship may be implicitly established by storing an owner's primary key in one of the members' domains. Cross references of n-tuples are obtained by the use of *foreign keys*. A domain (or domain combination) of a relation R is called a foreign key if it is not the primary key of the R, but its elements are values of the primary key of some relation S (the possibility $R = S$ not being excluded). In order not to burden the user with their ordering, domains are uniquely identified by distinct *role names*. The qualification of identical domains by distinct role names allows for the definitions of recursive structures.

In contrast to the network model, the relational model does not include predefined access paths, and hence there are no preferred request formats. However, this *symmetry of the data model* does not necessarily imply symmetry of the underlying physical data structure in which all accesses are provided by dictionaries. Thus, the creation of dictionaries for specific domains allows an optimization of the physical structure for actual requirements without affecting the user interface. Since the logical description of a relational data base model includes neither the physical structure nor the access paths, *physical data independence* is naturally guaranteed. *Logical data independence* is supported through a schema–subschema mechanism. Generally, the relational model is immune to modifications of the logical structure, as long as the normalization requirement does not affect existing domains.

The well-developed mathematical basis of the relational model allows the definition of relational completeness and the rigorous study of optimal data base design. However, the relational model has, for example, no equivalent form to the frequently used hierarchical record organization which violates the first normal form condition. In addition to that, the necessary existence of a primary key in every relation sometimes requires implementations of the relational model which violate the theoretical foundation [9]. Trade-offs between the relational model and the network model are discussed in [97].

2.4 LIST HANDLING

2.4.1 Implementation of Linear Lists

The most elementary data organization is the sequential file.[†] An entry in a file will be called a *record*. One data item in a record may play a distinguished role as the element selected to identify a record. This element is called a *key*. Records may be of constant or of variable length. If each entry in a file consists of only one elementary data item, the simplest way of setting up a sequential file obviously is to declare it as a one-dimensional array. If the entries are records of constant length,

[†]In the USASI Standard COBOL definitions, the term *list* is reserved for files that are organized by pointers, whereas the sequentially allocated storage of information items is called a *sequential file*.

the file may be declared as a two-dimensional array, either in the form FILE(M,N) or in the form FILE(N, M), if M is the number of records and N is the number of elements in each record. There is a substantial difference between the two forms in view of the fact that the elements of a two-dimensional array are internally stored as a one-dimensional array. If we assume that the internal storage happens in row-major order, we find in FILE(M, N) all records consecutively stored, whereas in FILE(N, M) first all first elements (e.g., the keys) of all records are stored, followed by the sequence of second elements, and so on. The first form is more efficient with respect to an insertion or deletion of records, whereas the second form is more advantageous for a key search as the keys now form a contiguous subarray.

In the case that the records are of variable length, the file cannot be represented by a two-dimensional rectangular array. Rectangularity of the data type array, however, is required in all high-level languages considered here. In this

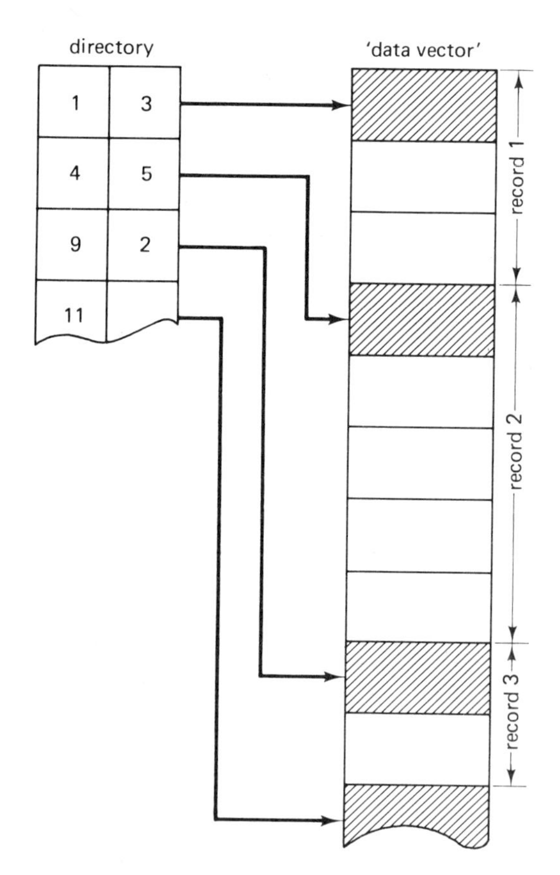

Figure 2.10 Sequential storage of records of variable length, accessed through a directory (hatching indicates a key).

case, we may store the records sequentially in a one-dimensional array and provide for a *directory* in the form of an (M, 2)-array, the first column containing pointers, pointing to the beginning of the respective records, and the second column containing integers specifying the length of the records (Fig. 2.10). A useful modification of this simple scheme may be to store the record keys as a third entry in the directory, thus facilitating the search for a key. Hence, we obtain a list whose records are accessed through pointers. The final step toward a linked list is the inclusion of pointers as elementary data items in the records.

As an example of a multilinked list representing a List structure, we return to the example of Figure 2.6, this time modified, however, such that the vertex coordinates are not stored as data items in the respective *polygon rings* but are entries of a separate *point list*. Consequently, the *line lists* contain only the appropriate pointers to the point list [31]. Thus, we obtain the structure depicted in Figure 2.11(A), which is a combination of the unidirectional linkage of two linear lists with the construction principle of hierarchical Lists. In terms of practical programming, such lists can be represented by arrays of an appropriate dimension (in our example, we reserve 1000 words each for the point list and the line list and 90 for the *picture lists*. The three lists have the format shown in Figure 2.11(b).

The implementation of such a picture file in the form of linked lists rather than in the form of sequential lists offers the easy possibility of adding or deleting graphic objects. The addition of new elements is aided by a *cursor list* which contains pointers pointing to the first free cell in each of the three lists. Deletion of an element is achieved by replacing its identifier by a pointer indicating the first free cell of that list (in the example, this occurs in location 1016). Of course, the cursor list must be updated accordingly. Such a measure adds the freed record to the reservoir of available records. It requires, however, that the program can distinguish between identifiers and pointers (e.g., this can be accomplished by using negative integers as identifiers). A special symbol (Λ) indicates that, from its occurrence on, the list is empty.

Such a solution is general and flexible but also quite wasteful, as even a simple picture requires the storage of a relatively large number of identifiers and pointers. This is aggravated in a language like FORTRAN with a static array structure, where such lists must be dimensioned for the maximal case, thus wasting in the average storage space.

Instead of having a point list and a line list, we could set up a point list only that has four data entries in each record: the coordinates of the two endpoints of a line segment. This still implies a considerable redundancy, as in most cases the points may occur twice, whenever the endpoint of one segment is the start point of the next segment. Finally (and this will be our favorite representation of graphic data), we may store the coordinates of each point only once, adding to each coordinate tuple a control parameter whose value is 0 or 1. The coordinate tuple specifies the terminal point of a segment, and the control parameter specifies the segment as invisible (0) or visible (1). Hence, the data represent the vertices of a general polygon which may represent any arbitrary line drawing composed of

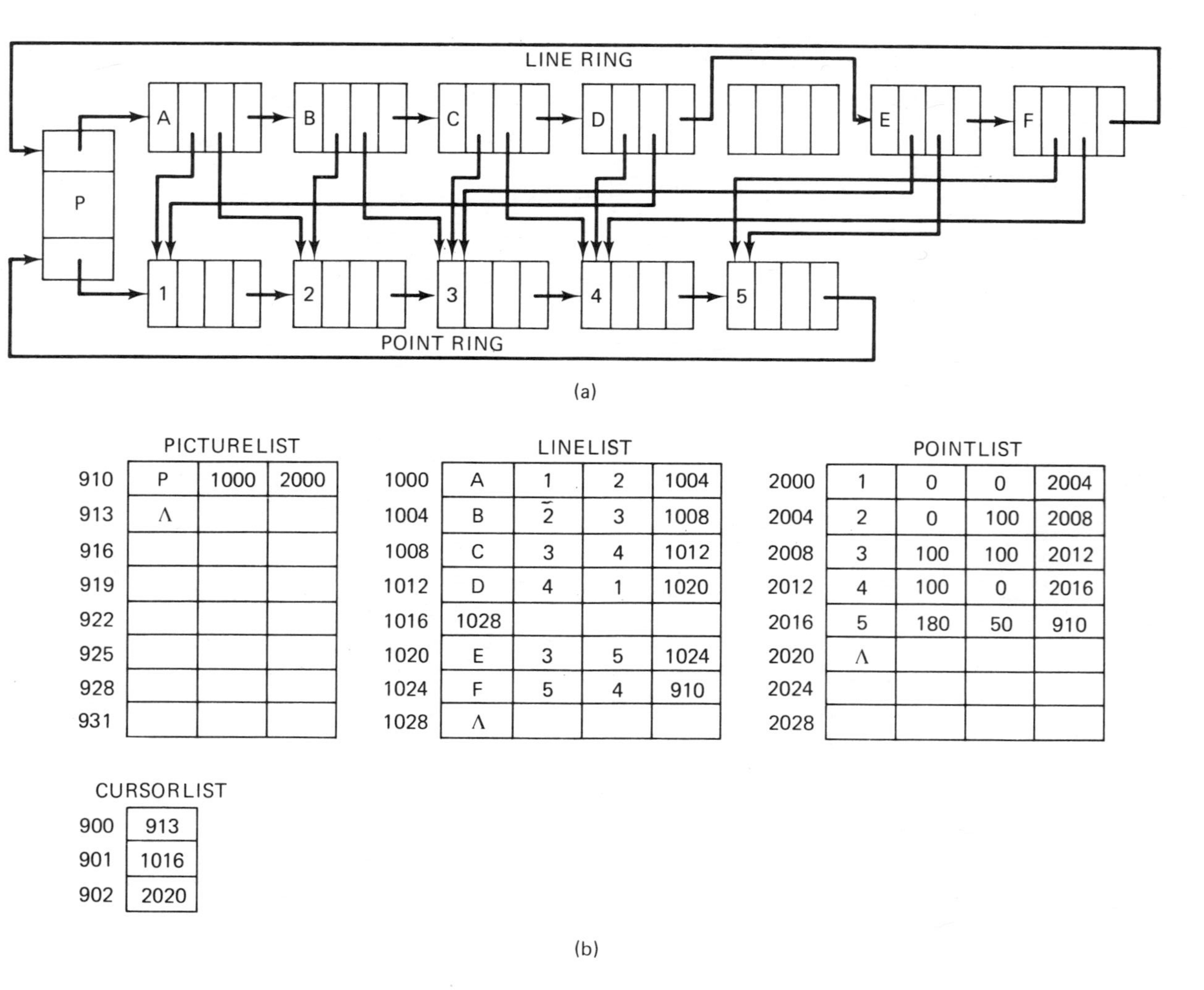

PICTURELIST

910	P	1000	2000
913	Λ		
916			
919			
922			
925			
928			
931			

LINELIST

1000	A	1	2	1004
1004	B	2	3	1008
1008	C	3	4	1012
1012	D	4	1	1020
1016	1028			
1020	E	3	5	1024
1024	F	5	4	910
1028	Λ			

POINTLIST

2000	1	0	0	2004
2004	2	0	100	2008
2008	3	100	100	2012
2012	4	100	0	2016
2016	5	180	50	910
2020	Λ			
2024				
2028				

CURSORLIST

900	913
901	1016
902	2020

(b)

Figure 2.11 (a) Structure; (b) realization.

straight-line segments. These three methods are illustrated in the following for the example of a three-sided polygon in 2-space.

Point list + line list

x_1	y_1
x_2	y_2
x_3	y_3
x_4	y_4

1	2
2	3
3	4

Point list only

x_1	y_1	x_2	y_2
x_2	y_2	x_3	y_3
x_3	y_3	x_4	y_4

Points + control parameter

x_1	y_1	0
x_2	y_2	1
x_3	y_3	1
x_4	y_4	1

If m denotes the number of lines, n the number of points, and D the dimension of the space ($D=2,3$), then we have for the required storage space (we assume that one word is needed for each value, even for the control parameter) [81]

$$\left.\begin{array}{ll}\text{Point list + line list:} & Dm+2n \\ \text{Point list only:} & 2Dn \\ \text{Points + parameter:} & (D+1)n\end{array}\right\} \text{ words for one polygon.}$$

Examples for the implementation of more complex list structures are given in Section 2.2.

2.4.2 List Construction and Manipulation

The manipulation of linked lists requires special operators. A general model of a set of low-level operators is given in [84]. In this model, it is assumed for the sake of simplicity that each node of a linked list has only two fields, named VALUE and LINK. VALUE contains the data item stored in the node, and LINK contains the pointer to the next node. Let the node be addressed by a pointer P; then LINK(P) and VALUE(P) are the references to the contents of the respective fields. The reservoir of available nodes may be organized as a special list by linking all unused nodes. A link variable AVAIL may refer to the first element of this "available space." AVAIL=0 indicates that the reservoir is exhausted. Two functions, GET(X) and PUT(X), may be employed in order to fetch a node from the available space or to put a node back in the available space, respectively.

A forward-linked list can be constructed by adding nodes to the given list head, named LH. A pointer LAST may be used for pointing to the last node of a list. Initially, we have LAST=LH; LINK(LAST)=LAST. The "program" for constructing a circular list can be written as follows (it is assumed that the data items to be stored in the nodes are originally found as the elements of an array called DATA) [84]:

```
DO I=1 TO N
GET(P); LINK(LAST)=P; VALUE(P)=DATA(I); LINK(P)=LH; LAST=P;
END;
```

Such an approach has its merits as a descriptive tool [84], but it certainly is too low-level to be taken as a blueprint for a practical programming system. The reader will readily realize that if he writes a procedure for the insertion of a node into the existing list or the deletion of a node from it; for an arbitrary node can only be accessed by starting with the head and sequencing from node to node through the link fields. This method of *pointer concatenation* (sometimes also called "pointer chasing") would be too awkward for a practical use. Therefore, a practical programming system would provide for more powerful operators. For example, such a set of operators, which could be implemented as a subroutine package, might be [30], for the purpose of building a linked list and retrieving data from it:

- INITIALIZE (⟨listname⟩): Creates a list head;
- START (⟨listname⟩): Returns the first data item of the list, and a list cursor is set to identify this node;
- NEXT/PRIOR (⟨listname⟩): Returns the data item from the node succeeding/preceding the node identified by the current value of the cursor;
- LAST/FIRST (⟨listname⟩): Returns a truth value indicating whether the item identified by the cursor is the last/first node in the list;
- INSERT/INSERTPRIOR (⟨listname⟩, ⟨data item⟩) Inserts the data item into a new node succeeding/preceding the node identified by the cursor;
- DELETE (⟨listname⟩): Deletes the node identified by the current cursor value.

However, such an approach would still have certain shortcomings, for it cannot handle multilinked lists where the number of links connecting a node with other nodes is arbitrary. Even if we further extended such a system, we would be restricted to the use of rigidly structured building blocks (e.g., the CORAL system briefly discussed in Section 2.2.5). Therefore, it is a better approach to devise a system where the user can declare and name nodes of arbitrary structure as well as individual pointer fields in such nodes. A structure then can be built by concatenating names of nodes and pointer variables. Of course, it is a desirable feature of a high-level programming language to directly provide for all facilities required for the construction of file management systems.

As an example of a language that does have data structure handling capabilities, let us briefly discuss the devices PL/I provides for this purpose. PL/I has four attributes defining storage classes [76]: STATIC, AUTOMATIC, CON-

TROLLED, and BASED. Variables that have the STATIC attribute are allocated storage before the program execution begins; and such allocations remain unchanged for the duration of the program (as in FORTRAN). Variables that have the AUTOMATIC attribute are allocated storage upon activation of the block in which that variable is declared; and the variable is freed when the block is terminated (as in ALGOL). Once a variable is freed, its value is lost. Variables that have the CONTROLLED attribute are allocated storage upon execution of an ALLOCATE statement; and the variable is freed upon execution of a FREE statement. Therefore, it is now the programmer who determines the instant at which a variable is allocated or freed. Storage may be allocated for a variable several times without freeing the variable before each new allocation. In this case, the allocations are stacked; i.e., they are not lost. However, only the current value of the variable is available at any given time. BASED storage is similiar to CONTROLLED storage, but now all the allocations are available for use at any time. Whenever a based variable is allocated, a pointer variable is set to a value specifying the address of the allocation. By including this pointer variable in a reference to the based variable, the programmer can choose between the different allocations of a based variable. Such a pointer variable is called a *locator*.

Reference to a specific value of a based variable B through a pointer variable P is given by the notation P–>B. The based variable B may be a structure containing a locator for another allocation, and so on. Thus, different allocations can be chained together. This is the simple concept underlying PL/I list processing. Two built-in functions can be employed in the process of building data structures:

- ADDR(X) returns as its value a pointer that identifies its argument (which can be a scalar, an array, an element of an array, a structure, an element of a structure, etc.)
- NULL returns a *null pointer*, i.e., a special pointer that is guaranteed not to identify any address in storage. NULL can be used for recognizing the beginning or the end of a chain by setting the pointer in the respective node to the value of NULL. Furthermore, if NULL is assigned to a pointer at the start of a program, a later test of the pointer will show whether a based variable qualified by that pointer has been allocated or not.

As an example, we can now write a PL/I procedure for the construction of a unidirectionally linked list, performed by adding to the list each new argument that is received. We assume again that each node has two variables: COORD (representing a pair of coordinate values) and LINK. The value of LINK of the last node in the chain shall be NULL (the list is not circular). A variable CURSOR shall retain the address of the most recently referenced node. NODE is declared as a based structure, as it contains data items of different type (an array of integer coordinates and a pointer).

```
FORWARD_LINK: /*CONSTRUCTS A FORWARDLY LINKED LIST. P IDENTIFIES THE
              NODE TO BE ADDED TO THE LIST (P IS CONTEXTUALLY DECLARED
              TO BE A POINTER VARIABLE). EACH ARGUMENT P WILL REFER TO
              A TWO-COMPONENT STRUCTURE CONSISTING OF COORD AND
              LINK. CURSOR IS ON THE FIRST INVOCATION OF THE PROCEDURE
              INITIALIZED TO NULL AND POINTS UPON EACH NEW INVOCATION
              TO THE NODE PREVIOUSLY ADDED. P->LINK IS SET TO NULL AS
              THE CURRENT NODE IS ALWAYS THE LAST NODE IN THE LIST*/
              PROCEDURE(P):
              DECLARE 1 NODE BASED(P),
                        2 COORD(2) FIXED DECIMAL(3,0),
                        2 LINK POINTER;
              DECLARE CURSOR POINTER STATIC INITIAL(NULL); /*INITIALIZA-
              TION*/
              P->LINK=NULL;
              IF CURSOR=NULL THEN CURSOR=ADDR(NODE); /*ENTER FIRST
              NODE*/
              ELSE DO; CURSOR->LINK=ADDR(NODE); /*LINK NEW NODE*/
                        CURSOR=ADDR(NODE); /*UPDATE CURSOR*/
              END:
              END FORWARD_LINK;
```

There are more aspects of PL/I list processing which cannot all be discussed here. What our simple example may demonstrate is the fact that the list processing features of PL/I are extremely general and simple. The trade-off for these advantages is that the features are given at a low level, providing rather building blocks than sophisticated structures. Of course, one can always add to a system a data management module which consists exactly of such sophisticated, "prefabricated" structures that may be custom-tailored for the specific applications. Such an approach becomes mandatory in cases where the available general-purpose programming language practically has no data structure handling capabilities.

2.4.3 Searching and Sorting of Lists

Searching is the operation most frequently performed on a list. The problem is: Given a list L and a key word X, find the index I such that $L(I)=X$. If no X can be found in the list, $I=0$ may be returned. The simplest (and most time-consuming) solution to this problem is to compare each key word in L with X. If the number of key words is N, this takes in maximum N and in the average about $N/2$ comparisons (provided that the probability for finding a key is uniformly distributed). This simple scheme has the advantage that the list need not be sorted, but it is only appropriate for short lists.

In certain applications it may happen that certain keys are often searched for, others less frequently, and others again almost never. In such a case it may pay to move after each successful search the respective element toward the top of the list, e.g., by interchanging it with its predecessor in the list. Thus, the elements frequently looked up will, after a while, have been ascended toward the top, and the average search time will decrease. This is, in effect, a sorting according to the relative frequency of a desired event. An example of such a sorting technique can be found in the scan grid hidden-line algorithm discussed in Chapter 5.

In the case of relatively long lists, the search time can be drastically reduced by performing a binary search, provided that the list is ordered. Let us assume that the keys in a list are numbers sorted into an ascending order of their magnitude. The core of the procedure is the comparison of the middle element with the given key. For N being odd, the index of the middle element is $(N+1)/2$; for N being even, we may take the element whose index is $M=\lfloor(N+1)/2\rfloor$.† If $L(M)=X$, the procedure can be terminated; else we have to iterate it on one of the two partitions, into which L has been bisected by taking out the middle element, according to the criterion: Take the partition with the indices $I\in[1:M-1]$ for $L(I)<X$ and $I\in[M+1:N]$ for $L(I)>X$. This procedure is iterated until either an element $L(I)=X$ has been found, or a partition with $I\in[I_{\min}:I_{\max}]$ is encountered that is empty, indicated by $I_{\min}>I_{\max}$. Hence, we have a procedure as described by the flow chart, Fig. 2.12. The maximal number of tests is $\lceil\log_2(N+1)\rceil$; the average number is, under the assumption that all search arguments are equally probable, approximately $\log_2 N-1$. No search method based on comparisons can do better than this [85]. Note that a test consists of two comparison operations, $L(I)=X$ and $L(I)<X$. Other, equally important search techniques exist which cannot all be discussed here (e.g., the Fibonacci search and the various tree search methods [85].

The binary search requires a sorted list. If this is not the case, the list must first be sorted. The simplest sorting scheme is the *sorting by insertion*, which works as follows. First, key K_2 of the second record is compared with key K_1 of the first record and, if $K_2<K_1$, both records are interchanged. Next, key K_3 of the third record is compared with K_2 and K_1. As a result of the comparison, this record either remains where it is or it has to be inserted between the first and the second record, or it has to be moved to the beginning of the list (shifting the other two records up one space). In general, let us assume that we have a list of length N and consider a record R_j, $1\leqslant j\leqslant N$. Let us furthermore assume that all the preceding records $R_1,\ldots,R_{j-1}$ have already been rearranged such that their keys $K_1,\ldots,K_{j-1}$ form the linear order

$$K_1\leqslant K_2\leqslant\ldots\leqslant K_{j-1}.$$

The key K_j of record R_j is compared with the keys $K_{j-1},K_{j-2},\ldots$ until discovering that R_j should be inserted between the records R_i and R_{i+1}. In order to put R_j into position $i+1$, the records R_{i+1}, $R_{i+2},\ldots,R_{j-1}$ must be moved up one space. Figure 2.13 presents a flow chart for this algorithm [110].

†We use the common notation $\lfloor X\rfloor$ for the next lower integer and $\lceil X\rceil$ for the next higher integer of a number X.

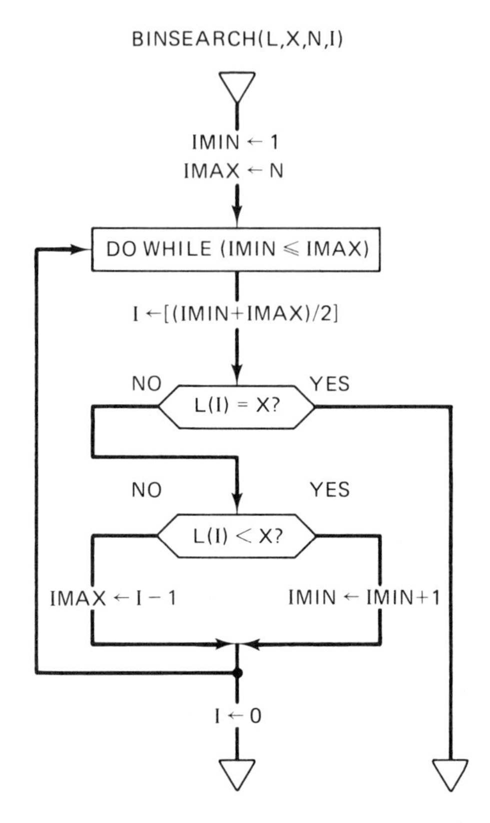

Figure 2.12 Flow chart of a binary search procedure.

The number of operations is in the worst case (if L were inversely sorted) $n(n-1)/2$ comparisons and moves; in the best case (if all records were already in the right place) $n-1$ comparisons [85]. The simple method can be recommended for short lists.

For longer lists, one of the recommendable algorithms is the *partition exchange* or *quicksort* algorithm. This algorithm applies the following comparison/exchange scheme [128]: The cursors, M and N, are kept with initially $M=1$ and N being the length of the list. $K(M)$ and $K(N)$ are compared. If no exchange is necessary, N is decreased by 1 and the comparison is repeated, and so on, until the first exchange occurs. From now on, M is increased by 1 and comparing and increasing is repeated until the next change occurs. Now, N is decreased again, and so on, until $M=N$. Note that all comparisons involve the initial key $K(1)$. When $M=N$, the associated record $R(1)$ will have moved into its final position, say I. This record separates two parts, $R(1),\ldots,R(1-1)$ and $R(I+1)$, $\ldots,R(N)$, of the original list. The same technique is now applied to the two parts, resulting in a refinement of the partition, and so on. This process is repeated until every record is in its right place.

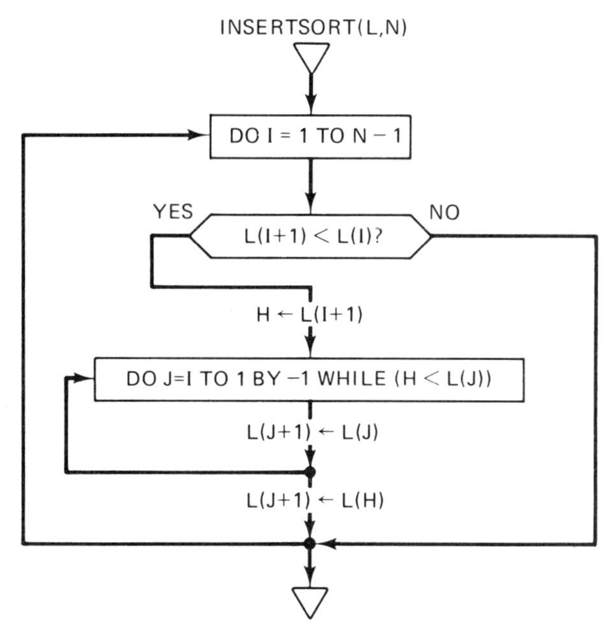

Figure 2.13 Sorting by insertion.

Thus, we have a recursive procedure as illustrated by the flow chart, Fig. 2.14. Two modifications of the scheme described above increase the efficiency: first, a check is built in which guarantees that only lists or partitions of lists with length >10 will be sorted. Shorter lists will be sorted by the INSERTSORT procedure. Second, a third record $L(M+N)/2$ is introduced into the comparison process, thus reducing the total number of comparisons [110]. The interesting part of the

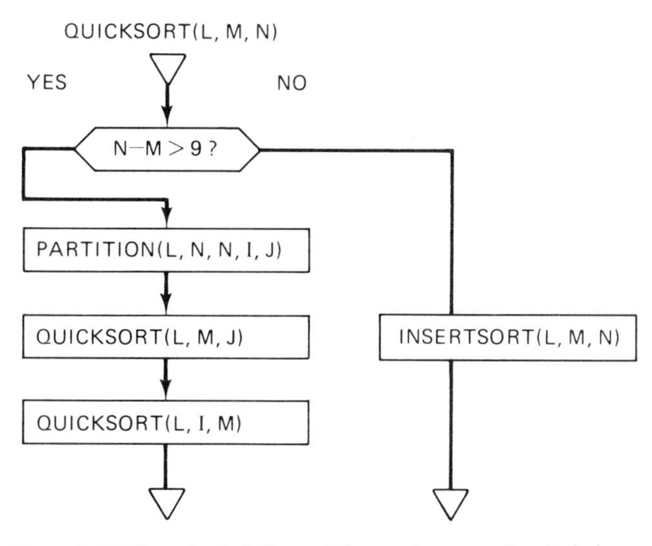

Figure 2.14 Flow chart of the partition–exchange sorting technique.

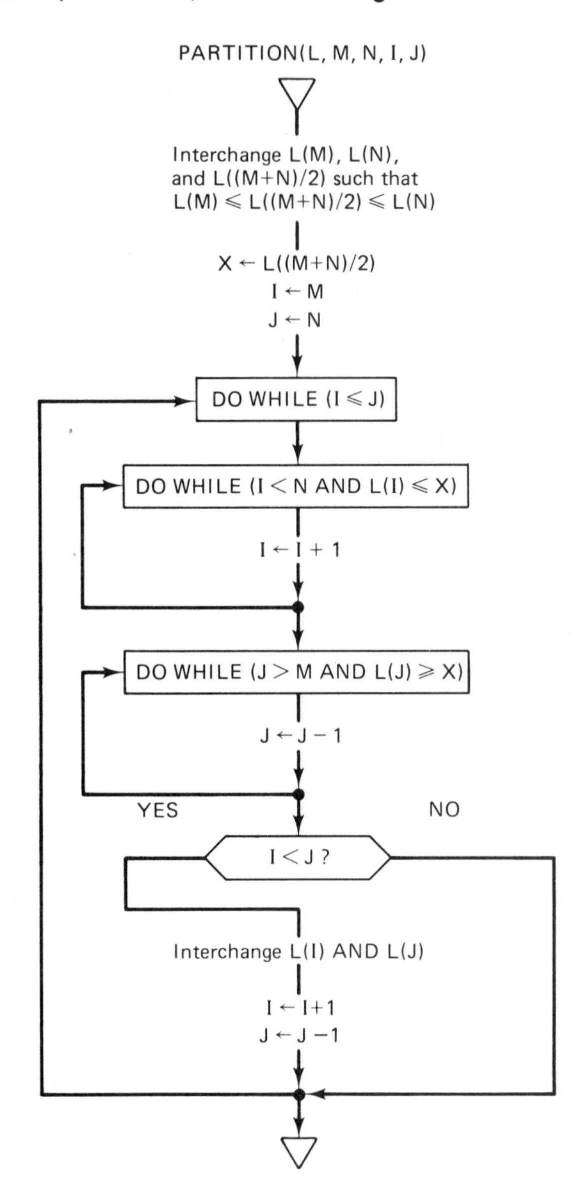

Figure 2.15 Procedure PARTITION.

scheme is the partition, which is accomplished by the procedure PARTITION flow-charted in Figure 2.15. The number of comparisons is in this case proportional to $N \cdot \log N$.

Other sorting techniques shall only be mentioned: Radix sort, Shell sort, Bubble sort, and Tree sort, all exhibiting certain advantages and disadvantages. For details, see [85]. Whereas the sophisticated system programmer may want to select the method tailored best to the particular application, the casual user may get along with one standard sorting procedure. Therefore, the quicksort algorithm with its ($N \log N$)-complexity, which is available as an ACM algorithm [128], certainly is a sensible choice.

2.4.4 Hash Coding

Searching can be avoided by storing data records at addresses derived from their respective keys by a hashing operation (see Section 2.2.5). Such an approach has two advantages: (1) the lists need not to be sorted, and (2) access to a data record can be found even faster than through a binary search. The problem with hash addressing is the following. The symbol string forming the key usually allows for a number of possible combinations much greater than the number of combinations that may actually occur. Hence, the mapping from the set of possible key words to the set of addresses need not be injective, meaning that we may have equivalence classes of key words known as *collision classes*. Since it is not known which of the possible key words will actually be used and which will not, it is not possible to optimize a hashing function such that the frequency of occurrence and the cardinality of collision classes is minimized.

Generally speaking, one can state that a hashing function is the better, the closer to a uniform distribution the distribution of the generated addresses will come. Thus, the problem of finding a good hashing transformation is similar to the problem of finding a good transformation for the generation of random numbers [85]. However, as stated above, a particular transformation is hard to evaluate, if the domain on which it will be applied is rather undefined—as is the case with binary-coded symbol strings interpreted as numbers.

Therefore, the evaluation of hashing operations can in this case only be empirically accomplished, usually on the basis of elaborate simulations. For example, such a comparison of a number of frequently used methods was published in [90]. Some of these more frequently used operations are the following.

1. *Division:* In this method the key is divided by a positive integer, for which usually the prime number that comes closest to the number of available addresses M is used. The remainder of the division becomes the address associated with the key.

2. *Midsquare:* This method was adopted from an old scheme for the generation of pseudo random numbers (developed by von Neumann): a key is multiplied by itself and its address is obtained by truncating digits at both ends of the product until the number of digits left equals the desired address length.

3. *Folding:* A key is partitioned into a number of parts, each of which, except the last, has the same length as the address. The parts are added together, either digitwise modulo 10 or partwise modulo M (M being the number of available addresses).

4. *Lin's method:* A key is expressed in radix p and the result taken modulo q^m, where p and q are relatively prime and m is a positive integer.

5. *Algebraic coding:* This method was adopted from the theory of linear cyclic codes: each digit of a key is considered to be a polynomial which is univariant for all the keys in a file. The coefficients of the remainder polynomial form the address. From the theory of error-correcting codes it is known that—provided that the divisor polynomial is properly chosen—all key words that have a certain minimum Hamming distance will have different addresses.

Lum, Yuen, and Dodd [91] found that the best performance was obtained by the division method. This result is in compliance with the recommendation of the division method by other authors (see [85]).

Collisions are handled by storing all records of a collision class, including their keys, in a linear list to the beginning of which the hash address gives access. In order to find the individual record, a search must be performed through this list for the given key. A "good" hashing function, although it cannot prevent collisions, will at least keep such a collision list small so that normally a simple sequential search is adequate. Note that these collision lists require additional memory space. The same space may remain unoccupied at some other places in the memory, as certain addresses may never occur. Hence, a hash-addressed file inevitably has "holes" (therefore, it is also called a "scatter memory")—a small price to pay for the advantages of the hash addressing scheme.

EXERCISES

1. Show why a reflexive relation is also domain-total and range-total.

2. (a) Given the set $S=\{a,b,c,d,e\}$ and a relation $R=\{(a,b),(b,c),(d,e)\}\subseteq S\times S$.

Is R reflexive? If not, define a minimal set $R_r\subseteq S\times S$ such that $R\cap R_r=\varnothing$ and $R\cup R_r$ is a reflexive relation. ($\varnothing$=empty set.)

Is R symmetric? If not, define a minimal set $R_s\subseteq S\times S$ such that $R\cap R_s=\varnothing$ and $R\cup R_s$ is a symmetric relation.

Is R antisymmetric? If not, define a minimal set $R_a\subseteq S\times S$ such that $R\cap R_a=\varnothing$ and $R\cup R_a$ is an antisymmetric relation.

Is R transitive? If not, define a minimal set $R_t\subseteq S\times S$ such that $R\cap R_t=\varnothing$ and $R\cup R_t$ is a transitive relation.

Is R irreflexive? If not, define a minimal set $R_i\subseteq S\times S$ such that $R\cap R_i=\varnothing$ and $R\cup R_i$ is an irreflexive relation.

(b) Define a minimal set $E\subseteq S\times S$ such that $R\cap E=\varnothing$ and $R\cup E$ is an equivalence relation. Try to use the sets R_r, R_s, R_a, R_t, and R_i for the definition of E.

(c) Define a minimal set $P\subseteq S\times S$ such that $R\cap P=\varnothing$ and $R\cup P$ is a partial ordering. Try to use the sets R_r, R_s, R_a, R_t, and R_i for the definition of P.

3. Given a set of points, P, forming the picture as shown in Fig. 2.6(a). This picture consists of 2 polygons, S and T (or, in other words, the set P consists of 2 subsets, S and T). Each polygon consists of a number of edges, and each edge has 2 vertices, as depicted in Fig. 2.6(b). Hence, we have a set of relations $\rho = \{\text{R1}, \text{R2}, \text{R3}\}$ with R2, R2, R3 being defined by the propositions p1: "to be a polygon of," p2: "to be an edge of," and p3: "to be a vertex of," respectively. Then we can easily read from Fig. 2.6 that $\text{R1} = \{(S,P),\ (T,P)\}$; $\text{R2} = \{(A,S),\ (B,S),\ (C,S),\ (D,S),\ (C,T),\ (E,T),\ (F,T)\}$; and $\text{R3} = \{(1,A),\ (2,A),\ (2,B),\ (3,B),\ (3,C),\ (4,C),\ (4,D),\ (1,D),\ (3,E),\ (5,E),\ (5,F),\ (4,F)\}$. Assume that an associative data structure has been established into which the pairs of $\text{R1} \cup \text{R2} \cup \text{R3}$ have been entered in the form of triads $(S,\text{p1},P)$, $(T,\text{p1},P)$, $(A,\text{p2},S), \ldots, (F,\text{p2},T)$, $(1,\text{p3},A), \ldots, (4,\text{p3},F)$. What are the components A, B, and π, of the "universe" (A,π,B) thus established? List all subsets of (A,π,B) that can be obtained as answers to the associative queries Q1–Q6 as defined in the table given in Section 2.1.6.

4. Read Section 2.3.1, "Traversing Binary Trees," and Section 2.3.2, "Binary Tree Representation of Trees" in Vol. 1, *Fundamental Algorithms*, of Knuth's *The Art of Computer Programming*.

5. How many ordered trees can you form with three nodes, A, B, and C? How many oriented trees are there? What is the number of nodes of a complete, n-level binary tree? (From [84].)

6. Construct a memory representation for the tree-structured picture of Fig. 3.3(a). To this end, assume that each node of the tree Fig. 3.3(a) is represented by a record consisting of two cells as depicted below. One cell contains a node identifier, ID, the other contains a data link, DL. The data link, in turn, consists of a base-bound pair, (BA, BO), through which a certain number of records of a linear list, the data list, are associated with a node (separation of structure and data).

ID, BA, and BO thus have integer values. If needed, additional cells can be added to each record to accommodate pointers through which the record is linked to other records of the structure. These records are the building blocks for the following representations.

6.1 Represent the tree-structured picture of Fig. 3.3(a):
   (a) By a linear list.
   (b) By a matrix.
   (c) By a linked list.

6.2 Compare the obtained representations with respect to
   (a) Memory space required.
   (b) Procedural steps required for the retrieval of a record.
   (c) Complexity of the addressing function.

6.3 Represent the tree-structured picture of Fig. 3.3(a) by a binary tree and represent that binary tree:
 (a) By a linear list.
 (b) By a matrix.
 (c) By a linked list.

6.4 Compare the obtained representations with respect to the three criteria listed in Exercise 6.2.

6.5 Compare the tree representation to the binary tree representation with respect to the three criteria of Exercise 6.2.

7. (a) Write a HLL program for the creation of a name table for N records. The name table will be represented by a matrix such that each record corresponds with a row of the matrix. A record is a character string of length $L+D$, in which the first L characters represent a name and the last D characters represent a pointer, i.e., a string of digits forming a nonnegative integer number, or a special symbol followed by D-1 blanks. Assume that the programming language provides an operator through which a string of digits can be converted into the corresponding number. The pointers are used to establish throughout the name table unidirectionally linked lists of records (in any arbitrary order). The character T in lieu of a pointer indicates that its record is the last one of a list; and the character E in lieu of a pointer marks a record as empty. Note that the pointers are row indices of the matrix. Initially, when the matrix is created, all records are empty.

(b) Write a function with the calling sequence NEXT(I) that receives as argument the index I of a record and returns as value the pointer it finds in this record (as a number, not as a character string). If no pointer can be found (but one of the special symbols, E or T), the value -1 is returned.

(c) Write a procedure with the calling sequence LAST(I, J), where I is an input parameter denoting a record. The procedure performs a "pointer chase" from record I on through the linked list record I is part of until the last record has been found, whose index is returned as value of the output parameter J. (*Note*: Use a high-level programming language of your choice. The recursiveness of the "pointer chase" suggests that LAST may be written as a recursive procedure. What difficulty will we encounter if the chosen language is FORTRAN?)

8. Someone has the idea to implement a bijective hash transformation in the following way. Names consist of four characters (two letters denoting the object type, followed by two digits). Each character is coded by 6 bits. The binary code words for all four characters of a name are interpreted as one bit string of length 24, and this bit string, after conversion into the corresponding decimal number, is the hash index. As this transformation is one to one, no collisions can occur. Is this method practically feasible? If not, could it be modified so that it becomes practically feasible?

3

Picture Structure and Picture Transformations

3.1 PICTURE STRUCTURE

A picture may be defined as a set of graphic primitives. Such a definition, however, would not suffice for the purpose of *picture editing*, for in this case the relationships between the elements of a picture must be considered as well. This leads to the definition of a picture as a set of primitives, together with a set of relations defined on this set. The reader will recognize that this is the same definition as was introduced in Chapter 2 for data structures; and in fact, a picture structure is determined by the data structure imposed on the set of graphic data of that picture.

This fact will be illustrated by a simple example. A polygon, for instance, can be defined as a set of line segments called *edges*. According to the definition of a polygon, we have a linear ordering among its edges. The same holds for a character string, for each character of the string has a successor or a predecessor (or both). Actually, all pictures generated on a refresh display are inherently ordered, namely, by the order in which its primitives are drawn. This, in turn, is the order of the primitive-generating instructions in the DPC program. It will be explained below how this fact can be exploited for the lightpen "pick" of primitives. The inherently given linear order of graphic primitives, however, may not be adequate. For example, the user may want the ability to pick an element of a figure and tell the system to delete the whole figure. In more general terms, the user may want to form equivalence classes for which each of its members can function as a representative. We know from the discussion in Chapter 2 that this leads to tree structures. Finally, the user may want to define certain sets of primitives with nonempty intersections, leading to a structure called "acyclic graph" or "generalized List." We call such a structuring of the graphic data of a picture the *picture structure*.

Hence, in the data base of an application program, graphic data may be structured in many ways, as discussed in Chapter 2, ranging from linear lists to trees or Lists or even to associative structures. Between such a data base representation of graphic objects and their representation by a sequence of primitive-generating display processor instructions, there exists a third level of representation given by the high-level GPL program. In such a program the user usually wants to refer to entities more complex than single primitives. Therefore, we introduced in Chapter 1 the definition of a graphic entity as a set of primitives which are displayed with a uniform appearance and status (see Section 1.3) and which are identified by a name. Consequently, a graphic programming system should offer facilities for grouping primitives into entities, thus creating a hierarchical structure. This can be accomplished by providing GPL constructs not only for the generation of primitives but also for the grouping of these primitives into more complex entities. However, it is more efficient and more in compliance with the notion of a high-level language to have constructs which directly generate more complex entities. Therefore, we introduce such an entity *as a set of primitives of the same type* and call it an *item*. Of course, an item may contain exactly one primitive. The restriction of items to homogeneous sets (all elements are of the same type) guarantees that all associated data are of the same type. Hence, the structure of the variables referenced in item-generating constructs may be the *homogeneous, rectangular array*, a structure type that is found in almost all high-level programming languages. Of course, pictures may consist of primitives of various types. We take this into account *by defining a picture as a set of segments, where a segment is defined as a set of items*. Hence, we have a second type of entity given in the form of an inhomogeneous set of items. The appropriate high-level language construct for a segment is the *block* or the *procedure*.

In the thus-defined hierarchical structure, the user has the choice of declaring either a whole segment or certain items of a segment as an entity. Any such entity may have its own origin (that may become, for instance, the origin of rotation), its individual attributes (*boldness*, *color*, *line style*, *blink status*), and its individual lightpen pick status. Any such entity is identifiable by its name, e.g., segments by a segment name and items by a pair (segment name, item name).

The efficiency of a program increases with increasing complexity of the entities in the picture structure. However, the actual structuring of a picture is usually determined by semantic considerations. Thus, we may say that items are a syntactic entity (the rule that all primitives of an item must be of the same type certainly is a syntactic rule), whereas segments are defined as semantic entities (e.g., a house, a floor plan, a molecular structure, the drawing of a machine part, a transistor pattern in a mask layout). Single primitives can be treated as an entity if declared as an item (with the primitive as its sole element).

Items, as defined as sets of primitives of the same type, are ordered sets. The ordering is given by the order in which the primitive-generating statements occur in the program or by which data are assigned to an item-generating statement. This means that, for such a set of primitives $P=\{p_1,p_2,\ldots,p_n\}$, there is a bijective mapping $\leqslant:[1:n]\rightarrow P$. Thus, primitives in an item may be identified by their ordinal numbers (rather than by name).

Primitives may be categorized as being either graphic primitives (dots, lines, etc.) or characters (alphanumeric or other characters for which there is a hardware character generator). In the former case the data type is that of real or integer numbers representing coordinates, and in the latter case the data type is that of character code. To define primitives through their data structures, we introduce the *coordinate point* for graphic primitives and the *code word* for characters. A coordinate point is a pair of numbers in the two-dimensional case, a triple in the three-dimensional case with regular coordinates, a quadruple in the three-dimensional case with homogeneous coordinates, etc. (see Section 3.4). In general, a point is an n-tuple of numbers (real or integer).

A dot is the pictorial representation of a point. A vector connects two points. For the sake of consistency and economy, a vector is defined by specifying explicitly only one point, its end point, whereas its start point is given by the current beam position (which may be the end point of the previously drawn vector). Vectors connected in such a way are said to be *concatenated.* A set of concatenated vectors forms a polygon. Figures composed of vectors, but in a more general fashion than are polygons, can be generated in the same way through vector concatenation if we include *blank* (invisible) *vectors*. Therefore, the item type *polygon* may be used to generate any kind of figure composed of straight lines, if for each line a control parameter is added that specifies whether the line is visible or blank. These control parameters may be ordered in the form of a boolean vector. Since this vector has the same dimension (length) as the vector of points, the point coordinates and the control parameters may be combined into one rectangular array, provided the programming language permits that. Similar considerations lead to rectangular arrays as the data structure of items consisting of circles or surface patches. Of course, in either case we need more than one point for the primitive specification, but this only affects the dimension of the respective array.

The ordering of the graphic data in such an array corresponds with the order in which the associated primitives are generated. However, the user normally need not know this order when he or she selects (e.g., by a lightpen pick) a primitive in order to have a certain operation performed on it. For example, a command to the system may be: "Identify the object to which I am pointing and delete it." The primitive identifier is in this case only internally used, whereas the user identifies it in the same manner as an infant would denote things without yet having names for them, simply by pointing to them [38].

In modern systems, the display processor is often a "dedicated" minicomputer. Therefore, it is an important design objective to keep the display file organization as simple as possible in order to make it manageable by the small minicomputer, which may have only limited computing power and memory capacity—of course, without sacrificing the possibility of local entity identification. This goal is achieved if we define the picture structure to be a tree rather than a (more general) graph by imposing the additional constraint that items be *disjoint* sets of primitives. This constraint requires that any two items must not have any common primitive.

Even if two items cannot share the same primitives, we can still connect two items by allowing them to share the same *points*. The difference may be illustrated by the following example. Given two items, a square $S = \{L1, L2, L3, L4\}$ and a triangle $T = \{L5, L6, L7\}$ (Fig. 3.1), which are connected by setting P1 = P5 and P4 = P6. In the picture segment $S \cup T$, either line L4 or line L5 must be eliminated from the respective item in order to maintain the rule that the intersection of items is empty. If we violated that rule, the following two undesirable features would occur:

1. The line $\overline{P1P4}$ would be drawn twice (once as L4 and a second time as L5), resulting in a higher intensity than that of all other lines of the picture.

2. A lightpen pick of $\overline{P1P4}$ would lead to an ambiguity as to which item has been picked.

Therefore, the imposed condition that items are partitions of the set of primitives of a picture not only facilitates the display file management but has additional practical merit.

Our discussion may be summarized at this point by the following definitions.

Definition: *Point*

A point is an n-tuple of coordinate values (integer or real).

Definition: *Primitive*

A primitive is either a graphic primitive or a character. A graphic primitive is specified either by a point (for dot or vector) or by an ordered set of points (for circle or surface patch). A character is specified by a code word.

Definition: *Item*

An item is an ordered set of primitives of the same type. The data structure of an item is the homogeneous, rectangular array.

Definition: *Segment*

A segment is a set of disjoint items. A segment may also contain symbol (subpicture) invocations. The data structure of a segment as a set of items is a tree formed by a partition, i.e., a set of equivalence classes.

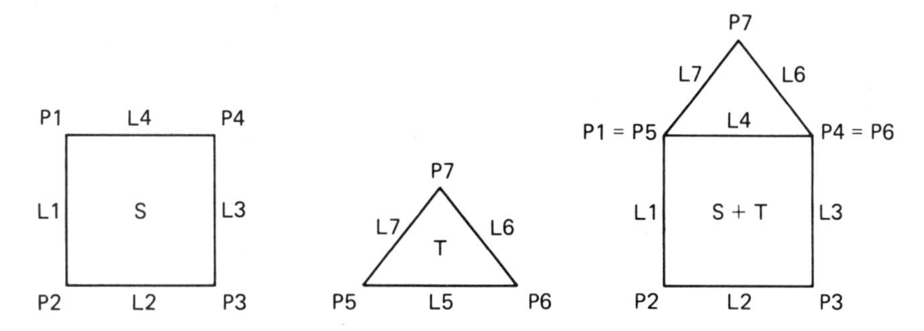

Figure 3.1 Segment $S \cup T$ consisting of two connected items (S and T).

Definition: *Symbol and Symbol Instance*

A symbol is an identifiable display processor program segment that has no appearance and status attributes of its own. Rather, these attributes are assigned to the symbol instances which are created by the invocations of a symbol (subroutine technique). A symbol may consist of primitives, items, or symbol instances. The items in a symbol are not identifiable and have no attributes of their own.

Hence, we find three types of entities in our picture structure: segments, symbol instances, and items. If we want to declare a single primitive as an entity, we declare it as an item. Note the possibility of nested symbol invocations, which somehow distorts the tree structure of a segment that contains symbols. The segment structure resulting from these definitions is depicted in Figure 3.2.

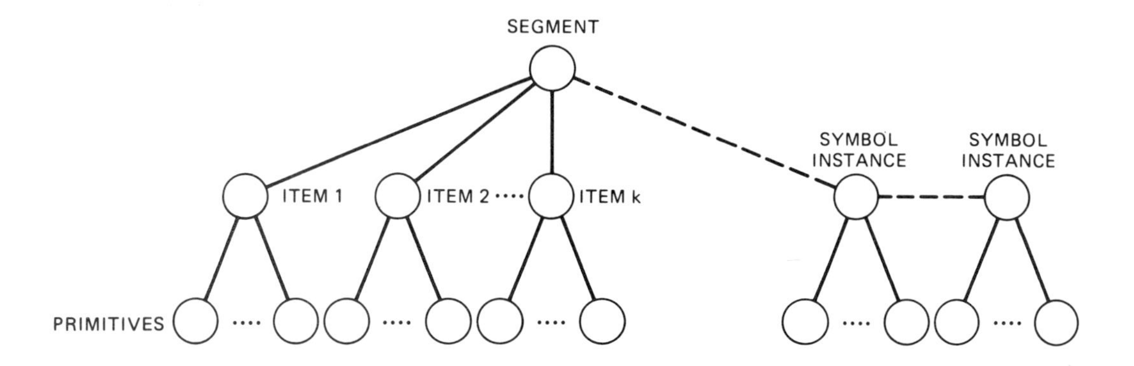

Figure 3.2 Tree structure of a segment as a set of items. If the segment also contains symbol instances, the tree structure is distorted because of the possibility of nested symbol invocations. Items in symbols are not individually identifiable.

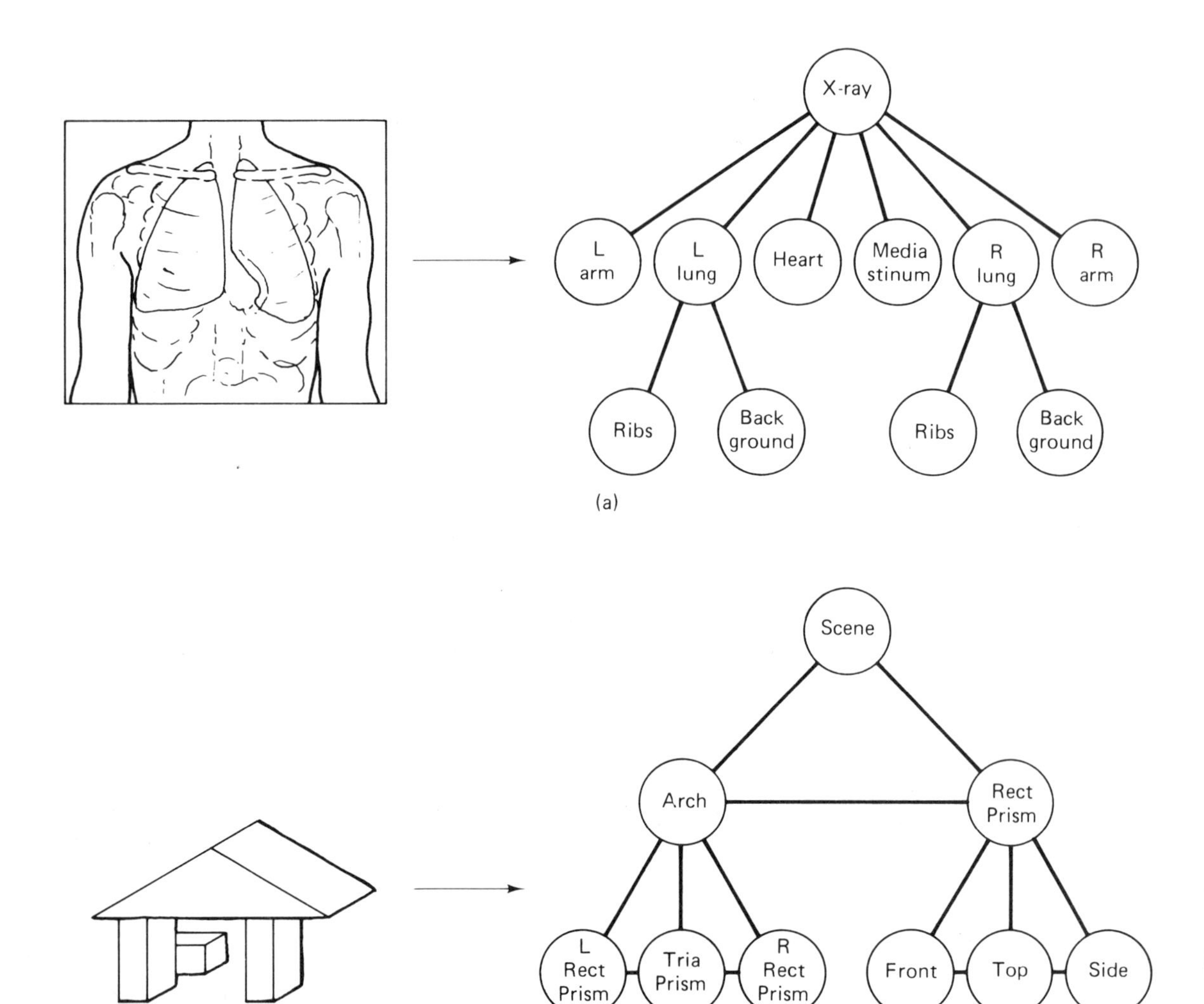

Figure 3.3 Examples of picture structures: (a) a tree; (b) a graph. (From "Some Properties of Web Grammars," Purdue University used by permission.)

The notion of a graphic entity suggests that a portion of a program written for the generation of an entity be conceptually divided into two parts:

1. The *entity initialization*, i.e., the specification of name and attributes of the entity and—if desired—of an individual entity origin.

2. The *entity specification*, i.e., the definition of the primitives, items, or symbol instances of an entity.

An entity specification consists of the assignment of data (points or code words) to an entity-generating procedure or program block. The usefulness of separating conceptually the notions of entity initialization and specification becomes apparent when we examine symbols. Symbols are prespecified entities to which an initialization must be added on an invocation. A refinement of this concept is presented in Chapter 10.

We pointed out the difference between items as "syntactic" entities and segments as "semantic" entities. The collection of all segments, in turn, forms a higher entity which we call *picture* or *image* and which encompasses everything visible on the display screen. Hence, two structures are involved: (1) the semantic structure linking segments in a picture, and (2) the syntactic structure of the segments. Whereas the segment structure exists in the display file as well as in the data base, the picture structure exists only in the data base and is certainly irrelevant for the picture-generating parts of a program. Whereas, for the sake of simplicity, the segment structure is basically a simple tree with the segment as root and the primitives as the terminal nodes, the (semantic) picture structure may be a more general hierarchical structure (e.g., an acyclic graph or a List) or may even consist in a plex of "associations," as discussed in Chapter 2.

Figure 3.3 depicts two illustrating examples [11]. In either case, the picture structure is represented by a graph such that the nodes represent segments or segment items, whereas the edges represent the relationship between these entities. The relationship may be of the parent–offspring type (as in a tree) or it may express the connectedness of equals. In either case, it is irreflexive. The picture shown in Figure 3.3(a) can be represented by a tree, whereas the picture Fig. 3.3(b) has the structure of a more general acyclic graph. More specificly, such a structure and a "picture grammar" defined on it is called a *node and edge labeled web* over an alphabet (or vocabulary) $V=(V_N \cup V_E)$ and defined [11] as a 4-tuple

$$W=(N_w, E_w, \nu_w, \varepsilon_w),$$

where N_w is a finite, nonempty set of nodes
E_w is a set or ordered pairs of distinct elements of N_w, called edges
$\nu_w: N_w \rightarrow V_N$ and $\varepsilon_w: E_w \rightarrow V_E$ are two mappings
$G_w=(N_w, E_w)$ is called the underlying graph of the web.

A grammar G defined on the alphabet $V=(V_N \cup V_E)$ is said to be a web grammar.

The construction of webs follows the rules of web grammars, analogously to the construction of sentences according to a string grammar (albeit web grammars are more complicated than string grammars). Such *picture grammars* are playing an important role in the field of pattern recognition in pictures and scene analysis; i.e., they are part of *cognitive computer graphics* rather than of *generative computer graphics*. As the examples demonstrate, the semantic picture structure depends strongly on the application. No general rules can be given.

3.2 DOMAIN TRANSFORMATIONS

A statement that generates a graphic object can be considered as being a function whose argument is an array of coordinate values and which results in the appearance of the object on the display screen. In principle, these functions can be defined on two different domains. These are:

1. The *screen domain* SD: The screen domain is measured in *raster units*, that is, it is given by the Cartesian product of a finite set of integer numbers. For example, if a given display has a raster of 1024 by 1024 addressable points, the screen domain may be

$$SD=[0:1023]\times[0:1023]. \tag{3.1}$$

Naturally, the display processor instructions are defined in SD; but even high-level language display programming systems often use the screen domain.†

2. The *world domain* WD: The world domain is the domain that is most convenient for the programmer, i.e., the domain of real numbers. Therefore, with $\mathbf{R}$ denoting the set of real numbers, we have

$$WD=\mathbf{R}\times\mathbf{R}. \tag{3.2}$$

(More precisely, in this case we can only use a finite domain given by the approximation of the Cartesian plane $\mathbf{R}\times\mathbf{R}$ by the floating-point-number representation used in the respective computer.)

In a good high-level language display programming system, the user should be offered the possibility of defining pictures in the world domain. In this case a special function must be provided as a mapping from an arbitrary area $W\subset\mathbf{R}\times\mathbf{R}$, specified by the user, to the screen domain. We call this subarea of $\mathbf{R}\times\mathbf{R}$ a *window* and the mapping

$$w: W\rightarrow SD \tag{3.3}$$

a *windowing function*. For the sake of simplicity, we shall assume that the window W is always a rectangle. If the user fails to specify a window, the system may set a *default window*. Many existing systems avoid the need for such transformations by

†However, this should not be advocated, as it is counterproductive to the desirable device independence of graphic programming.

forcing the programmer to specify his/her pictures solely in terms of screen raster units, thereby requiring only integer arithmetic. In stand-alone systems with a small-scale dedicated computer, the restriction to integer operations may be inevitable. However, even then it is a rather crude approach to use the number of addressable points on the screen which the hardware just happens to provide (usually some power of 2), as it hinders the portability of a programming system. Therefore, a *normalized screen domain* should be introduced. On systems with floating-point arithmetic, the user should be given the alternative of specifying coordinates as real numbers in a world domain, freeing her/him from the burden of having to worry about a domain overflow in the case of picture transformations.

If the selected window and the screen domain are not similar, different scale factors for abscissa and ordinate are required as well as possibly a rotation of the coordinate system. An additional complication arises if the world domain is three-dimensional, for a three-dimensional window must now be applied. Naturally, such a window cannot be injectively mapped into the two-dimensional screen area. The solution to this problem is to apply the same projectional transformation to the three-dimensional window that is applied to the objects in the world domain. Figure 3.4 illustrates this situation.

In computer graphics, two types of projection are mostly used (see Section 3.4), *orthographic projection* and *central projection*. In the case of an orthographic projection, the three-dimensional rectangular window or *viewbox* is not affected by the projectional transformation. If a central projection is applied, the viewbox changes into a *viewing pyramid* as depicted in Figure 3.4(b), i.e., a pyramid whose base is the projection plane and whose apex is the viewpoint.

Performing a perspective transformation alone may not suffice to create in the viewer's perception a realistic, unambiguous image of a three-dimensional line drawing. Therefore, additional steps must be taken to aid the viewer in the perception of depth. A feature that carries depth information is called a *depth cue*. A depth cue can be, for example, the suppression of hidden lines (see Chapter 5).

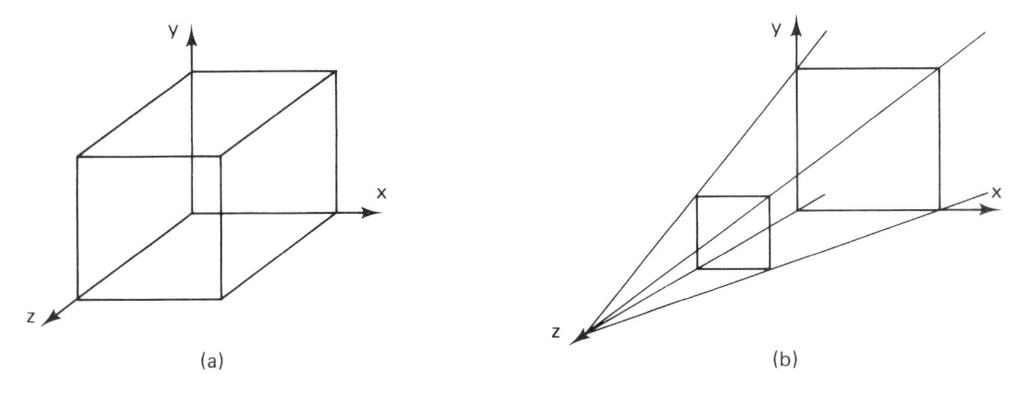

Figure 3.4 Windowing in the case of a three-dimensional world space: (a) three-dimensional window; (b) its transformation by a central projection (viewpoint on the z-axis).

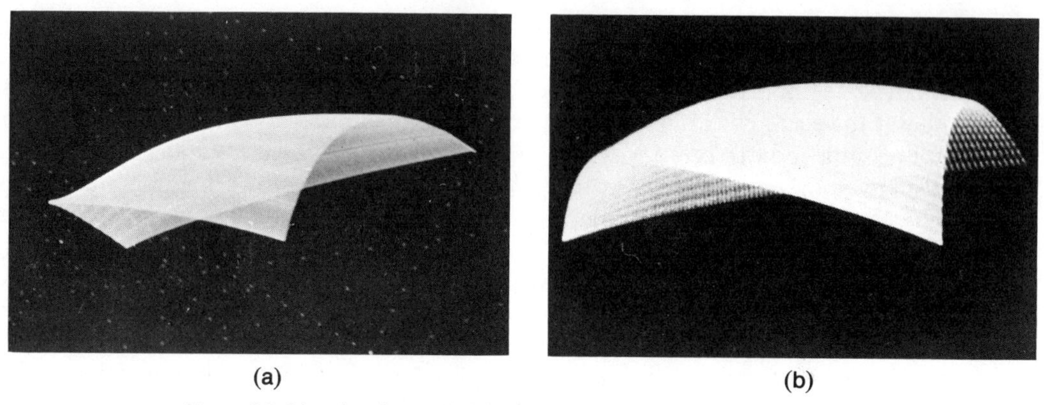

(a) (b)

Figure 3.5 Line drawing: (a) without depth cueing; (b) with depth cueing.

Another possibility is the modulation of the beam intensity as a function of depth: the farther away from the viewer a line is, the lower is its brightness and, eventually, a picture may fade away in the "yonder." Figure 3.5 depicts a line drawing rendered (a) without and (b) with an intensity cue. In the latter case, the viewport may be considered as a three-dimensional rather than a two-dimensional region, where two dimensions are given by the screen coordinates and the third by the beam intensity.

The problem of windowing not only arises with transformations from a world domain to the screen domain, but it may also occur if some of the geometric transformations are applied on objects already defined in the screen domain. This happens whenever a rotation, scaling, or translation moves part of an object outside a boundary of the screen area. Usually, this problem is solved by providing

	$x_0 x_1$	$\bar{x}_0 \bar{x}_1$	$\bar{x}_0 x_1$
$\bar{y}_0 y_1$	1101	0001	0101
$\bar{y}_0 \bar{y}_1$	1100	0000 screen area	0100
$y_0 y_1$	1111	0011	0111

Figure 3.6 Definition space, consisting of an array of nine tiles (with the screen area as a center tile) and tile coding in form of minterms $x_0x_1y_0y_1$.

a larger domain than the screen area for the definition of objects. The screen domain is then a subspace of this larger "image definition domain," and a clipping procedure must yield a display file representation of the image in which all portions of the objects lying outside the screen area are suppressed.

A logical way of extending the image definition space is to compose it out of nine subareas, each one having the size of the screen area, such that these nine rectangular areas are arranged in the form of a 3×3 array [133]. If we call such a subarea a *tile*, we have then a quadratic array of nine tiles with the screen area as the center tile (Fig. 3.6). This means that the screen domain must be tripled in each dimension. To this end, two more bits are necessary in each coordinate representation. These two bits can be used to provide a tile coding by interpreting these additional bits as boolean variables representing the propositions†

x_0x_1: tile is left of the screen area
$\bar{x}_0x_1$: tile is right of the screen area
y_0y_1: tile is below the screen area
$\bar{y}_0y_1$: tile is above the screen area.

The individual tiles are uniquely identified by 9 of the 16 possible *minterms* which can be formed with the four variables x_0, x_1, y_0, y_1, as indicated in Figure 3.6. Points lying in the screen area are easily recognized by the condition

$$c = x_0 \vee x_1 \vee y_0 \vee y_1 = 0. \tag{T1}$$

This particular coding scheme has the advantage that any two adjacent tiles have at least one 1 in their code words in common—except for the center tile. Let P_i and P_j be two points in the definition space. Let

$$x_0^i x_1^i y_0^i y_1^i \quad \text{and} \quad x_0^j x_1^j y_0^j y_1^j$$

be the code for the respective tiles. Let

$$c^i = x_0^i \vee x_1^i \vee y_0^i \vee y_1^i \quad \text{and} \quad c^j = x_0^j \vee x_1^j \vee y_0^j \vee y_1^j.$$

With c^i and c^j we may form the following propositions:

$$p_1 = \bar{c}^i \wedge \bar{c}^j \quad \text{and} \quad p_2 = c^i \wedge c^j. \tag{T2}$$

For $p_1 = 1$, both points lie inside the screen area; for $p_2 = 1$, both points lie outside the screen area; and for $\bar{p}_1 \wedge \bar{p}_2$, one point lies inside and one point lies outside the

†Here and in the following, $\bar{x}$ denotes the negation of x, $\wedge$ denotes the logical AND, and $\vee$ denotes the logical OR. In a conjunctive term, the symbol $\wedge$ is omitted; i.e., x_0x_1 is the same as $x_0 \wedge x_1$. In a disjunctive form, $\wedge$ has precedence over $\vee$.

screen area. Based on these simple tests, we can now approach the problem of windowing as follows.

Windowing of Points: Windowing performed on single dots is simplest to perform. All that is required is to perform test T1. If the dot is inside the screen area, it is displayed; otherwise it is suppressed.

Windowing of Lines: Naturally, it is more complicated to perform windowing on lines. Here, we must distinguish several cases.

1. CASE IN–IN: First, proposition p_1 is tested for the two endpoints of the line. If it is true (if $p_1 = 1$), the entire line is inside the screen area and thus visible. No action need be taken.

2. CASE IN–OUT: If p_1 is false ($p_1 = 0$), proposition p_2 is additionally tested. If p_2 is also false, part of the line is inside and part is outside the screen area. In this case, the next step is to calculate the intersection of the line, passing through p_i and p_j, with the respective boundary of the screen area, i.e., with one of the lines $x = x_{min}$, $x = x_{max}$, $y = y_{min}$, or $y = y_{max}$.

A test of the last two bits of the tile code, y_0 and y_1, of the point outside the screen area indicates whether the ray starting from the point inside points into the top row (01) or the center row (00) or the bottom row (11). If the vector points into the center row, x_0 determines which of the boundaries, $x = x_{min}$ or $x = x_{max}$, is intersected by the ray. In this case the appropriate value, x_{min} or x_{max}, must be inserted into the line equation

$$y = y_i + \frac{y_j - y_i}{x_j - x_i}(x - x_i), \tag{3.4}$$

yielding the y-coordinate of the intersection. If the ray points into the top row, $y = y_{max}$ is inserted into the equation

$$x = x_i + \frac{x_j - x_i}{y_j - y_i}(y - y_i); \tag{3.5}$$

if the ray points into the bottom row, $y = y_{min}$ is used instead. In either case it is not guaranteed, however, that the point of intersection is on the screen area boundary (Fig. 3.7). Suppose that we calculate in both cases of Fig. 3.7 the intersection with $y = y_{max}$ and subsequently perform test T1 again. In case A, the test will indicate that the point of intersection lies outside the center tile, and the next step is to calculate the point of intersection with $x = x_{min}$. In case B test T1 yields the information that the remaining line is already lying entirely within the center tile (the boundaries are assumed to be part of the screen area).

3. CASE CUT–OUT: If proposition p_2 is true (p_1 must then be false), both endpoints of the line lie outside the screen area. This does not necessarily mean

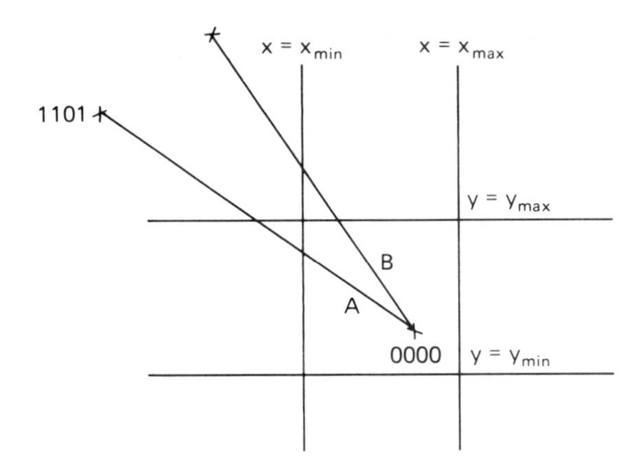

Figure 3.7 Two vectors, A and B, pointing from a point in the center tile (0000) into the same outer tile (1101). Line A intersect $x = x_{min}$; line B intersects $y = y_{max}$.

that the whole line is outside the screen area, as we may have a case OUT–IN–OUT. Which of the two possibilities applies is determined by an additional test, this time by the proposition

$$p_3 = x_0^i x_0^j \vee \bar{x}_0^i \bar{x}_0^j x_1^i x_1^j \vee \bar{y}_0^i \bar{y}_0^j y_1^i y_1^j \vee y_0^i y_0^j. \tag{T3}$$

If $p_3 = 1$, then we are sure that the entire line is outside and, thus, must be eliminated from the display file.

4. CASE OUT–IN–OUT: For $p_3 = 0$, the line may or may not lie partly inside the screen area. However, the tile coding provides in this case no criterion by which this question can be answered. Therefore, we have no other choice but to calculate the points of intersection with two of the lines that are bounding the window. This time, we may, for example, use the two propositions $p_4 = (x_0^i \vee x_0^j)$ and $p_5 = (y_0^i \vee y_0^j)$ in order to determine which coordinate values to insert into the equations for the points of intersection, (3.4) and (3.5). Since the maximum length of a line equals the screen area diagonal, we have only four possible cases which are listed in the following table.

p_4	p_5	$x=$	$y=$	p_4	p_5	$x=$	$y=$
0	0	x_{max}	y_{max}	1	0	x_{min}	y_{max}
0	1	x_{max}	y_{min}	1	1	x_{min}	y_{min}

Of course, the points of intersection calculated thus need not be on the screen area boundary if the line is entirely outside the screen area. This case, however, has been ruled out by test (T3).

As equations (3.4) and (3.5) show, the calculation of the point of intersection requires addition and subtraction as well as multiplication and division. On minicomputers without multiply/divide hardware, it may be more advantageous to find the point of intersection with the screen area boundary by a successive approximation procedure (incidentally, a division procedure is a similar kind of algorithm). To this end, we calculate at first the midpoint of the vector as given by the equations

$$x_m = \frac{x_i + x_j}{2} \quad \text{and} \quad y_m = \frac{y_i + y_j}{2}. \tag{3.6}$$

Subsequently, test T1 is performed, indicating which sector of the bisected vector must be considered further. This procedure is repeated until the successive bisectioning and testing results in two adjacent points, one which lies inside and the other which lies outside the screen area (adjacent points are adjacent with respect to the screen raster grid). If the original vector is k raster units long, this procedure requires $\log_2 k$ steps. The reader may easily verify that our tile code is chosen such that negative coordinates (those x-coordinates of the tiles left of the center and those y-coordinates of the tiles below the center) are represented in two's-complement. The first bit gives the sign (0 for positive and 1 for negative numbers), and the second bit doubles the range. This distinguishes our code from the coding as was originally suggested by Sproull and Sutherland (who are to be credited for having invented this very useful approach) [133]. The Sproull/Sutherland code leads to slightly simpler boolean expressions for the various criteria as discussed above. However, this gain in computational simplicity is negligible, whereas it is a decisive advantage of the tile coding presented here to be integrated into the two's-complement number representation.

It should be emphasized that the method of successive approximation combined with logical tests performed on the "tile code" bits assume:

1. Integer coordinates.
2. Windowing with respect to the screen domain.

In the more general case of windowing with respect to an arbitrarily specified subdomain, the computation-saving tile coding is not applicable, and we have to resort to the purely arithmetic operations of calculating points of intersection. Thus, the former method is adequate for machine-level implementations, whereas the latter method is appropriate for high-level language implementations in the host computer.

We shall now examine how windowing affects the above-defined picture structure. As an example, we consider a segment (Fig. 3.8) that consists of three items. Item (1) is a two-edge polygon, item (2) is an arc, and item (3) is a four-edge polygon. By introducing the windowing (indicated by dashed lines), item (1) is transformed into an item consisting of three line segments (of which one is blank). Item (2) has now three arcs as primitives (of which one is blank) and item (3) is reduced from four to three primitives. In Figure 3.8, blank primitives are marked

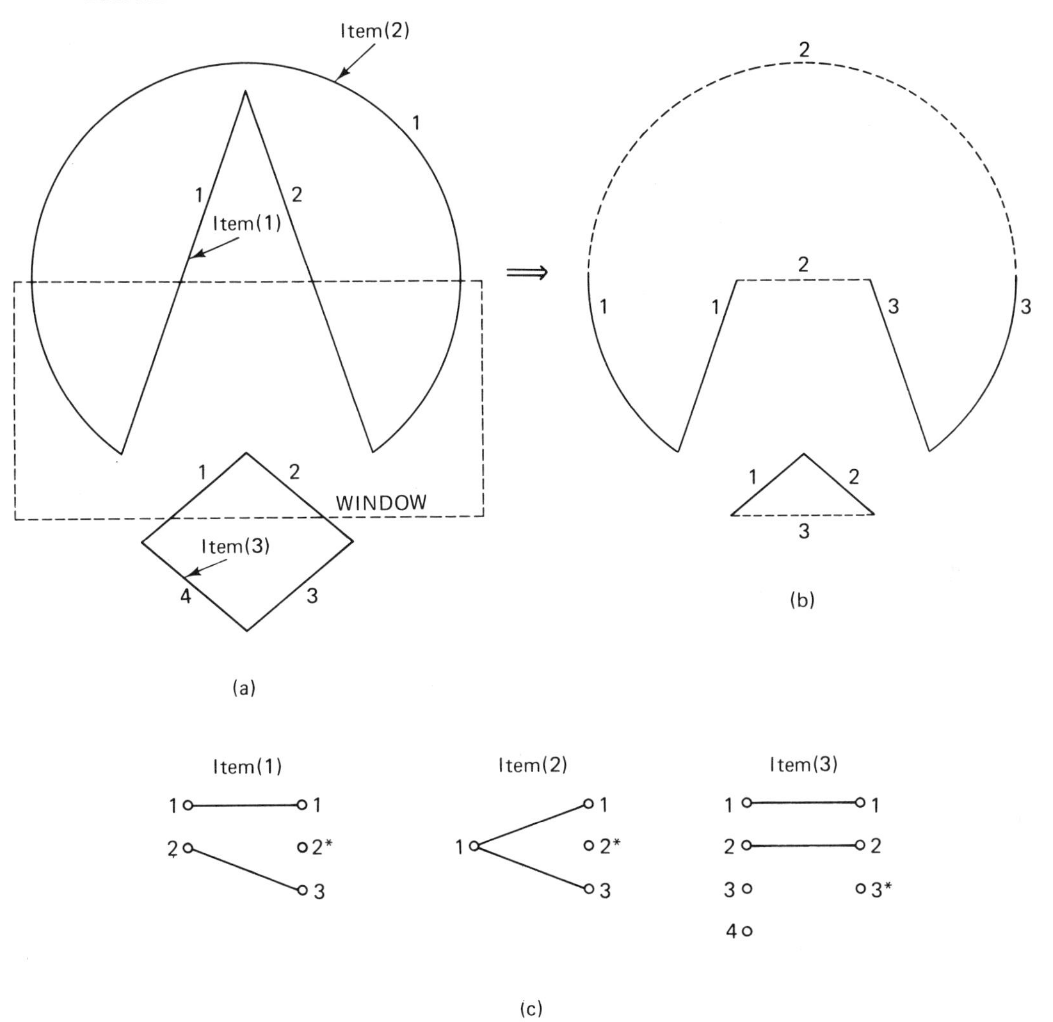

Figure 3.8 (a) Original segment; (b) segment after windowing (dashed lines are actually blank); (c) relations between the items of the original and the transformed segment (starred primitives are invisible and serve solely the purpose of a beam positioning).

by dashed lines. From this example, we draw the following important conclusions concerning the effect of windowing on the above:

1. The calculation of new data as implied by windowing, given by the intersections of the primitives of a windowed picture with the window boundaries, requires that the primitive type be known.

2. Windowing may eliminate certain primitives of an item or it may create new primitives for an item. Hence, windowing may affect the cardinality of items.

3. Windowing does not change the type of an item. Thus, it is ensured that the data structure of an item, the homogeneous, rectangular array, is maintained. However, with a change of the cardinality of an item, the dimension of the array that contains the data of the item is also changed.

4. As a measure for preserving an initially given picture structure, blank (invisible) primitives may be introduced in order to move the beam from the end of a visible portion of an item to the beginning of the next visible portion. This is accomplished by introducing a control vector that determines the visibility of each primitive of an item.

It is desirable for the user to perform windowing by the invocation of appropriate library subroutines rather than to have to program each windowing transformation individually. The fact that a windowing operation depends on the primitive type of an item necessitates different windowing routines for different primitive types. As the result of the fact that a domain transformation can be described only in the form of relations between original and transformed items and not in the form of one-to-one mappings, the new data obtained after windowing cannot simply be stored in the array that contained the original data of an item, if the programming language does not allow for dynamic array dimensions.

The latter problem, which fortunately is not insurmountable, does not exist if windowing is performed strictly at the primitive level, i.e., without a more complex structuring of graphic data. Beam moves between disconnected primitives must then be explicitly programmed by appropriate "move" commands. Whereas such a low-level operation is undesirable for high-level programming, it is a perfectly adequate solution if domain transformations are performed by hardware, as is the case in high-performance display systems.

3.3 GEOMETRIC TRANSFORMATIONS

In the following discussion we shall consider three coordinate transformations:

1. Rotation.
2. Translation.
3. Scaling.

These transformations shall be subsumed under the generic name *geometric transformation*. By definition, geometric transformations are bijective mappings from the coordinate space onto itself. The picture structure as defined previously is not affected at all, and the given incidence relations between the points of a picture [e.g., relations of the kind "being connected by a line (arc, surface patch), etc."] are preserved. Therefore, the primitive type is not pertinent to the execution of a geometric transformation and, consequently, such transformations are strictly performed on data (points); i.e., they do not affect the program. As a result, geometric transformations may be performed either before or after execution of the DPC generating procedures.

3.3.1 Rotation

At first, we consider the rotation of the XY-coordinate system in the Cartesian plane with the origin as the pivotal point. The geometric relations between the original XY-system and the system $X'Y'$ obtained after a rotation by the angle α (in the mathematically positive sense, i.e., counterclockwise) is depicted in Fig. 3.9.

From this picture, we read

$$\begin{aligned} x' &= x\cos\alpha + y\sin\alpha \\ y' &= -x\sin\alpha + y\cos\alpha. \end{aligned} \tag{3.7}$$

If we rotate a picture in the XY-coordinate system (counterclockwise), we must find the coordinates of any point in the $X'Y'$-system and then rotate the $X'Y'$-system back into the normal position. We obtain the same result in equations (3.7) if we replace α by $-\alpha$. Furthermore, we want to consider this rotation in the XY-plane as a special case of a rotation in the XYZ-space, namely, as a planar rotation with respect to the Z-axis (i.e., in Fig. 3.9 the Z-axis is perpendicular to the XY-plane, penetrating the pivotal point). Therefore, $z' = z$, and hence we may write for the coordinates x', y', z' of a point $P(x,y,z)$ after a rotation by the angle α (in a mathematically positive sense),[†]

$$\begin{aligned} x' &= x\cdot\cos\alpha - y\cdot\sin\alpha + 0\cdot z \\ y' &= x\cdot\sin\alpha + y\cdot\cos\alpha + 0\cdot z \\ z' &= x\cdot 0 + y\cdot 0 + z\cdot 1 \end{aligned} \tag{3.8}$$

or, in matrix notation,

$$[x' \quad y' \quad z'] = [x \quad y \quad z] \times \begin{bmatrix} \cos\alpha & \sin\alpha & 0 \\ -\sin\alpha & \cos\alpha & 0 \\ 0 & 0 & 1 \end{bmatrix}. \tag{3.9}$$

Analogously, we have for the transformed coordinates of a point after a rotation with respect to the Y-axis by the angle β

$$[x' \quad y' \quad z'] = [x \quad y \quad z] \times \begin{bmatrix} \cos\beta & 0 & -\sin\beta \\ 0 & 1 & 0 \\ \sin\beta & 0 & \cos\beta \end{bmatrix}, \tag{3.10}$$

[†]Note that in the case of the rotation of two-dimensional objects, we use, of course, only the first two equations in (3.8), setting z to zero.

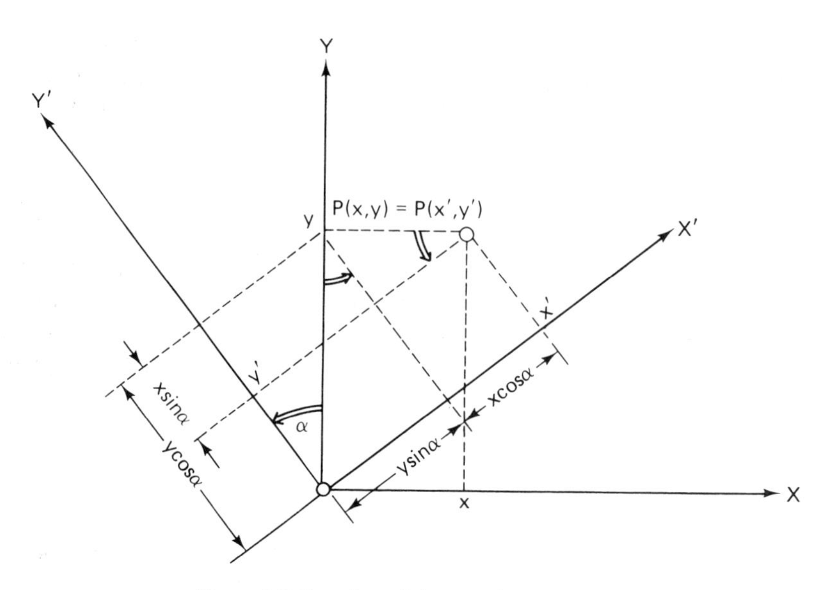

Figure 3.9 Rotation of the coordinate system.

and after a rotation with respect to the X-axis by the angle γ

$$\begin{bmatrix} x' & y' & z' \end{bmatrix} = \begin{bmatrix} x & y & z \end{bmatrix} \times \begin{bmatrix} 1 & 0 & 0 \\ 0 & \cos\gamma & \sin\gamma \\ 0 & -\sin\gamma & \cos\gamma \end{bmatrix}. \tag{3.11}$$

A general rotation in the three-dimensional XYZ-space with respect to all three orthogonal axes (and the origin as pivot) is obtained by a superposition of the three planar rotations. Because of the orthogonality, we obtain a general rotation simply by multiplying the three matrices (3.9), (3.10), and (3.11) with each other. Since a matrix multiplication is not commutative, however, we must specify a certain order in performing the rotation. This convention is arbitrary; although, after the order has been fixed, we must adhere to it.

If we denote the three matrices (3.9), (3.10), and (3.11) by $R(\alpha)$, $R(\beta)$, and $R(\gamma)$, respectively, and if we adopt the customary convention that the rotation with respect to the X-axis is performed first and the rotation with respect to the Z-axis is performed last, we obtain for the coordinates x_R, y_R, z_R of a point after rotation,

$$\begin{bmatrix} x_R & y_R & z_R \end{bmatrix} = \begin{bmatrix} x & y & z \end{bmatrix} \times R(\gamma) \times R(\beta) \times R(\alpha);$$

or with $p = [x \quad y \quad z]$, $p_R = [x_R \quad y_R \quad z_R]$, and

$$R(\alpha,\beta,\gamma) = R(\gamma) \times R(\beta) \times R(\alpha), \tag{3.12}$$

then

$$p_R = p \times R(\alpha,\beta,\gamma). \tag{3.13}$$

We call the points in the three-dimensional space, p and p_R, the original (point) and its image (with respect to a given rotation), and $R(\alpha,\beta,\gamma)$ is termed the 3×3 *rotation matrix*. Execution of the matrix multiplications $R(\alpha,\beta,\gamma)=R(\gamma)\times R(\beta)\times R(\alpha)$ yields the result given below [equation (3.14)].

Example: After a rotation by 90° with respect to all 3 axes, the new coordinates of the point $P(1,1,1)$ must be determined. By applying equation (3.12) with $\cos\alpha=\cos\beta=\cos\gamma=0$ and $\sin\alpha=\sin\beta=\sin\gamma=1$, we obtain

$$[1 \quad 1 \quad 1]=\begin{bmatrix} 0 & 0 & -1 \\ 0 & 1 & 0 \\ 1 & 0 & 0 \end{bmatrix}=[1 \quad 1 \quad -1].$$

If we wish to rotate the point by $-90°$ back to its original position, we must rotate in the reverse order. That is, we must calculate the expression

$$\begin{aligned}[x \quad y \quad z]&=[x_R \quad y_R \quad z_R]\times R(-\alpha)R(-\beta)R(-\gamma)\\ &=[x_R \quad y_R \quad z_R]\times R^{-1}(\alpha,\beta,\gamma)\end{aligned}$$

This results in

$$\begin{aligned}&[1 \quad 1 \quad -1]\times\begin{bmatrix} 0 & -1 & 0 \\ 1 & 0 & 0 \\ 0 & 0 & 1 \end{bmatrix}\times R(-\beta)\times R(-\gamma)\\ &=[1 \quad -1 \quad -1]\times\begin{bmatrix} 0 & 0 & 1 \\ 0 & 1 & 0 \\ -1 & 0 & 0 \end{bmatrix}\times R(-\gamma)\\ &=[1 \quad -1 \quad 1]\times\begin{bmatrix} 1 & 0 & 0 \\ 0 & 0 & -1 \\ 0 & 1 & 0 \end{bmatrix}=[1 \quad 1 \quad 1]\end{aligned}$$

(i.e., the original point). Had we performed the backward rotation in the same order as the forward rotation and not in the reverse order, we would end up with the point $[-1 \quad -1 \quad 1]$, which is clearly not the original.

Thus, for the rotation function

$$R(\alpha,\beta,\gamma)=R(\gamma)\times R(\beta)\times R(\alpha),$$

we have the inverse

$$R^{-1}(\alpha,\beta,\gamma)=R(-\alpha)\times R(-\beta)\times R(-\gamma),$$

which is the rotation that cancels the initial rotation $R(\alpha,\beta,\gamma)$. It follows directly from the definition of R and R^{-1} that $p\times R(\alpha,\beta,\gamma)\times R^{-1}(\alpha,\beta,\gamma)=p$ and hence

$R(\alpha,\beta,\gamma)\times R^{-1}(\alpha,\beta,\gamma)=I$, where I is the identity matrix,

$$I=\begin{bmatrix}1&0&0\\0&1&0\\0&0&1\end{bmatrix}.$$

Carrying out the operation equation (3.12) yields the *3 × 3 rotation matrix*

$$R(\alpha,\beta,\gamma)$$
$$=\begin{bmatrix}\cos\alpha\cdot\cos\beta & \sin\alpha\cdot\cos\beta & -\sin\beta\\ -\sin\alpha\cdot\cos\gamma+\cos\alpha\cdot\sin\beta\cdot\sin\gamma & \cos\alpha\cdot\cos\gamma+\sin\alpha\cdot\sin\beta\cdot\sin\gamma & \cos\beta\cdot\sin\gamma\\ \sin\alpha\cdot\sin\gamma+\cos\alpha\cdot\sin\beta\cdot\cos\gamma & -\cos\alpha\cdot\sin\gamma+\sin\alpha\cdot\sin\beta\cdot\cos\gamma & \cos\beta\cdot\cos\gamma\end{bmatrix} \tag{3.14}$$

For later reference, we abbreviate

$$R(\alpha,\beta,\gamma)=\begin{bmatrix}A&B&C\\D&E&F\\G&H&I\end{bmatrix}; \tag{3.15}$$

i.e., $A, B, C, D, E, F, G, H, I$ stand for the respective terms in the matrix above.

3.3.2 Translation

Translation of a point is performed by adding a positive or negative constant to each coordinate of the point. If these translation parameters are denoted by x_T (translation in x-direction), y_T (translation in y-direction), and z_T (translation in z-direction), we have the coordinates of a rotated and (subsequently) translated point

$$[x_{RT}\quad y_{RT}\quad z_{RT}]=[x_R\quad y_R\quad z_R]+[x_T\quad y_T\quad z_T]$$

or (3.16)

$$p_{RT}=p_R+p_T.$$

Multiplication (from the right-hand side) of equation (3.16) by R^{-1} yields, with $p_R\times R^{-1}=p$,

$$p_{RT}\times R^{-1}=p+p_T\times R^{-1}.$$

$p_{RT}\times R^{-1}$ is the original of p_{RT}; and hence, instead of rotating a set of points first and translating them later, we obtain the same result by translating the set of points first by $p_T\times R^{-1}$, followed by the rotation. It depends on the respective application which procedure is more straightforward; however, if translation is performed in order to relocate a rotated picture, it will certainly be simpler to perform the operation given by equation (3.16).

3.3.3 Scaling

Scaling is performed by the operation

$$[x_S \quad y_S \quad z_S]=[x \quad y \quad z]\times\begin{bmatrix} Sx & 0 & 0 \\ 0 & Sy & 0 \\ 0 & 0 & Sz \end{bmatrix} \quad \text{or} \quad p_S = p \times S. \qquad (3.17)$$

The components Sx, Sy, Sz of the scaling matrix S are the scaling factors in the X, Y, and Z direction, respectively. If the scaling factors are smaller than 1, the object to which scaling is applied is compressed; if the factors are greater than 1, the object is stretched. If $Sx = Sy = Sz$, the object is homogeneously compressed or stretched. Different scaling factors, for instance, must be applied if an object is to be connected to another one but at first does not meet the compatibility conditions. Homogeneous scaling is invariant with respect to rotation, whereas inhomogeneous scaling naturally is affected by rotation; i.e., in the latter case it makes a difference whether scaling is performed before or after rotation.

In the special case of homogeneous scaling, i.e., in the case $Sx = Sy = Sz = c$, the matrix S can be replaced by the scalar c, and instead of equation (3.17) we have the simpler expression

$$p_S = c \cdot p.$$

The rotation of a point p can be interpreted as a transformation of the vector space V of the original into the vector space W of the image. If such a transformation is described by the function $R(p) = p_R$, $p \in V$ and $p_R \in W$, we have for the combined operations of translation, homogeneous scaling, and rotation,

$$R(c \cdot p + p_T) = c \cdot R(p) + R(p_T),$$

i.e., a linear transformation. Note that this holds only for homogeneous scaling.

The combination of the three operations: rotation, translation (after rotation), and scaling (after rotation and translation) can be combined into one matrix multiplication, if we switch to a representation of the point set of an object in a four-dimensional vector space (Section 3.5).

3.4 THE PERSPECTIVE REPRESENTATION OF THREE-DIMENSIONAL OBJECTS

As the display screen is a two-dimensional space, we cannot display three-dimensional objects but only their *projections*. Computationally, projectional transformations are in general quite expensive. Since the generation of a perspective

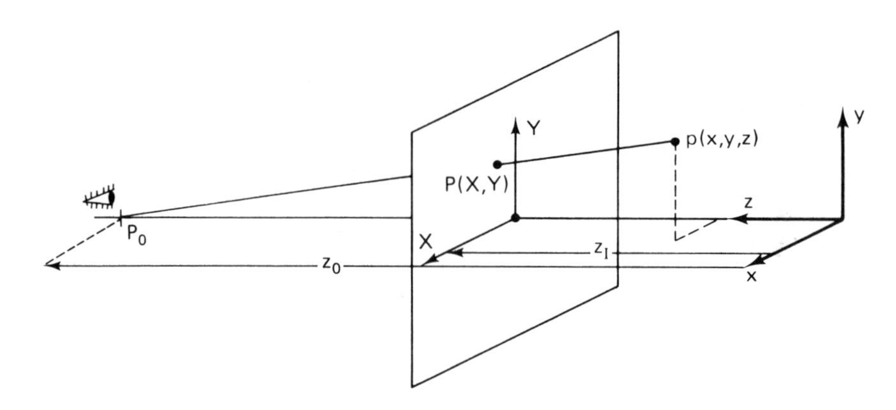

Figure 3.10 Central projection of a point $p(x,y,z)$ onto the image plane.

view of a given object may require the projectional transformation of a considerable number of points, the applied projection is usually restricted to the *central projection* and sometimes to the even simpler *parallel* or *orthographic* projection in order to keep the execution time for the generation of a perspective view within reasonable limits. But even if the time element were not the limiting factor, it would hardly pay to use very sophisticated projections, for the rather restricted resolution of the display screen allows only relatively crude representations and thus rules out the creation of precision artwork.

Figure 3.10 explains the central projection as it is usually applied in computer graphics. The problem is to determine the projection of an object point, located somewhere in a three-dimensional space, onto a plane in that space, called the *image plane*. This projection is called the *image point* of the corresponding object point. In a central projection, the *center of projection*, also called the *viewpoint*, is located on one of the axes of the three-dimensional orthogonal coordinate system. In Figure 3.8 the viewpoint is arbitrarily located on the z-axis. This fact can also be expressed by saying that the *optical axis* is aligned with the z-axis of the coordinate system. The image plane is perpendicular to the optical axis; i.e., in Figure 3.10 it is parallel to the xy-plane of the coordinate system. This fact accounts for the simplicity of a central projection.

Let the coordinates in the two-dimensional coordinate system of the image plane, which we may call the image coordinate system, be denoted by X and Y. Let the distance of the image plane to the origin of the spatial coordinate system be denoted by z_I and the distance of the viewpoint to the origin of the coordinate system by z_O. Then it follows from an elementary theorem of geometry that

$$X = \frac{z_0 - z_I}{z_0 - z} \cdot x$$

$$Y = \frac{z_0 - z_I}{z_0 - z} \cdot y. \tag{3.18}$$

A further simplification can be obtained by setting $z_I = 0$, i.e., by moving the image plane into the xy-plane of the spatial coordinate system. Then we have

$$X = \frac{z_0}{z_0 - z} \cdot x = \frac{1}{1-(z/z_0)} \cdot x$$

$$Y = \frac{z_0}{z_0 - z} \cdot y = \frac{1}{1-(z/z_0)} \cdot y. \tag{3.19}$$

For even further simplification, we may move the viewpoint C toward infinity. With $z_0 \rightarrow \infty$, $1/(1-z/z_0) \rightarrow 1$, and we have the trivial solution

$$X = x \quad \text{and} \quad Y = y.$$

This projection is called *orthographic*. The orthographic projection is a special form of the *parallel projection* by which parallel lines of the three-dimensional object are transformed into parallel lines of its image.

A disadvantage of the perspective transformation from an object point to its image point, as given by equations (3.19), is its nonlinearity. This can be avoided by representing the object points in *homogeneous coordinates*. Homogeneous coordinates play a central role in *projective geometry*. Although in computer graphics we are not concerned with projective geometry but only with the simpler Euclidean geometry, we may briefly explain the underlying notion of homogeneous coordinates, keeping in mind, however, that in our case homogeneous coordinates are only an artifact for linearizing the perspective transformations we may have to perform [111].

So far, a point in a two-dimensional space was defined by an ordered pair (x,y). A line, on the other hand, may be defined as the set of solutions (x,y) to a linear equation $ax + by + c = 0$, $a \neq 0$ or $b \neq 0$. In this equation of a line, $ax + by + c = 0$, we think normally of a,b,c as fixed constants and x,y as variables. For example, if we would ask for the lines passing through a given point (u,v), any such line would be represented by the linear equation above, and the answer to our question would be given by all triples (a,b,c) for which $au + bv + c = 0$. If a,b,c are understood to be constants, and x and y variables, the expression $ax + by + c = 0$ has two terms of first degree (x and y occur only to the first power) and one term of zero degree (the constant c). Therefore, this expression is called *inhomogeneous*. If, on the other hand, a,b, and c are considered as the variables, and x,y as constants, all terms have the same degree, and the expression is called *homogeneous*.

In projective geometry [64], two classes of objects are involved, the set of all points and the set of all lines; and no attempt is made to principally distinguish between these two sets. Furthermore, according to the principle of duality, any true proposition, such as "there is exactly one line on two distinct points," leads to another true proposition if the items "lines" and "points" are interchanged. In classic Euclidean geometry, there is the exception that two parallel lines fail to intersect. In projective geometry, this exception is eliminated by creating additional points which have no real existence, which are called "ideal" points. It is customary, for obvious reasons, to call an ideal point also a "point at infinity." Each line has exactly such an ideal point which is on no other line; and the ideal points in a plane constitute an ideal line, the line at infinity. Following this concept, the orthographic projection that we introduced in the preceding section as a special case of the central projection is a central projection with an ideal point as the viewpoint.

The additional ideal points can be easily introduced, and the yet existing discrepancy between points and lines—the first defined by an inhomogeneous expression and the latter by a homogeneous one—can be easily eliminated by modifying the defining linear equation into the form

$$au + bv + cw = 0,$$

which has the advantage of being homogeneous. The original form, $ax + by + c = 0$, can be easily restored. Suppose that $w \neq 0$; then u/w and v/w are a pair of numbers satisfying the original equation.

Conversely, if (x,y) is a pair of numbers satisfying the original equation, then $(x,y,1) = (1/w)(u,v,w)$ will be a triple satisfying $au + bv + cw = 0$, as will $(\lambda x, \lambda y, \lambda)$, as long as $x = u/w$, $y = v/w$, and $w \neq 0$. There are further solutions to $au + bv + cw = 0$, which do not correspond to solutions of $ax + by + c = 0$, namely, those for which $w = 0$. The triple $(0,0,0)$ is a trivial solution, but any other triple $(u, v, 0)$, $u \neq 0$ or $v \neq 0$, yields additional information on the coefficients a and b, namely, that a and b are such that $au + bv = 0$.

By the specification $x = u/w$, $y = v/w$, and $w \neq 0$, we associate uniquely a triple (u, v, w) with a pair (x,y). In geometry, the pair (x,y) represents (we may also say *is*) a point. Under the condition above, the triple (u, v, w), $u = wx$; $v = wy$; $w \neq 0$, represents this point, too. Therefore, the triple (u, v, w) is said to be the *homogeneous coordinates* of the point (x,y). Note that if (u, v, w) are homogeneous coordinates of a point, then all triples $(\lambda u, \lambda v, \lambda w)$ are also homogeneous coordinates of the same point, provided that $\lambda \neq 0$; i.e., any point forms an entire class of solutions to $au + bv + cw = 0$.

If we write the equation $au + bv + cw = 0$ in matrix form,

$$[u \quad v \quad w] \times \begin{bmatrix} a \\ b \\ c \end{bmatrix} = 0,$$

the matrix $[u \quad v \quad w]$ represents all points on a line, and the matrix $\begin{bmatrix} a \\ b \\ c \end{bmatrix}$ represents all lines on a point, and we have now the desired symmetry between points and lines. With $u = wx$ and $v = wy$, we now obtain the line equation in the form

$$w \cdot [x \quad y \quad 1] \times \begin{bmatrix} a \\ b \\ c \end{bmatrix} = 0.$$

As this satisfies the equation for any $w \neq 0$, we might as well say that

$$[x \quad y \quad 1]$$

represents all points lying on a line. Furthermore, it can be shown [64] that the equation $au + bv + cw = 0$ provides exactly one additional class of solution given by the case $z = 0$, and this additional solution specifies the ideal point on the line $ax + by + c = 0$. (Note that the same ideal point is assigned to parallel lines, whereas different ideal points are assigned to intersecting lines.)

In analogy to the two-dimensional case, we define the quartuple (r,s,t,u), $u \neq 0$, as homogeneous coordinates of a (real) point (x,y,z) in a three-dimensional space. The relation between the homogeneous and the inhomogeneous coordinates are now $x = r/u, y = s/u, z = t/u, u \neq 0$; and a point in the three-dimensional space may be represented by†

$$[x \quad y \quad z \quad 1] = \frac{1}{u}[r \quad s \quad t \quad u]. \tag{3.20}$$

Since the only condition for u is that $u \neq 0$ but is otherwise arbitrary, we may, for example, set $u = 1$, or we may as well use u as an arbitrary scaling factor [e.g., we could set $u = \max(r,s,t)$, thus scaling the given space into a unit space].

Any point, represented by homogeneous coordinates, is now a vector of dimension four; and the matrices for rotation, translation, and scaling are now square matrices of rank 4×4. Furthermore, we shall see that the perspective, as given by a central projection, can also be defined as a matrix operation applied to all points of a three-dimensional structure, if these points are represented in homogeneous coordinates.

We suggested above that the perspective transformation given by equations (3.19) is linear when expressed in terms of homogeneous coordinates. We now assert that the matrix

$$\mathrm{PT} = \begin{bmatrix} 1 & 0 & 0 & 0 \\ 0 & 1 & 0 & 0 \\ 0 & 0 & 0 & -1/z_0 \\ 0 & 0 & 0 & 1 \end{bmatrix} \tag{3.21}$$

is exactly such a linear transformation that maps an object point (given in homogeneous coordinates) to its image point (also given in homogeneous coordinates).

The proof for this assertion can easily be given by carrying out the operation. To this end, let $v = [wx \quad wy \quad wz \quad w]$ be the homogeneous coordinate vector representation of an object point $p(x,y,z)$. Multiplying v by PT yields

$$v \times \mathrm{PT} = [wx \quad wy \quad 0 \quad w(1 - (z/z_0))].$$

†An n-tuple can alternatively be represented as a vector of n components.

Assuming that the vector obtained by this operation is the homogeneous coordinate representation of the image point, we restore the normal Cartesian coordinates of the image point by dividing the first three components of this vector by its fourth component. Hence, we obtain the vector

$$\left[\frac{x}{1-(z/z_0)} \quad \frac{y}{1-(z/z_0)} \quad 0\right].$$

A comparison with equations (3.19) shows that this is, in fact, the coordinate vector of the image point.

3.5 THE "4×4 MATRIX" FOR ROTATION, SCALING, TRANSLATION, AND PERSPECTIVE

We can now show that the homogeneous coordinate representation of an object point allows us to perform the combined transformations of rotation, scaling, translation, perspective by one single matrix multiplication, i.e., one single linear transformation [111]. The result is the homogeneous coordinate representation of the corresponding image point. In order to obtain the coordinates of the image point in the XY-coordinate system in the image plane, a subsequent division of the first two components of the result vector by the fourth component must be performed (the third component of the result vector is identically zero). Hence, the division required in the perspective transformation cannot be avoided, but its place is now outside the sequence of linear transformations for rotation, scaling, translation, and perspective. In the following, we denote again an object point by the (homogeneous) coordinate vector $[wx \quad wy \quad wz \quad w]$.

For the sake of simplicity, we set for the moment $w=1$, that is, a point is given by $[x \quad y \quad z \quad 1]$[†]. Instead of the 3×3 rotation matrix, we have now a 4×4 rotation matrix which is of the form

$$\tilde{R}(\alpha,\beta,\gamma)=\begin{bmatrix} A & B & C & 0 \\ D & E & F & 0 \\ G & H & I & 0 \\ 0 & 0 & 0 & 1 \end{bmatrix}, \qquad A,B,\ldots,I \text{ being the components of the } 3\times3 \text{ matrix given in equations (3.15)}$$

(3.22)

Proof: Let $[x_R \quad y_R \quad z_R \quad 1]=[x \quad y \quad z \quad 1]\times\tilde{R}$ be the image of point $[x \quad y \quad z \quad 1]$, both points represented in homogeneous coordinates. Execution of the matrix multiplication yields $[Ax+Dy+Gz \quad Bx+Ey+Hz \quad Cx+Fy+Iz \quad 1]$; i.e., $x_R=Ax+Dy+Gz$, $y_R=Bx+Ey+Hz$, $z_R=Cx+Fy+Iz$, and we have the same result, $p_R=p\times R$, as in Section 3.3.1.

[†]If we choose to set $w\neq1$, the only difference is that we eventually have to divide all coordinate values by w.

Rotation and scaling may be combined by multiplying the 3×3 rotation matrix with the (diagonal) scaling matrix given in equation (3.17) and expanding subsequently the result into a 4×4 matrix. Consequently, if we use homogeneous coordinates, rotation and scaling can be performed by multiplying a point with the matrix

$$\widetilde{RS} = \begin{bmatrix} A\cdot Sx & B\cdot Sy & C\cdot Sz & 0 \\ D\cdot Sx & E\cdot Sy & F\cdot Sz & 0 \\ G\cdot Sx & H\cdot Sy & I\cdot Sz & 0 \\ 0 & 0 & 0 & 1 \end{bmatrix}; \tag{3.23}$$

that is, $[x_{RS} \quad y_{RS} \quad z_{RS} \quad 1]=[x \quad y \quad z 1]\times \widetilde{RS}$.

In a consistent way, translation can be performed by multiplying a point with the matrix

$$\tilde{T} = \begin{bmatrix} 1 & 0 & 0 & 0 \\ 0 & 1 & 0 & 0 \\ 0 & 0 & 1 & 0 \\ x_t & y_T & z_T & 1 \end{bmatrix}. \tag{3.24}$$

Proof: $[x \quad y \quad z \quad 1]\times \tilde{T}=[x+x_T \quad y+y_T \quad z+z_T \quad 1]$.

Rotation, scaling, and translation may be performed simultaneously by one matrix multiplication, if we construct a matrix $\widetilde{RST}$ that takes care of all three operations. Because of the particular structure of $\widetilde{RS}$ and $\tilde{T}$, we obtain simply

$$\widetilde{RS} \times \tilde{T} = \widetilde{RST} = \begin{bmatrix} A\cdot Sx & B\cdot Sy & C\cdot Sz & 0 \\ D\cdot Sx & E\cdot Sy & F\cdot Sz & 0 \\ G\cdot Sx & H\cdot Sy & I\cdot Sz & 0 \\ x_T & y_T & z_T & 1 \end{bmatrix}, \tag{3.25}$$

and hence

$$[x_{RST} \quad y_{RST} \quad z_{RST} \quad 1]=[x \quad y \quad z \quad 1]\times \widetilde{RST}.$$

Hence, by multiplying a point $p=[x \quad y \quad z \quad 1]$ with the matrix $\widetilde{RST}$, we obtain the rotated, scaled, and translated image of $p, p_{RST}=p\times \widetilde{RST}$. The projection of an object point thus transformed onto the image plane of perspective view is, according to the preceding section, obtained by the operation $p_{RST}\times \text{PT}$. Thus, the total transformation from an object point p to its perspective image p_I is given by

$$p_I = p \times \widetilde{RST} \times \mathrm{PT} = p \times R4, \qquad R4 = \widetilde{RST} \times \mathrm{PT}. \tag{3.26a}$$

Carrying out the matrix multiplication $\widetilde{RST} \times \mathrm{PT}$ yields

$$R4 = \begin{bmatrix} A\cdot Sx & B\cdot Sy & 0 & -C\cdot Sz/z_0 \\ D\cdot Sx & E\cdot Sy & 0 & -F\cdot Sz/z_0 \\ G\cdot Sx & H\cdot Sy & 0 & -I\cdot Sz/z_0 \\ x_T & y_T & 0 & 1-z_T/z_0 \end{bmatrix}. \tag{3.26b}$$

The matrix $R4$ is termed the "4×4 matrix of rotation, scaling, translation, and perspective." Thus, the entire process of rotation, scaling, translation, and calculation of the coordinates of a perspective representation requires the following steps:

1. All points of a structure in the three-dimensional space must be represented in homogeneous coordinates, i.e., a point is given by

$$p = [x \quad y \quad z \quad 1].$$

2. A point is transformed by applying the 4×4 matrix, resulting in

$$p \times R4 = \left[x_{RST} \quad y_{RST} \quad 0 \quad 1 - \frac{z_{RST}}{z_0} \right].$$

3. The first two components of the resulting vector must be divided by the fourth component of the same vector [which is $(1 - z_{RST}/z_0)$], and the obtained result are the coordinates of the perspective representation of the point†

The whole complex of transformation may thus be integrated into a software module (a subroutine) that has the coordinate values (x,y,z) of a point in the three-dimensional space as input, together with 10 transformation parameters,

- The three angles of rotation α, β, γ.
- The three scaling factors Sx, Sy, Sz.
- The three translation parameters x_T, y_T, z_T.
- The parameter z_0, defining the center of perspective.

Output variables are the coordinate values x_p and y_p of the perspective representation. This module can be depicted as a transfer system as shown in Figure 3.11. In some computer display systems, the combined transformation has

†The division is avoided if orthographic projection is performed.

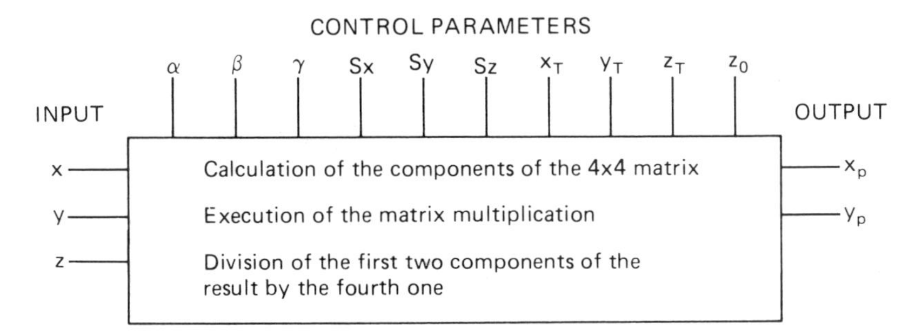

Figure 3.11 Transfer system view of a module for rotation, scaling, translation, and perspective.

been partially or totally realized by a special hardware module [147, 130, 133, 32, 131, 11, 80]. In that case, the combined transformations can be performed fast enough to be executed in real time.

A natural default for the control parameters of a geometric transformation is the identity element of the operation. That is, we apply the following sets of default parameters (written in vector form)

$$\begin{bmatrix}0 & 0 & 0\end{bmatrix} \quad \text{for rotation}$$
$$\begin{bmatrix}1 & 1 & 1\end{bmatrix} \quad \text{for scaling}$$
$$\begin{bmatrix}0 & 0 & 0\end{bmatrix} \quad \text{for translation.}$$

Some Topics of Elementary Geometry in Homogeneous Coordinates: We conclude this section by presenting for practical use some topics of elementary geometry in homogeneous coordinates [111]. We restrict ourself to the two-dimensional case. The homogeneous coordinate representation of a point in 2-space is the row vector

$$p = \begin{bmatrix} wx & wy & w \end{bmatrix}, \tag{3.27}$$

and a line is represented by a column vector

$$\lambda = \begin{vmatrix} a \\ b \\ c \end{vmatrix}. \tag{3.28}$$

The equation of a line is given by $p \times \lambda = 0$. The location of a point p with respect to a line can thus be tested by forming the inner product

$$p \times \lambda = D.$$

For $D = 0$, the point is on the line. For $D \neq 0$, the sign of D indicates on which side of the line the point lies. In this case, the perpendicular distance is

$$\text{distance} = \left(\frac{p}{w} \times \lambda\right)(a^2 + b^2)^{-1/2}. \tag{3.29}$$

A line λ_{pq} going through two points, $p=[p_1 \quad p_2 \quad p_3]$ and $q=[q_1 \quad q_2 \quad q_3]$, is represented by the column vector

$$\lambda_{pq}=\begin{vmatrix} q_3p_2-q_2p_3 \\ q_1p_3-q_3p_1 \\ q_2p_1-q_1p_2 \end{vmatrix}. \tag{3.30}$$

The point of intersection $p_{\lambda\mu}$ of two lines,

$$\lambda=\begin{vmatrix} a \\ b \\ c \end{vmatrix} \quad \text{and} \quad \mu=\begin{vmatrix} d \\ e \\ f \end{vmatrix} \tag{3.31a}$$

is given by the vector

$$p_{\lambda\mu}=\left[(fb-ec) \quad (cd-af) \quad (ea-db)\right]. \tag{3.31b}$$

In 3-space, a point is represented by the vector

$$p=\left[wx \quad wy \quad wz \quad w\right]. \tag{3.32}$$

A line λ can be represented in parametric form (see Chapter 4) as

$$\lambda=[t \quad 1]\times\begin{bmatrix} a & b & c & d \\ e & f & g & h \end{bmatrix}. \tag{3.33}$$

A line passing through the points $p=[p_1 \quad p_2 \quad p_3 \quad p_4]$ and $q=[q_1 \quad q_2 \quad q_3 \quad q_4]$, with the (arbitrarily chosen) values of the parameter t being $t=t_p$ in point p and $t=t_q$ in point q, must satisfy the matrix equation

$$\begin{bmatrix} t_p & 1 \\ t_q & 1 \end{bmatrix}\times\begin{bmatrix} a & b & c & d \\ e & f & g & h \end{bmatrix}=\begin{bmatrix} p_1 & p_2 & p_3 & p_4 \\ q_1 & q_2 & q_3 & q_4 \end{bmatrix},$$

and hence

$$\lambda=[t \quad 1]\times\begin{bmatrix} t_p & 1 \\ t_q & 1 \end{bmatrix}^{-1}\times\begin{bmatrix} p_1 & p_2 & p_3 & p_4 \\ q_1 & q_2 & q_3 & q_4 \end{bmatrix}. \tag{3.34}$$

A plane is represented by a column vector

$$\pi=\begin{bmatrix} a \\ b \\ c \\ d \end{bmatrix}. \tag{3.35a}$$

The equation of a plane (condition that a point p be on the plane π) is

$$p \times \pi = 0. \tag{3.35b}$$

The perpendicular distance from p to π is

$$\text{distance} = \left(\frac{p}{w} \times \pi\right)(a^2 + b^2 + c^2)^{-1/2}. \tag{3.36}$$

Three planes π_1, π_2, and π_3 intersect in one point p such that $p \times \pi_1 = 0$, $p \times \pi_2 = 0$, and $p \times \pi_3 = 0$. If the three planes are represented by

$$\pi_1 = \begin{bmatrix} a_1 \\ b_1 \\ c_1 \\ d_1 \end{bmatrix}, \quad \pi_2 = \begin{bmatrix} a_2 \\ b_2 \\ c_2 \\ d_2 \end{bmatrix}, \quad \pi_3 = \begin{bmatrix} a_3 \\ b_3 \\ c_3 \\ d_3 \end{bmatrix}, \tag{3.37a}$$

and if M denotes the matrix

$$M = \begin{bmatrix} a_1 & a_2 & a_3 & 0 \\ b_1 & b_2 & b_3 & 0 \\ c_1 & c_2 & c_3 & 0 \\ d_1 & d_2 & d_3 & 1 \end{bmatrix}, \tag{3.37b}$$

then the point p is given by the bottom row of the inverse matrix M^{-1} [5]. The three planes do not intersect in one point if M has no inverse.

Three points determine a plane if they are not collinear. The condition for the three points, $p = [p_1 \quad p_2 \quad p_3 \quad p_4]$, $q = [q_1 \quad q_2 \quad q_3 \quad q_4]$, and $r = [r_1 \quad r_2 \quad r_3 \quad r_4]$ not to be collinear is that the matrix

$$p = \begin{bmatrix} p_1 & p_2 & p_3 & p_4 \\ q_1 & q_2 & q_3 & q_4 \\ r_1 & r_2 & r_3 & r_4 \\ 0 & 0 & 0 & 1 \end{bmatrix}$$

is nonsingular, i.e., has an inverse.

3.6 A STANDARD TRANSFORMATION SYSTEM

In Section 1.6 we discussed briefly the problem of deciding in which order to perform the clipping operation and the geometric picture transformations. Further, the discussion in the preceding sections of this chapter exhibits a certain dichotomy between two-dimensional and three-dimensional objects. A transformation system

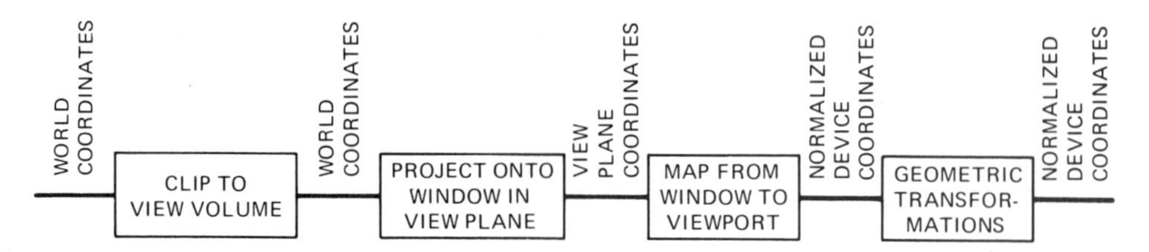

Figure 3.12 Process of image generation of the CORE systems.

that eliminates this dichotomy and answers the question concerning the order of operations has been proposed as part of the CORE system†. In this system the dichotomy between two-dimensional and three-dimensional objects is eliminated by defining two-dimensional objects as a special case of three-dimensional objects, and the question of the order of picture transformations has been resolved, after elaborate deliberations, by proposing a standardized transformation system (Fig. 3.12). As mentioned above, graphic objects are defined always as three-dimensional objects. Consequently, the "world" is a three-dimensional, Cartesian space, and the window in that space is a three-dimensional window, called *view volume.* The form of the view volume depends on the projection type applied.

The general process of image creation is conceptualized as a four-step procedure. The first step consists in the clipping of world coordinate objects to the specified view volume; i.e., this step defines the portion of the world coordinate space which is to be projected onto a projection plane or *view plane* in step 2 of the process. The view plane is a plane in the world. On the view plane, a second coordinate system is defined, called view-plane coordinates. The intersection of the view volume and the view plane defines a window in the view plane, onto which the world coordinate objects in the view volume are projected. Conversely, a window in the view plane, together with a given center of projection (for perspective projection) or direction of projection (for parallel projection), determines the view volume. In step 3 of the process, the view-plane window is mapped to a *viewport* that is defined in a *normalized device coordinate system.* Step 4 of the process finally consists of a mapping from the (two-dimensional) normalized device coordinates into physical device coordinates.

The four operations are specified by a set of *viewing parameters* or their appropriate default values. There is no ordering on the set of parameters, and only one operation can be in effect at any one time. Changes of the viewing parameters cannot take place during the construction of a segment. The view plane is used both to serve as the projection plane and to allow the specification of a window (and hence of a view volume) for clipping. The view plane is defined by specifying a "view reference point," a "view plane normal," and a "view plane distance." The

†CORE is the standard graphic software system proposed by the Graphic Standards Planning Committee of ACM/SIGGRAPH. See *Computer Graphics 11*, 3, Fall 1977.

view reference point is usually a point on or near the object to be viewed. Its default is the origin of the view-plane coordinate system. The *view-plane normal* is a vector perpendicular on the view plane and beginning in the view reference point, thus specifying the orientation of the view plane in the three-dimensional world space. The *view-plane distance* is a signed quantity which determines the distance between the reference point and the view plane, measured along the view-plane normal. The default for the view-plane orientation is a plane $Z=d$, where d is the view-plane distance. The default for the view-plane distance is zero, in which case the reference point is in the view plane.

Having specified a view plane through the viewing parameters, the specification of a window in the view plane and the type of projection determines both the view volume (for clipping) and the projection of a (clipped) object onto the window. If a perspective projection is selected, the center of projection must be specified relative to the view reference point. If a parallel projection is selected, the end point of a vector with start point in the view reference point must be given. This vector defines the *direction of projection*. If the projection direction is normal to the view plane, the projection is *orthographic*. If no projection type is selected, the default is the orthographic projection.

The edges of the window in the view plane are parallel to the view-plane coordinate system, in short the *UV system*. The origin of the *UV* system is the point where the line collinear with the view-plane normal pierces the view plane. The orientation of the *UV* system is determined by a *view-up vector*. The view reference point is the start point of the view-up vector, whereas the end point is specified by the programmer relative to the start point (in world coordinates). The view-up vector is projected onto the view plane, with a projection parallel to the view-plane normal. This projection determines the positive direction of the V-axis as shown in Figure 3.13. The *UV* system is a left-hand coordinate system. The window size is given in terms of maximum and minimum U and V values. Figures 3.14 and 3.15 present examples of view volumes, determined by a window specification and a projection selection.

The view volumes shown in Figures 3.14 and 3.15 are semiinfinite and infinite, respectively. In either case, they may be truncated to a finite-view volume by specifying a *front clipping plane* and a *back clipping plane*. Both clipping planes are parallel to the view plane; and both are specified by a distance along the view-plane normal from the view reference point (Figs. 3.16 and 3.17). This clipping scheme is called *depth clipping*.

Geometric transformations are separate transformations unrelated to the viewing operations. Geometric transformations are applied on the object representation in normalized device space, that is, *after* the viewing operations have been applied to the object in world space. The insertion of geometric transformations between step 3 and 4 of the image-generating process makes this process and the geometric transformations mutually independent. Of course, this rules out the use of a 4×4 transformation matrix.

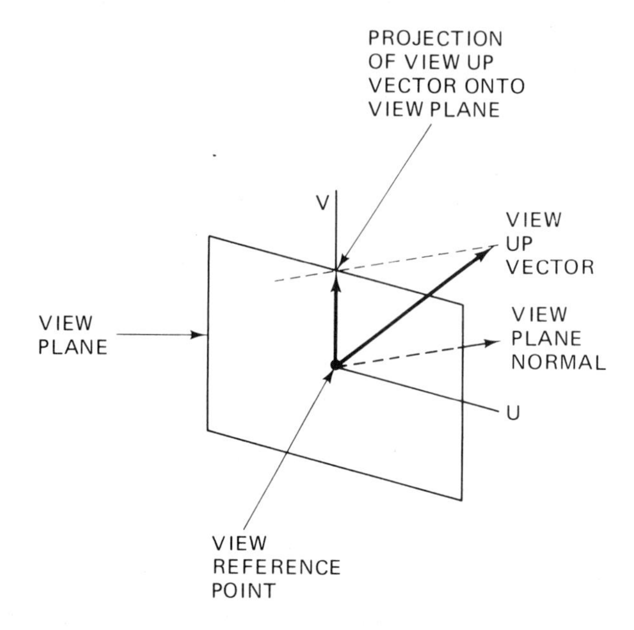

Figure 3.13 Definition of the *UV* system in the view plane.

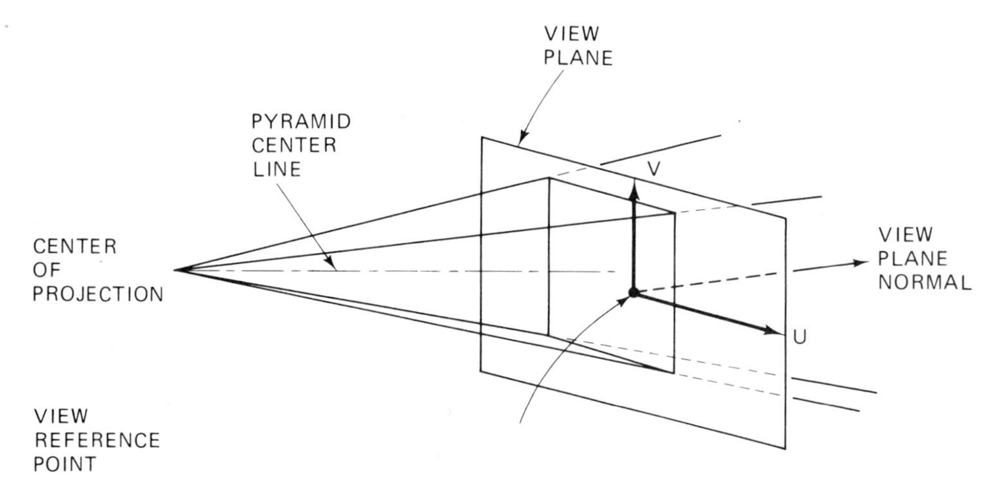

Figure 3.14 View volume for a perspective projection ("viewing pyramid").

The CORE concept differs in two major aspects from the conventional approach as discussed in the preceding sections. First, the viewing specification is based on a certain part of the object (selected by the choice of a reference point) rather than on a "viewpoint" or "eye point" in the world system. Furthermore, the center line of the view volume need not be normal to the view plane. Second, all objects are conceptualized as three-dimensional objects; of which two-dimensional

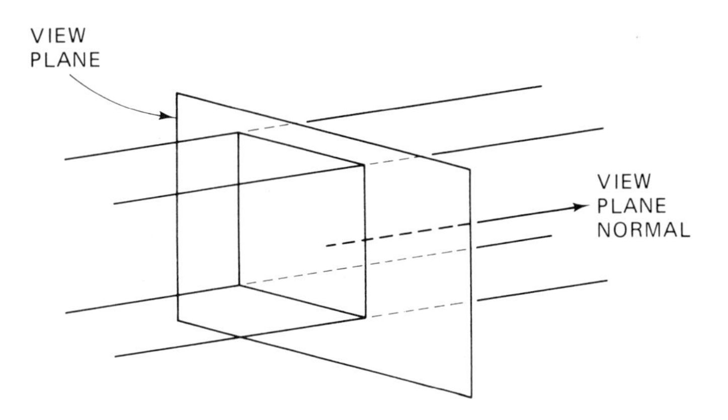

Figure 3.15 View volume for an orthographic projection ("viewing parallelepiped").

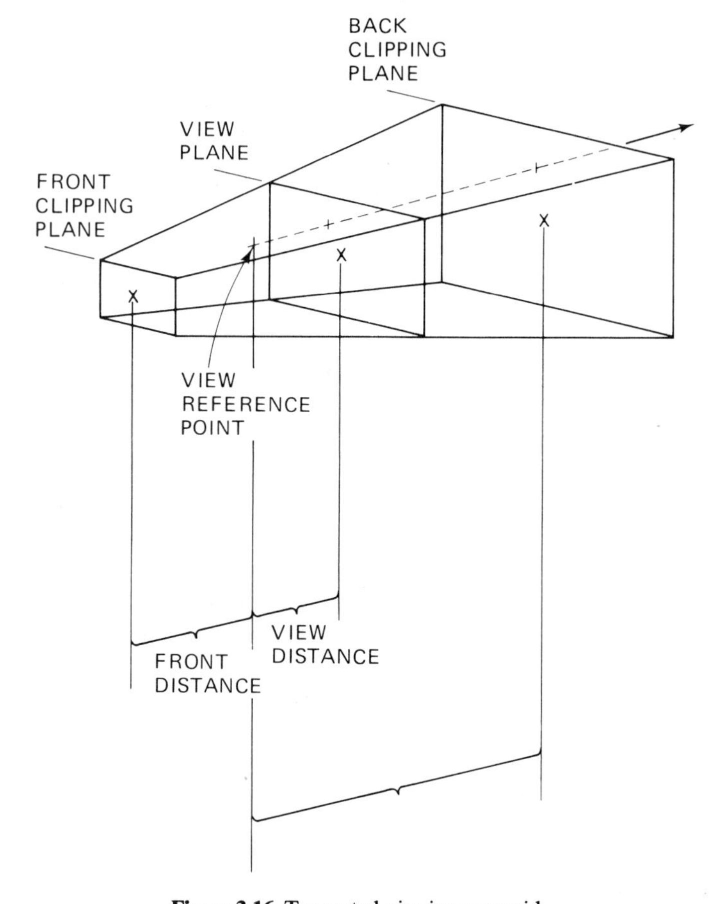

Figure 3.16 Truncated viewing pyramid.

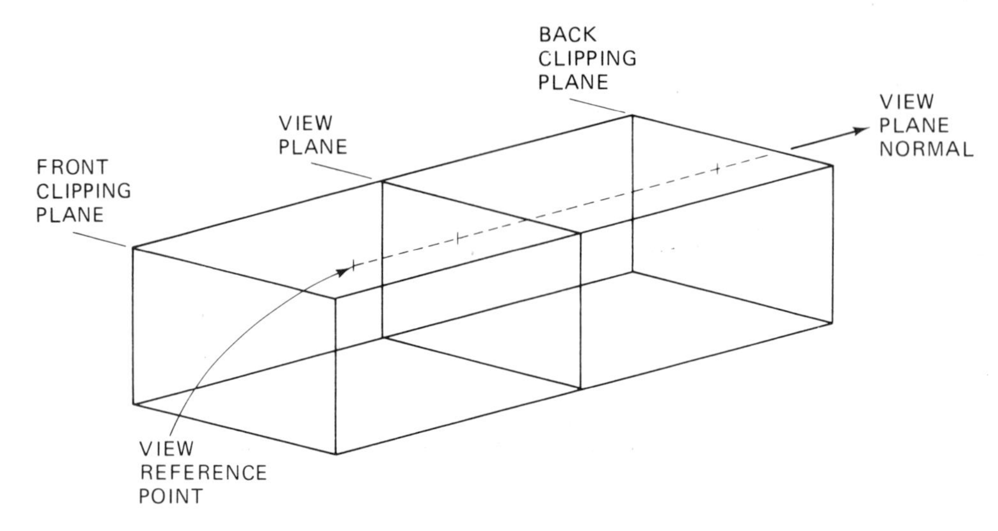

Figure 3.17 Truncated viewing parallelepiped.

objects are a special case. CORE is conceived as a universal system that encompasses the entire realm of computer graphics activities from plotting to interactive graphics.

EXERCISES

1. Given a three-dimensional picture definition space D×D×D with D= [0 : 1000] and four points P1=(100, 100, 100), P2=(300, 100, 100), P3= (100, 400, 100), P4=(100, 100, 200).
   (a) Augment the set of points {P1, P2, P3, P4} by the appropriate number of additional points so that the resulting set of points forms the vertices of a rectangular prism.
   (b) Organize a picture file for the rectangular prism
      (b1) In the form of a point list only.
      (b2) In the form of a point list and an associated line list.
      (b3) In the form of a list that contains the point coordinates together with a control parameter for the visibility of a line segment.
   (c) Write a procedure for an efficient conversion of the picture file representation (b2) into the representation (b3). By "efficient" we mean that certain redundancies may be tolerated for the sake of a simpler conversion scheme; however, they should be kept small. Use a language of your choice. The parameters are: M=number of lines; N=number of points; D=dimension of the space; PLIST=point list; LLIST=line list; POCO= list containing points and control parameters.

2. (a) Given the following picture:

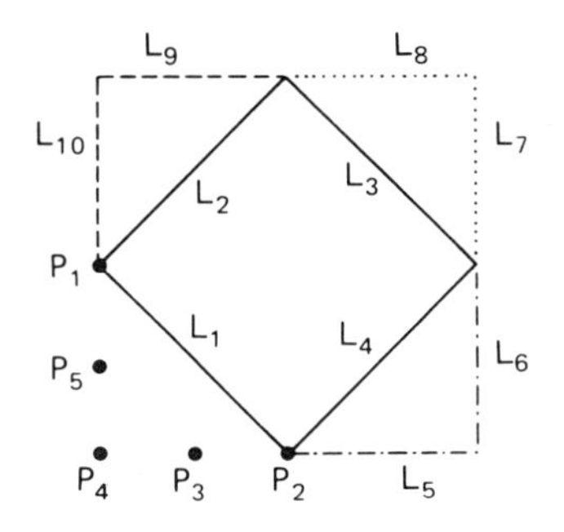

Define its structure according to the hierarchical structuring of pictures into segments, items, and primitives as introduced in this chapter. Your definition should allow for the deletion of the blinking inner square in response to a lightpen pick of either one of the primitives L_1, L_2, L_3, or L_4. Construct a table that lists the segments of the picture; the items of each segment; and the primitives, the mode, the style, and the status of each item.

(b) Define a structure of the picture in (a) which allows for the deletion of each one of the triangles in response to a lightpen pick of the appropriate included (blinking) primitive L_1, L_2, L_3, or L_4. Construct a table that lists the segments of the picture; the items of the segments; and the primitives, the mode, the style, and the status of each item.

3. Given the picture in a virtual domain as diagrammed below. This picture shall be transformed into the picture in the screen domain also depicted below.

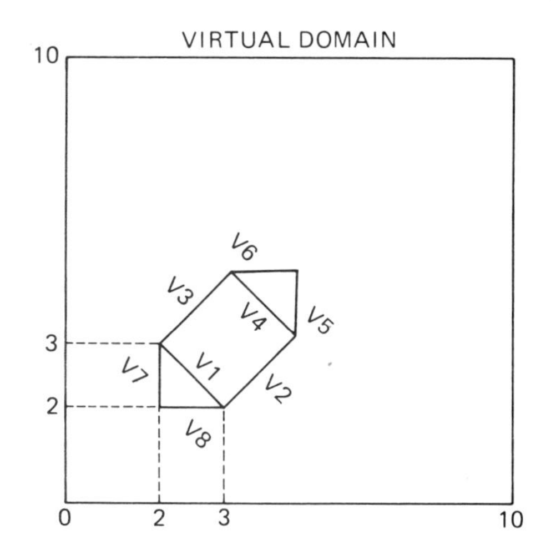

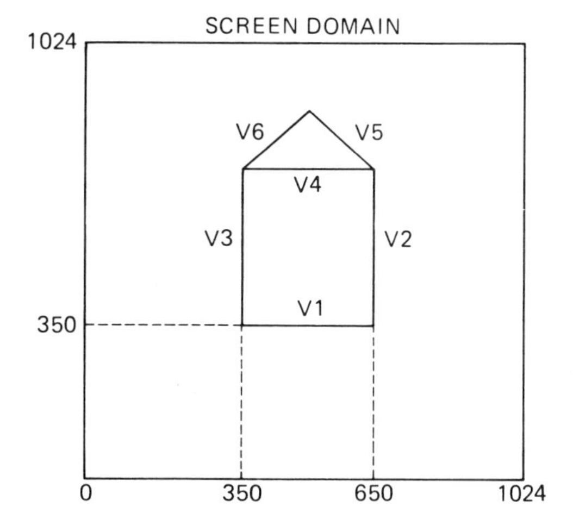

(a) Determine the scale factor for the scaling from the virtual domain to the screen domain.

(b) Draw diagrams in the virtual domain and in the screen domain, which illustrate the procedural steps in the picture transformation, for the two alternative sequences of transformation steps as discussed in Section 1.6.

4. Suppose that your graphical objects are defined in the screen domain SD= [0:1023]×[0:1023]. Write a FORTRAN subroutine for the clipping of a line

CLIP(IX1, IX2, IY1, IY2, IXS, IXT, IYS, IYT).

IX1, IX2, IY1, IY2 are input parameters specifying the window boundaries (which are given by the four equations: IX1=constant, IX2=constant, IY1= constant, IY2=constant). IXS, IXT, IYS, IYT are transition parameters; i.e., they function as input as well as output parameters. (IXS, IYS) and (IXT, IYT) are the end-point coordinates of a line segment before and after windowing. The subroutine changes the original coordinates of the end points of a line segment so that, after windowing, it is scissored to the part that is within the window boundaries.

Make sure that you handle the possible case "out–in–out" as illustrated in the figure.

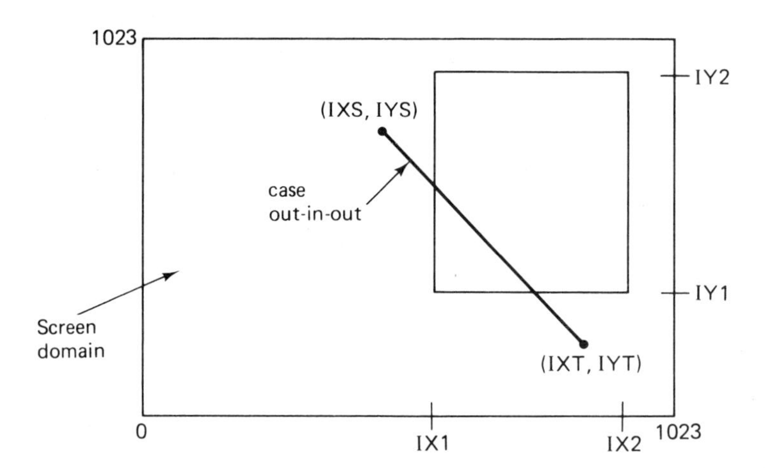

5. Assume a picture definition domain consisting of a 3×3 array of tiles with the screen area as the center tile. Assume a tile coding as introduced in Section 3.2. Formulate the following propositions A to I as boolean expressions (allowed operatores are NOT, AND, OR) of the propositions p_1, p_2, p_3, p_4, p_5, as defined in the text, as well as of the bits x_0, x_1, y_0, y_1 of the tile code (if needed).

A: Two given points, P_i and P_j, lie inside the screen area.

B: Two given points, P_i and P_j, lie outside the screen area.

C: One of the two points, P_i and P_j, lies inside and the other lies outside the screen area.

D: Point P_i lies inside the screen area, and point P_j lies in the top row of tiles.

E: Point P_i lies inside the screen area, and point P_j lies in the bottom row of tiles.

F: Point P_i lies inside the screen area, and point P_j lies in the left column of tiles.

G: Point P_i lies inside the screen area, and point P_j lies in the right column of tiles.

H: Point P_i lies inside tile T_i and point P_j lies inside tile T_j such that T_i and T_j are adjacent and have at least one 1 in their code words in common. Define all possible cases.

I: Point P_i lies inside tile T_i and point P_j lies inside tile T_j such that the code words of T_i and T_j have no 1 in common. Define all possible cases.

6. Assume a picture definition domain consisting of nine tiles as in Exercise 5, with a normalized screen domain $SD=[0:1000]\times[0:1000]$. Assume the origin of the picture definition space in the lower-left corner of the screen domain. Given the points $P_1=(100,800)$, $P_2=(800,100)$, $P_3=(-800,500)$, $P_4=(200,1500)$, $P_5=(1700,1600)$, $P_6=(200,-800)$, $P_7=(1540,680)$.

(a) Construct a table that lists the seven points, $P_1,\ldots,P_7$, together with the respective values of the four tile coding variables, x_0,x_1,y_0,y_1, and a tag bit t that indicates whether $(t=1)$ or not $(t=0)$ the point is in the screen domain.

(b) Perform the clipping of all objects that are not completely inside the screen area for the following lines. Complete the following table.

Start point	End point	Case	Propositions tested	Intersected boundary of SD	Point of intersection
P_1	P_2				
P_1	P_6				
P_1	P_3				
P_3	P_4				
P_6	P_5				

(c) Calculate the intersection of the line with start point P_1 and end point P_7 with the appropriate boundary of SD by using the successive approximation procedure. Specify the sequence of results obtained by a repetitive application of test T1.

7. Given the point $P=(10,20,30)$ in a three-dimensional space.

(a) Calculate the coordinates of the point P_z which is defined by the rotation of P with respect to the z-axis by the angle $\alpha=30°$.

(b) Calculate the coordinates of the point P_y which is defined by the rotation of P with respect to the y-axis by the angle $\beta=60°$.
(c) Calculate the coordinates of the point P_x which is defined by the rotation of P with respect to the x-axis by the angle $\gamma=90°$.
(d) Perform a rotation of P with respect to all three axes. Assume the angles of rotation, α, β, γ, as specified above. Calculate the coordinates of the resulting point P_{xyz}.
(e) Rotate the point P_{xyz} back to its original position. How do you efficiently determine the corresponding 3×3 rotation matrix?
(f) Calculate the coordinates of the point P_t which results from P by a translation with the parameters $x_t=5$, $y_t=-5$, and $z_t=-15$.
(g) Calculate the coordinates of the point P_s which results from P by scaling with the scaling factors $S_x=1.5$, $S_y=0.75$, and $S_z=0.5$.
(h) Calculate the X- and Y-coordinates of the point P_{pp} which is obtained from P by the orthographic projection.
(i) Calculate the X- and Y-coordinates of the point P_{cp} which is obtained from P by the central projection with the parameters $Z_I=0$ and $Z_O=200$.
(j) Determine the 4×4 matrix for rotation, scaling, translation, and perspective which defines the transformations of (d), (f), (g), and (i).
(k) Calculate the coordinates of the point P_I which is obtained from P by applying all transformation defined by the 4×4 matrix in (j). Assume that P has the homogeneous coordinates $(10, 20, 30, 1)$.

8. Given the following four items of type polygon.

Item	Line	Start point	End point	Item	Line	Start point	End point
I1	L1	P1	P2	I3	L7	P7	P8
	L2	P2	P3		L8	P8	P9
	L3	P3	P1		L9	P9	P7
I2	L4	P4	P5	I4	L10	P10	P11
	L5	P5	P6		L11	P11	P12
	L6	P6	P4		L12	P12	P10

(a) Draw the picture in a two-dimensional space which is defined by the following specification of the start points and end points $P_i=(x_i, y_i)$: P1 = P11 = (100, 200), P2 = P4 = (500, 200), P5 = P7 = (500, 600), P8 = P10 = (100, 600), P3 = P6 = P9 = P12 = (300, 400).
(b) Define a segment for the object drawn under (a) which consists of one single item whose primitives are taken from the set {L1, L2, ..., L12}. List the primitives of the item, together with a control parameter that determines their visibility.
(c) Define a segment for the object drawn under (a) as the set {I1, I2, I3, I4} [with I1, I2, I3, I4 as defined under (a)]. List the primitives for each item, together with a control parameter that determines their visibility.

(d) Define a segment for the object drawn under (a) such that a lightpen pick of the primitive L1 or L7 will lead to the erasure of the item I1 or I3, respectively. Items I2 and I4 cannot be erased. List the primitives of each item together with their status and control parameters.

(e) Diagram the picture structure of point (d) as a tree.

9. Given a surface FACE = {L1, L2, L3} in a three-dimensional space. FACE is defined by the edges $L1=\overline{P1P2}$, $L2=\overline{P2P3}$, $L3=\overline{P3P1}$, with P1 = (300, 500, 100), P2 = (700, 300, 200), P3 = (700, 700, 300), given in absolute coordinates.

(a) Assume a projection plane consisting of nine tiles with the screen area as the center tile. Let the tile coding be defined as in the text. The screen area is normalized to SD = [0 : 1000] × [0 : 1000]. Assume the origin of the picture definition space in the lower-left corner of the screen area. Draw the orthographic projection of the surface in the above-defined projection plane.

(b) Perform the following transformations on the given surface. Represent the points P1, P2, and P3 in homogeneous coordinates $(x, y, z, 1)$.

(1) Rotation with respect to the z-axis by the angle $\alpha = 45°$, rotation with respect to the y-axis by the angle $\beta = 0°$, rotation with respect to the x-axis by the angle $\gamma = 0°$. The pivotal point is the point P1 = (300, 500, 200).

(2) Translation with the parameters $x_t = 500$, $y_t = 600$, $z_t = 100$.

(3) Scaling with the scaling factors $S_x = 1$, $S_y = 1$, $S_z = 2$.

(4) Parallel projection onto the xy-plane.

Determine the value of the projection parameters z_I and z_O.

Determine the value of the 4×4 matrix for rotation, scaling, translation, and perspective that defines the transformations (1) to (4).

Determine the homogeneous coordinates of the points P1′, P2′, P3′ which are to be multiplied in order to perform the transformations above.

Calculate the coordinates of the image points.

Depict the projection of the transformed surface FACE = {L1, L2, L3} in the projection plane defined above. Scissor all lines which are not completely inside the screen area. Generate a picture file (points and control parameters) for the transformed and scissored surface.

10. Given the homogeneous coordinate representation of two points, P_1 and P_2, in the two-dimensional space.

$$P_1 = [wx \quad wy \quad w] = [3 \quad 4 \quad 2]$$

$$P_2 = [wx \quad wy \quad w] = [6 \quad 5 \quad 3].$$

Also given the homogeneous coordinate representation of a line λ_1 in the two-dimensional space.

$$\lambda_1 = \begin{bmatrix} 2 \\ -2 \\ 1 \end{bmatrix}.$$

(a) Calculate the ordinary representation of P_1 and P_2 in the two-dimensional space, i.e., $P_1' = (x_1, y_1)$, $P_2' = (x_2, y_2)$.
(b) Does the point P_1 lie on the line λ_1? If no, calculate the perpendicular distance d_1 of P_1 from the line λ_1.
(c) Does the point P_2 lie on the line λ_1? If no, calculate the perpendicular distance d_2 of P_2 from the line λ_1.
(d) Calculate the homogeneous coordinate representation of a line λ_2 which goes through P_1 and P_2.
(e) Calculate the point of intersection of the lines λ_1 and λ_2. Represent the point of intersection in homogeneous coordinates and in normal coordinates of the two-dimensional space.

11. Which of the two propositions, A: "The projection is orthographic" and B: "The projection is a central projection" holds true for the following five cases. Note that both propositions, or none, may be true.
(a) A projection in which the viewpoint is located on one of the axes of the three-dimensional coordinate system.
(b) A projection in which the "line of sight" (the line connecting a point in 3-space with the viewpoint) is contained in the image plane.
(c) A projection in which the viewpoint lies on one of the coordinate axes and is moved toward infinity.
(d) A projection in which the viewpoint is a finite point on one of the coordinate axes and the image plane is moved along this axis toward infinity.
(e) A projection in which the viewpoint is on a line perpendicular to the image plane.

12. Assume a "viewing pyramid" as depicted in Figure 3.5(B), against which all lines of a three-dimensional picture must be clipped. Extend the Sutherland clipping algorithm, as described in Section 3.2 for a two-dimensional window, for this case.

4

Interpolation and Approximation of Curves and Surfaces

4.1 INTRODUCTORY REMARKS

An important area in computer graphics is concerned with the design of curves and surfaces. Frequently, this leads to a *data-fitting problem*: given a number of points through which a curve or a surface is to be fitted. Hence, we have in this case the classical interpolation problem.

Interpolation is a special case of the more general approximation problem [109]. Given a function $f(t)$ which shall be approximated by a finite sum

$$g(t)=\sum_{i=1}^{n} c_i\psi_i(t)$$

of simpler functions, $\psi_i(t)$, such that a certain set of constraints on $g(t)$ is satisfied. As n unknown constants, $c_1,\ldots,c_n$, must be determined, at least n conditions (or constraints) must be specified for $g(t)$. Usually, these constraints are imposed on $g(t)$ such that $g(t)$ is a "good" approximation to $f(t)$. Of course, this leaves yet to define what a good approximation is. If $f(t)$ is continuous on the approximation interval $[a,b]$, then it is certainly a good choice if the functions $\psi_1(t),\ldots,\psi_n(t)$ are also continuous on $[a,b]$. If we introduce the linear space $C[a,b]$ as the set of all functions continuous on $[a,b]$, we may thus postulate that $\psi_i(t)\in C[a,b]$. A good approximation $g(t)$ to $f(t)$ may then, for example, be defined in such a way as to:

1. Make $\|f-g\|$ small, and (hopefully)
2. Ensure a unique solution for the set of coefficients c_i.

Here, $\|\ldots\|$ denotes the Tchebyscheff norm on $C[a,b]$.†

Such a condition still leaves a great number of choices as to the constraints to be imposed on $g(t)$. Constraints very frequently encountered in approximation theory are, for example:

1. Interpolatory constraints

$$g(t_i)=f(t_i), \qquad \forall i\in[1:n],$$

i.e., both functions shall have the same values at a distinct number of points in $[a,b]$.

2. Mixture of interpolatory and smoothness constraints
   a. $g(t_i)=f(t_i)$, $i\in[1:k<n]$.
   b. $g'(t_1)=f'(t_1)$ and $g'(t_k)=f'(t_k)$.
   c. $g(t)\in C^2$, i.e., $g(t)$ is twice continuously differentiable.

3. Orthogonality constraints

$$(f-g,\psi_i)=\int_a^b [f(t)-g(t)]\psi_i(t)\,dt=0,$$

i.e., the constants $c_1,\ldots,c_n$ are chosen such that the orthogonality condition above is satisfied and the functions $\psi_1,\ldots,\psi_n$ thus are orthogonal.

4. Variational constraint

$$\|f-g\|=\min\{\|f-h\|\,|\,h\in\text{span}\{\psi_1,\ldots,\psi_n\}\},$$

i.e., the coefficients $c_1,\ldots,c_n$ are chosen such that of all functions $\|f-h\|$ from the set of all possible linear combinations, $h=c_1\psi_1+c_2\psi_2+\ldots+c_n\psi_n$, the minimum is obtained.

†The Tchebyscheff norm of a function $f(t)\in C[a,b]$ is defined as the real-valued function

$$\|f\|=\max_{a\leqslant t\leqslant b}|f(t)|.$$

In general, $C^k[a,b]$ denotes the set of functions continuous in the first k derivatives (or, in other words, k-differentiable) on the interval $[a,b]$. $\forall$ means "for all".

In computer graphics, we are not primarily concerned with the quality of approximation, e.g., as measured in terms of some estimate of the approximation error, but with certain qualities which are defined in terms of the appearance of a curve or a surface to the human observer. Forrest [41] calls this quality the *property of shape*. As we shall see, the representation of shape information may require certain modifications of the usual approach taken in approximation theory. Another concern is to avoid computationally ill-behaved problems, e.g., such as having to invert ill-conditioned matrices.

In general, shapes in computer-aided design problems cannot be equated with ordinary, single-valued functions in the $y=f(x)$ sense. A first reason for this is that the shape of most engineering objects is intrinsically independent of the coordinate system; i.e., if we have to fit a curve or a surface through a set of selected points, which may have been obtained by taking measurements of a model, then the significant factor in determining the shape of the object is the relationship between these points themselves and not between the points and some arbitrarily chosen coordinate system. Therefore, in many applications it is a requirement that the choice of a coordinate system should not affect the shape. Moreover, shapes of engineering objects may have vertical tangents. If such a shape were represented by an ordinary function of the $y=f(x)$ type, vertical tangents would rule out a function approximation by analytic functions (e.g., polynomials). Last but not least, the curves and surfaces one has to deal with in computer-aided design very frequently are not planar and/or closed and, therefore, cannot be represented by an ordinary function at all.

For all these reasons, the dominant representation of shapes in computer-aided design is in the *parametric form*, that is, a 2-space curve is represented not by a function $y=f(x)$ but rather by a set of two functions

$$x=x(t) \quad \text{and} \quad y=y(t)$$

of a parameter t. Hence, we can say that a point on such a curve is represented by the vector

$$p_c=\left[x(t) \quad y(t)\right].$$

Likewise, a point on a space curve is given by the vector

$$p_c=\left[x(t) \quad y(t) \quad z(t)\right],$$

and a point on a surface is represented by the vector

$$p_s=\left[x(u,v) \quad y(u,v) \quad z(u,v)\right].$$

Such a parametric representation of shapes not only avoids the above-mentioned mathematical problems but is, moreover, the most adequate description of the way in which curves are drawn by a plotter or on the CRT screen: Here, two time

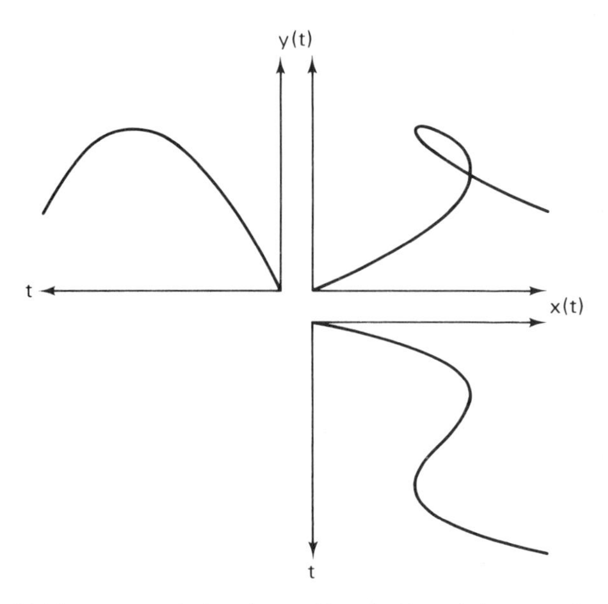

Figure 4.1 Two parametric functions, $x(t)$ and $y(t)$, and the cross-plot y vs. x.

functions, $x(t)$ and $y(t)$, are actually applied as a "driving function" to the servo system of the plotter or to the deflection system of the cathode ray tube, causing the pen or the electron beam, respectively, to move on the appropriate curve. This manner of curve generation is illustrated in Figure 4.1 [60]. In this example, the resulting curve cannot be represented by an ordinary function $y=f(x)$.

With curves or surfaces being represented in parametric form, the problem of interpolation or approximation of such an object is now the problem of determining by interpolation or approximation the components of the vector representation of the object.

At this point, the reader may ask why we should bother about sophisticated approximation and interpolation methods when the only graphical primitives a display can record are dots and straight lines. However, this does not prevent us from displaying satisfactorily smooth curves by approximating them by a polygon consisting of a sufficiently great number of small line segments. In order to obtain the data for such an approximation, first an interpolation between the given points (whose number is usually much smaller) must be carried out.

We anticipate for the future that a better display system will not only contain the now ubiquitous vector and character generators but also curve and surface generators. Some early steps in this direction consisted of adding special hardware for curve generation to a display. However, this approach was restricted to either the generation of Lissajous figures or conic sections. Because of their limited applicability, such systems have not found a widespread use. A general approach to a curve and surface generation by hardware leads exactly to the classical approximation problem, namely, the approximation of arbitrary curves or surfaces by a sum of simpler functions which can be generated by hardware function generators. A class of functions that are very appropriate for hardware generation are polynomials. This opens a whole new field for high-speed curve and surface generation [131].

We mentioned already that the properties by which we judge the qualities of generated shapes may differ from the properties by which the quality of an interpolation or approximation is mathematically measured. In computer-aided design, the mathematical notion of approximation accuracy usually is of secondary importance, whereas other criteria, such as, for example, the smoothness of a curve or a surface, may be very significant. Especially in the realm of an interactive design of objects, the notion of approximation accuracy becomes irrelevant as the success of a design activity may be solely evaluated in terms of the appearance of the pictorial representation of an object. On the other hand, other aspects, which may bear no relevance in the classical approximation problem, may now become very important. One such aspect is the desired *locality* of an operation that will enable the user to change certain details of a shape without affecting the shape as a whole.

For all these reasons, the classical approaches to interpolation and curve fitting, i.e., Lagrange and Hermite interpolation, are rather limited in their usefulness for computer-aided design. If we, nevertheless, discuss them briefly in the following section, it is primarily for the sake of completeness.

4.2 CLASSICAL METHODS: LAGRANGE AND HERMITE INTERPOLATION

Lagrange and Hermite interpolation both belong to the class of polynomial interpolations. The Lagrange polynomials are among the simplest of the interpolating polynomials.

Given $n+1$ distinct real numbers, $x_0<x_1<\ldots<x_n$, and a second set of $n+1$ corresponding real numbers y_i, $i\in[0:n]$. Then the Lagrange polynomial of degree n associated with these two sets, $\{x_i\}$ and $\{y_i\}$, is a polynomial $p(x)$ of degree n solving the interpolation problem

$$p(x_i)=y_i, \qquad \forall i\in[0:n].$$

The x_i are called the *knots* and the y_i are called the *constraints* of the interpolating function. Such polynomials always exist and are unique. Specifically, if $f(x)$ is a function defined at $x_0,x_1,\ldots,x_n$ such that $f(x_i)=y_i$, $\forall i\in[0:n]$, the Lagrange *interpolate* of degree n to $f(x)$ with knots at x_i, $i\in[0:n]$, is the polynomial

$$p_n(x)=\sum_{i=0}^{n} f(x_i)L_{i,n}(x) \tag{4.1}$$

with

$$L_{i,n}=\frac{(x-x_0)\ldots(x-x_{i-1})(x-x_{i+1})\ldots(x-x_n)}{(x_i-x_0)\ldots(x_i-x_{i-1})(x_i-x_{i+1})\ldots(x_i-x_n)}. \tag{4.2}$$

It follows from the definition of $L_{i,n+1}(x)$,

$$L_{i,n}(x_j)=\delta_{ij}, \qquad i,j\in[0:n],$$

where d_{ij} is the Kronecker's delta,

$$\delta_{ij}=\begin{cases}1 & \text{if } i=j\\ 0 & \text{else.}\end{cases}$$

As the simplest possible example, we choose $n=1$. Hence, we have

$$p_1(x)=\frac{x-x_1}{x_0-x_1}\cdot y_0+\frac{x-x_0}{x_1-x_0}\cdot y_1=y_0+(y_1-y_0)\cdot\frac{x_0-x}{x_0-x_1}.$$

The result is the equation of the straight-line segment connecting $y_0=f(x_0)$ and $y_1=f(x_1)$. Thus, the polygonal interpolation passing through a number of data points, so common in computer graphics, is a piecewise Lagrange polynomial interpolation of degree 1.

Of course, the Lagrange polynomials are not the only polynomials solving the interpolation problem $p(x_i)=f(x_i)$, $\forall i\in[0:n]$. As a matter of fact, there exist infinitely many polynomials that may be used, but the Lagrange polynomial is the unique polynomial *of degree n* solving the problem. However, the Lagrange interpolation over the full interval $[x_0,x_n]$ has a serious drawback: As the degree n of the Lagrange polynomial corresponds to the number of knots (plus 1), any attempt to increase the approximation accuracy by increasing the number of knots results also in an increase of the degree of the polynomial. However, polynomials of a higher degree, say five or more, will increasingly cause ripples in the curve (Runge–Méray phenomenon). Thus, it is not possible to let the curve $p_n(x)$ converge pointwise to $f(x)$ by increasing the number of evenly spaced knots. This problem could be mitigated by tailoring the spacing of the knots to the particular function $f(x)$; however, this would be a very impractical procedure.

One possible solution to the problem of instability and ripples, which are usually unacceptable for engineering purposes, is to subdivide the interval $[x_0,x_n]$ into a number of subintervals and to piece together a number of low-degree Lagrange polynomials, each approximating the function over one of the subintervals. Hence, any arbitrary approximation accuracy in the Tchebyscheff norm sense may be achieved; however, at the price of ending up with an overall approximation that may not be differentiable at some or all of the knots, i.e., may exhibit sharp corners at these points. Thus, we trade corners in the curve for ripples, both equally unacceptable in computer-aided design.

One possible solution is to piece together Hermite polynomials rather than Lagrange polynomials. Hermite interpolation fits a polynomial to a function f such that, at a given number of knots, not only f is interpolated but also a given number of consecutive derivatives of f: e.g., there is exactly one polynomial of degree 3

$$p_3(t)=a_3t^3+a_2t^2+a_1t+a_0,$$

which satisfies the constraints

$$p(a)=f(a), \quad p'(a)=f'(a)$$
$$p(b)=f(b), \quad p'(b)=f'(b).$$

This polynomial is called the *cubic Hermite interpolate* to f.

Given a function $f(t)$ and given $n+1$ knots $a=t_0<t_1<\ldots<t_n=b$, dividing $[a,b]$ into n subintervals $[t_{i-1},t_i]$, $i\in[1:n]$. Then we may construct the cubic Hermite polynomials $p_i(t)$ satisfying the constraints

$$p_i(t_{i-1})=f(t_{i-1}), \quad p_i'(t_{i-1})=f'(t_{i-1})$$
$$p_i(t_i)=f(t_i), \quad p_i'(t_i)=f'(t_i),$$

and form subsequently the piecewise cubic polynomial

$$s(t)=p_i(t), \quad \forall i:t\in[t_{i-1},t_i].$$

In each of the intervals (t_{i-1},t_i), $s(t)$ is a polynomial and thus infinitely differentiable,† At all interior knots t_i, $i\in[1:n-1]$, we have

$$f(t_i)=p_i(t_i)=p_{i+1}(t_i)=s(t_i)$$

and

$$p_{i+1}'(t_i)=s'(t_i)=p_i'(t_i)=f'(t_i).$$

Therefore, $s(t)$ is at least once differentiable at the interior knots, or

$$s(t)\in C^1[a,b].$$

If one wants to ensure an even greater smoothness, one must turn to piecewise Hermite polynomials of a degree greater than 3. However, as the same can be accomplished in a simpler way by employing cubic polynomial splines, we shall restrict our discussion to cubic Hermite interpolates.

Computationally, the interpolation of a function by piecewise cubic Hermite polynomials is quite straightforward. It is given by

$$s(t)=\sum_{i=0}^{n} f(t_i)\cdot I_{i0}(t)+\sum_{i=0}^{n} f'(t_i)\cdot I_{i1}(t) \tag{4.3}$$

†Note that $[a,b]$ denotes the closed interval and (a,b) denotes the open interval.

with

$$I_{i0}(t)=\begin{cases}\dfrac{(t-t_{i-1})^2}{(t_i-t_{i-1})^3}\left[2(t_i-t)+(t_i-t_{i-1})\right] & \text{if } t\in[t_{i-1},t_i]\\[2ex] \dfrac{(t_{i+1}-t)^2}{(t_{i+1}-t_i)^3}\left[2(t-t_i)+(t_{i+1}-t_i)\right] & \text{if } t\in[t_i,t_{i+1}]\\[2ex] 0 \quad \text{otherwise} \end{cases} \tag{4.4a}$$

$$I_{i1}(t)=\begin{cases}\dfrac{(t-t_{i-1})^2(t-t_i)}{(t_i-t_{i-1})^2} & \text{if } t\in[t_{i-1},t_i]\\[2ex] \dfrac{(t-t_{i+1})^2(t-t_i)}{(t_{i+1}-t_i)^2} & \text{if } t\in[t_i,t_{i+1}]\\[2ex] 0 \quad \text{otherwise} \end{cases} \tag{4.4b}$$

$$I_{00}(t)=\begin{cases}\dfrac{(t_1-t)^2}{(t_1-t_0)^3}\left[2(t-t_0)+(t_1-t_0)\right] & \text{if } t\in[t_0,t_1]\\[2ex] 0 \quad \text{otherwise} \end{cases} \tag{4.4c}$$

$$I_{n0}(t)=\begin{cases}\dfrac{(t-t_{n-1})^2}{(t_n-t_{n-1})^3}\left[2(t_n-t)+(t_n-t_{n-1})\right] & \text{if } t\in[t_{n-1},t_n]\\[2ex] 0 \quad \text{otherwise} \end{cases} \tag{4.4d}$$

$$I_{01}(t)=\begin{cases}\dfrac{(t-t_1)^2(t-t_0)}{(t_1-t_0)^3} & \text{if } t\in[t_0,t_1]\\[2ex] 0 \quad \text{otherwise} \end{cases} \tag{4.4e}$$

$$I_{n1}(t)=\begin{cases}\dfrac{(t-t_{n-1})^2(t-t_n)}{(t_n-t_{n-1})^3} & \text{if } t\in[t_{n-1},t_n]\\[2ex] 0 \quad \text{otherwise.} \end{cases} \tag{4.4f}$$

These functions have the properties

$$\begin{cases} I_{i0}(t)=\delta_{ij} \\ I'_{i0}(t)=0, \qquad i,j\in[0:n] \end{cases} \tag{4.5}$$

and

$$\begin{cases} I_{i1}(t)=0, & i,j\in[0:n] \\ I'_{i1}(t)=\delta_{ij}, \end{cases} \tag{4.6}$$

The question arises what to do if the derivatives of $f(t)$ at the interior knots are unknown. There are several ways to overcome this difficulty. As a case in point, we discuss the following Hermite interpolation, known as *Coons' approximation.*

A particular case of cubic Hermite interpolates is the approximation by rational cubic polynomials as introduced by Coons [20,4]. This method has found a certain notoriety in computer-aided design. We shall present this method already in the form appropriate for computer graphics applications.

Let a point on a space curve be given in homogeneous coordinates (as introduced in Chapter 3), that is, by the vector

$$\tilde{P}(t)=[wx(t) \quad wy(t) \quad wz(t) \quad w(t)]. \tag{4.7}$$

Let the functions $wx(t)$, $wy(t)$, $wz(t)$, and $w(t)$ be parametrized in the form of cubic polynomials

$$\begin{aligned} wx(t)&=a_3t^3+a_2t^2+a_1t+a_0 \\ wy(t)&=b_3t^3+b_2t^2+b_1t+b_0 \\ wz(t)&=c_3t^3+c_2t^2+c_1t+c_0 \\ w(t)&=d_3t^3+d_2t^2+d_1t+d_0. \end{aligned} \tag{4.8}$$

The ordinary coordinates $x=wx/w$, $y=wy/w$, and $z=wz/w$ are then given in the form of rational cubic polynomials. With equations (4.7) and (4.8), we can write

$$\tilde{P}(t)=[t^3 \quad t^2 \quad t \quad 1]\times\begin{bmatrix} a_3 & b_3 & c_3 & d_3 \\ a_2 & b_2 & c_2 & d_2 \\ a_1 & b_1 & c_1 & d_1 \\ a_0 & b_0 & c_0 & d_0 \end{bmatrix}=[t^3 \quad t^2 \quad t \quad 1]\times A. \tag{4.9}$$

The first derivative of the curve in point $\tilde{P}$ is

$$\tilde{P}'(t)=[3t^2 \quad 2t \quad 1 \quad 0]\times A,$$

and the second derivative is

$$\tilde{P}''(t)=[6t \quad 2 \quad 0 \quad 0]\times A.$$

The advantage of a parametric representation is that we may select any arbitrary spacing of the parameter values on the curve. Hence, for computational simplicity we insist that the parameter t assumes values from 0 to 1. The problem is now to determine the coefficient matrix A such that the function $\tilde{P}(t)$ satisfies the following constraints

$$\begin{aligned} \tilde{P}(t)|_{t=0} &= p_0 \quad \text{and} \quad \tilde{P}'(t)|_{t=0} = p_0' \\ \tilde{P}(t)|_{t=1} &= p_1 \quad \text{and} \quad \tilde{P}'(t)|_{t=1} = p_1'. \end{aligned} \tag{4.10}$$

Thus, we have

$$\begin{bmatrix} p_0 \\ p_1 \\ p_0' \\ p_1' \end{bmatrix} = \begin{bmatrix} 0 & 0 & 0 & 1 \\ 1 & 1 & 1 & 1 \\ 0 & 0 & 1 & 0 \\ 3 & 2 & 1 & 0 \end{bmatrix} \times A.$$

Solving this matrix equation for A yields

$$A = M \times \begin{bmatrix} p_0 \\ p_1 \\ p_0' \\ p_1' \end{bmatrix}, \tag{4.11a}$$

with

$$M = \begin{bmatrix} 0 & 0 & 0 & 1 \\ 1 & 1 & 1 & 1 \\ 0 & 0 & 1 & 0 \\ 3 & 2 & 1 & 0 \end{bmatrix}^{-1} = \begin{bmatrix} 2 & -2 & 1 & 1 \\ -3 & 3 & -2 & -1 \\ 0 & 0 & 1 & 0 \\ 1 & 0 & 0 & 0 \end{bmatrix}. \tag{4.11b}$$

(The matrix M has sometimes been dubbed the "magic matrix," although there is nothing magic about it.) From equations (4.9) and (4.11) we obtain

$$\tilde{P}(t) = [2t^3 - 3t^2 + 1 \quad -2t^3 + 3t^2 \quad t^3 - 2t^2 + t \quad t^3 - t^2] \times \begin{bmatrix} p_0 \\ p_1 \\ p_0' \\ p_1' \end{bmatrix}. \tag{4.12}$$

On the other hand, the cubic Hermite interpolation of a function $f(t)$, $t \in [0, 1]$ with the constraints $f(0) = p_0$, $f'(0) = p_0'$, $f(1) = p_1$, and $f'(1) = p_1'$ is, according to equation (4.3),

$$s(t) = I_{00} \cdot p_0 + I_{10} \cdot p_1 + I_{01} \cdot p_0' + I_{11} \cdot p_1'.$$

The comparison of equations (4.4) and (4.12) confirms that the curve $\tilde{P}(t)$ is the cubic Hermite interpolate for the case $n = 1, t_0 = 0, t_n = 1$.

The problem of fitting a smooth curve through a given $(n+1)$-tuple (an ordered set) of points $(p_0, p_1, \ldots, p_n)$ may now be attacked by piecing such cubic "splines" together [20, 4]. Since at the interior points usually only values are prescribed but not slopes, the given 4 degrees of freedom for determining the coefficients may be used to satisfy the constraints for two adjacent splines, $\tilde{P}_i$ and $\tilde{P}_j$,

$$\tilde{P}_i(1) = \tilde{P}_j(0) = p_{j0}, \qquad \tilde{P}_i'(1) = \tilde{P}_j'(0), \qquad \tilde{P}_i''(1) = \tilde{P}_j''(0),$$

thus obtaining an approximation $\tilde{P}(t) \in C^2$. At the two end points of the curve, the values and the tangent vectors shall be specified as before. However, the calculational procedure for such an approximation is quite tedious (if not clumsy as compared with the B-spline approach discussed in the following section). Moreover, the resulting matrices may be ill-conditioned, thus further complicating the design process. For all these reasons, we shall not elaborate here on this method. Applications of Coons' curves and surfaces are, for example, discussed in [107].

4.3 INTERPOLATION WITH B-SPLINES

Polynomials have been so widely used to approximate other functions because of their simple mathematical properties. However, as mentioned before, the problem of polynomials in general is that they tend to exhibit undesirable undulations when fitted through more than a few data points. As long as the objective is a best fit in the Tchebyscheff sense, these undulations may not bother us as long as their amplitude is small. However, in engineering curves and surfaces they usually cannot be tolerated. On the other hand, it is common experience that curves drawn with the aid of a spline or a French curve are much smoother. Therefore, it is very desirable to have functions that combine the simplicity of polynomials with a "built-in" smoothness.

Spline functions have been studied very intensively during the 1960s. The literature on spline functions is numerous and includes a number of monographs and textbooks. It is quite obvious that we cannot discuss in this text the theory and application of spline functions at any depth. Therefore, we shall restrict our discussion to listing briefly the most interesting properties of polynomial spline functions, the *B-splines*, which are extraordinarily well suited for computer graphics use.

Given a sequence of real numbers $x_0 < x_1 < \ldots < x_n$. A spline function $S(x)$ of degree m with the knots $x_0, x_1, \ldots, x_n$ is a function defined on the entire real line having the following two properties:

1. In each interval (x_i, x_{i+1}) for $i = 0, 1, \ldots, n$ with $x_0 = -\infty$ and $x_{n+1} = +\infty$, $S(x)$ is given by some polynomial of degree m or less.

2. $S(x)$ and its derivatives of orders $1, 2, \ldots, m-1$ are continuous everywhere.

Thus, a spline function is a piecewise polynomial function (e.g., the Coons' function discussed in the preceding section is a spline function). For $m=0$, condition (2) is not valid, as a spline function of degree 0 is a step function. A spline function of degree 1 is a polygon. For $m>0$, a spline function of degree m may be as well defined as a function in C^{m-1} which is obtained as the mth-order indefinite integral of a step function.

In general, $S(x)$ is given by different polynomials in adjoining intervals (x_{i-1}, x_i) and (x_i, x_{i+1}). However, the definition above does not require this and, as a special case, $S(x)$ may be a single polynomial on the entire real line. Hence, the general definition of spline functions includes *all* polynomials of degree m or less.

Cubic splines have proved to be one of the most useful tools for the solution of practical interpolation problems. Unfortunately, they occasionally exhibit a flaw, the production of extraneous inflection points in the curve. On these occasions it would be desirable to have an interpolant that is not only constrained to pass through the given values but also able to respond to a virtual "tension," which might be thought of as being produced by pulling on its ends. By applying such a tension, the unwanted inflection points could be removed.

The notion of "splines under tension" can be analytically approximated, leading to the interpolatory or approximatory construction of curves and surfaces under tension. A set of algorithms for the interpolatory construction of curves, in their ordinary as well as in their parametric representation, has been published and is available as ACM algorithm No. 476.†

For practical purposes, the restriction to a particular simple type of spline function has proved to be particularly useful. This type of spline is given by the truncated power function

$$x_+^m = \begin{cases} x^m & \text{if} \quad x>0 \\ 0 & \text{if} \quad x \leqslant 0. \end{cases} \tag{4.13}$$

A spline $S(x)$ of *odd* degree $2k-1$ with the knots $x_1, x_2, \ldots, x_n$ is called a *natural spline* if it is given in each of the two intervals $(-\infty, x_1)$ and $(x_n, +\infty)$ by some polynomial of degree $<k$. It can be shown that natural splines provide a unique solution to the interpolation (the data-fitting) problem. Furthermore, such an interpolation $S(t)$ is the smoothest interpolating function for the n data points, i.e., a function that minimizes the integral

$$\int_a^b \left[s^{(k)}(x) \right]^2 dx \qquad \text{for } 0<k<n.$$

We shall now come back to the interpolation problem,

$$\begin{aligned} s'(t_0) &= f'(t_0) \\ s(t_i) &= f(t_i), \qquad t_i = i \cdot h, \quad i \in [0:n] \\ s'(t_n) &= f'(t_n). \end{aligned} \tag{4.14}$$

†CACM 17, 4, Apr. 1974, 218–223.

Computationally, this problem would be greatly simplified if it were possible to obtain $s(t)$ by piecing together spline functions into which the desired smoothness is built a priori. For example, if we use cubic splines, the resulting curve $s(t)$ should have the property $s \in C^2[t_0, t_n]$, so that we do not have to enforce this condition explicitly (see the discussion of Coons' approximation in the preceding section). Such a reduction of an intrinsicly mixed-constraint problem to a pure interpolatory problem can be accomplished by using a particular class of spline functions called B-splines. In the following, we introduce B-splines as presented in [89]. We shall restrict our discussion to natural B-splines.

We define first a B-spline of odd degree $r = 2k - 1$ with uniform knot spacing of unity as the function

$$\beta_r(\tau) = \frac{1}{r!} \sum_{j=-k}^{k} (-1)^{j+k} \binom{2k}{j+k} (j-\tau)_+^r. \tag{4.15}$$

The function $\beta_r(\tau)$ is symmetric about $\tau = 0$, bell-shaped, and nonnegative on the interval $[-k, k]$. Outside this interval it is identically zero. More specifically, it has the properties

$$\beta(\tau) \begin{cases} > & 0 \quad \text{if } |\tau| < k \\ = & 0 \quad \text{if } |\tau| = k \\ \equiv & 0 \quad \text{if } |\tau| > k \end{cases} \tag{4.16a}$$

$$\beta_r(-\tau) = \beta_r(\tau) \tag{4.16b}$$

$$\beta_r(0) > \beta_r(\tau) \qquad \text{if } \tau \neq 0 \tag{4.16c}$$

$$\sum_{i=-\infty}^{\infty} |\beta_r(i)| = \sum_{i=1-k}^{k-1} \beta_r(i) = 1. \tag{4.16d}$$

The derivatives $\beta_r^{(p)}(\tau)$, $p = 1, 2, \ldots, r$, of $\beta_r(\tau)$ are given by

$$\beta_r^{(p)}(\tau) = \frac{1}{(r-p)!} \sum_{j=-k}^{k} (-1)^{j+k+p} \binom{2k}{j+k} (j-\tau)_+^{r-p}. \tag{4.17}$$

It follows from equation (4.17) that

$$\beta_r(\tau) \in C^{r-1} \tag{4.18}$$

$$\beta_r^{(p)}(\pm k) = 0, \qquad \forall p \in [0 : r-1]. \tag{4.19}$$

The graph $\beta_3(\tau)$ is depicted in Figure 4.2.

Now let us consider the approximation interval $[0, 1]$ and a uniform knot spacing $h = 1/q$. It can be shown that the $(q + r)$ functions $\beta_r(qt - i)$, $i \in [1 - k : q +$

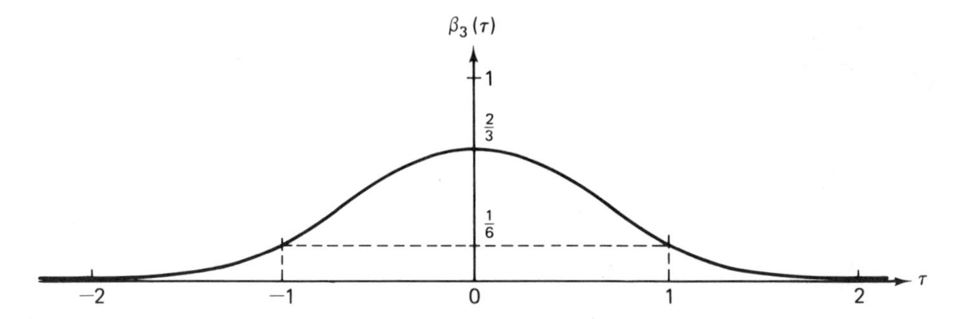

Figure 4.2 Graph of the B-spline $\beta_3(\tau)$.

$k-1]$, are linearly independent on $[0,1]$. Therefore, an arbitrary spline of degree r on $[0,1]$, with $(q+1)$ knots at $t_j = j \cdot h, j \in [0:q]$, can be represented by

$$s_r(t) = \sum_{i=1-k}^{q+k-1} a_i \beta_r(qt-i). \tag{4.20}$$

Because of the compact support for each $\beta_r(qt-i)$ [the interval over which $\beta_r(qt-i)$ is nonzero], there are at most $(r+1)$ nonzero terms in this summation for any fixed $t \in [0,1]$. Furthermore, we have

$$\sum_{i=1-k}^{q+k-1} \beta_r(qt-i) = 1, \qquad \forall t \in [0,1], \tag{4.21}$$

and therefore

$$\|s_r(t)\| \leqslant \sup_i |a_i|, \qquad \forall t \in [0,1]. \tag{4.22}$$

The derivative of $s_r(t)$ is

$$\frac{d^p}{dt^p} s_r(t) = q^p \sum_{i=1-k}^{q+k-1} a_i \beta_r^{(p)}(qt-i). \tag{4.23}$$

The solution of the interpolation problem (4.14) (or any similar problem) is now quite straightforward. To this end, we choose the cubic B-spline

$$s_3(t) = \sum_{i=-1}^{n+1} a_i \beta_3(nt-i). \tag{4.24}$$

The first derivative is

$$s_3'(t) = n \sum_{i=-1}^{n+1} a_i(-1)\beta_3'(nt-i).$$

From equation (4.14), we have

$$s_3'(t_0) = a_{-1} n\beta_3'(nt_0 + 1) + a_0 n\beta_3'(nt_0) + \ldots + a_{n+1} n\beta_3'(nt_0 - n - 1) = f'(t_0)$$

. . . .

$$s_3(t_i) = a_{-1}\,\beta_3(nt_i + 1) + a_0\,\beta_3(nt_i) + \ldots + a_{n+1}\,\beta_3(nt_i - n - 1) = f(t_i)$$

. . . .

$$s_3'(t_n) = a_{-1} n\beta_3'(nt_n + 1) + a_0 n\beta_3'(nt_n) + \ldots + a_{n+1} n\beta_3'(nt_n - n - 1) = f'(t_n),$$

$\forall i \in [0:n]$.

This constitutes a set of $n+3$ linear equations, i.e., a system

$$B \times a = c, \tag{4.25a}$$

where $a = [a_{-1}, a_0, \ldots, a_{n+1}]^T$ is the vector of the (yet unknown) coefficients and $c = [f'(t_0), f(t_0), f(t_1), \ldots, f(t_i), \ldots, f(t_n), f'(t_n)]^T$ is the vector of constraints. The matrix B is determined by values of $\beta_3(\tau)$ and $\beta_3'(\tau)$ at the knots as listed in the following table:

$\tau=$	-2	-1	0	$+1$	$+2$
$\beta_3(\tau)$	0	$\frac{1}{6}$	$\frac{2}{3}$	$\frac{1}{6}$	0
$\beta_3'(\tau)$	0	$\frac{1}{2}n$	0	$-\frac{1}{2}n$	0
$\beta_3''(\tau)$	0	n^2	$-2n^2$	n^2	0

which results in

$$B = \begin{bmatrix} -\frac{n}{2} & 0 & \frac{n}{2} & 0 & 0 & 0 & \cdots & & & & \\ \frac{1}{6} & \frac{2}{3} & \frac{1}{6} & 0 & 0 & 0 & \cdots & & & & \\ 0 & \frac{1}{6} & \frac{2}{3} & \frac{1}{6} & 0 & 0 & \cdots & & & & \\ 0 & 0 & \frac{1}{6} & \frac{2}{3} & \frac{1}{6} & 0 & \cdots & & & & \\ \vdots & & & & & & & & & & \\ 0 & 0 & 0 & & \cdots & & & 0 & \frac{1}{6} & \frac{2}{3} & \frac{1}{6} \\ 0 & 0 & 0 & & \cdots & & & 0 & -\frac{n}{2} & 0 & \frac{n}{2} \end{bmatrix}. \tag{4.25b}$$

B must be inverted in order to obtain the vector of coefficients $a = B^{-1} \times c$.

Since the matrix B is diagonally dominant with strict diagonal dominance on all but the first and last rows, it is nonsingular. Thus, the very small support of the

B-splines leads to well-conditioned matrices with a special structure which simplifies the computation very strongly. Furthermore, such a B-spline approximation exhibits the desirable locality. If we change one of the values c_i, it affects at most 4 of the coefficients a_i. Schoenberg—who was the first to study B-splines [118, 119] —proved that they are the unique nonzero splines with the smallest support.

The reader will have noticed that any appropriately chosen set of $n+r$ constraints suffices for the determination of the coefficients of an approximation. Moreover, it is not mandatory to select data points as interpolatory constraints which correspond with the knots of the B-splines, although the computation is considerably simplified if the knots are chosen as the data point coordinates.

4.4 BÉZIER APPROXIMATION OF CURVES

So far we have considered curve definitions that *interpolate* given data. Another approach is to provide a good, smooth representation of a surface that *approximates* given data. Such an approach is more appropriate whenever "the properties of the approximation in the large are of more importance than the closeness of fit" [26, p. 116] and whenever design constraints must be satisfied which cannot be measured in a simple mathematical manner (e.g., in the sense of the minimization of a given norm, as considered previously).

A very typical case in point is the problem in the automobile design process where a mathematical representation shall be found for a sketch or a clay model furnished by the designer. In such a case there is no definable "best" fit, but the quality of a fit depends primarily on the designer's judgment. It is thus logical to use an interactive technique in which the user can experiment with a variety of shapes without having to know anything about the mathematical principles involved. However, certain smoothness conditions should a priori be built into the class of curves the designer will experiment with. Such a design procedure is greatly facilitated if the shape of a curve can be controlled in a predictable manner by the alteration of a few parameters, even more so if these parameters can also be specified in a graphical form.

Over the past decade, the automobile, aircraft, and ship-building industries have developed various practical schemes for the solution of this problem, the most interesting approach probably being that developed by Bézier. Bézier based his approximation method on a classical approximation method, known as the *Bernstein polynomial approximation* and defined as follows. Let $f(t)$ be an arbitrary real-valued function continuous on the interval $[0,1]$. The Bernstein polynomial approximation of degree n to f is [60]

$$B_n(f(t))=\sum_{k=0}^{n} f\left(\frac{k}{n}\right)\phi_{k,n}(t), \tag{4.26}$$

with

$$\phi_{k,n}(t)=\binom{n}{k}t^k(1-t)^{n-k}, \qquad k\in[0:n]. \tag{4.27}$$

Readers familiar with probability theory will recognize that the functions equation (4.27) are identical with the binomial probability density functions. Recalling some of the elementary properties of these functions, we can state:

1. $\forall k\in[0:n]$: $\phi_{k,n}\geqslant 0$; $\forall t\in[0:1]$: $\sum_{k=0}^{n}\phi_{k,n}(t)=1$.
2. $\forall k\in[1:n-1]$: $\phi_{0,n}(0)=1$; $\phi_{k,n}^{(k)}(0)=\dfrac{n!}{(n-k)!}$.

   $\forall p\in[0:k-1]$: $\phi_{k,n}^{(p)}(0)=0$; $\forall q\in[0:n-k-1]$: $\phi_{k,n}^{(q)}(1)=0$.
3. $\forall p\in[0:n-1]$: $\phi_{n,n}^{(p)}=0$; $\phi_{n,n}(1)=1$.
4. $\phi_{k,n}^{(n-k)}(1)=\dfrac{(-1)^{n-k}n!}{(n-k)!}$.
5. $\phi_{k,n}\left(\dfrac{k}{n}\right)=\binom{n}{k}k^k(n-k)^{n-k}>\phi_{k,n}(t)$ if $t\neq\dfrac{k}{n}$.

Conditions 2 and 3 imply that the two end-point values, $f(0)$ and $f(1)$, are in general the only values that are interpolated by the Bernstein polynomial B_n. From the conditions for $\phi_{k,n}(t)$ listed above, the end-point derivatives of $B_n(t)$ can be obtained as follows:

$$\frac{d^p}{dt^p}B_n(f(t))|_{t=0}=\frac{n!}{(n-p)!}\sum_{k=0}^{p}(-1)^{p-k}\binom{p}{k}f\left(\frac{k}{n}\right)$$

$$\frac{d^p}{dt^p}B_n(f(t))|_{t=1}=\frac{n!}{(n-p)!}\sum_{k=0}^{p}(-1)^{k}\binom{p}{k}f\left(\frac{n-k}{n}\right). \tag{4.28}$$

Hence, the pth derivative at the end-points, $t=0$ and $t=1$, is determined by the values of $f(t)$ at the respective end-point and at the p points nearest to that end-point. Specifically, the first derivatives are equal to the slope of the straight line joining the end-point and the adjacent interior point.

Bernstein polynomials satisfy the Weierstrass approximation theorem; i.e., they converge with increasing n uniformly to the function they approximate. Furthermore, the approximation $B_n(f(t))$ is smoother than f itself if smoothness is measured in terms of the number of oscillations about a given straight line [120]. The reason why, despite all these desirable features, the Bernstein approximation has never been widely used in the minimal norm approximation discussed above is that they converge very slowly in the uniform norm. However, as proved by Bézier, they are very well suited to the problem of an interactive design of smooth free-form curves and surfaces [7].

Bézier's approach is to specify a well-ordered set of points, say $n+1$,

$$P=\left\{\left(\frac{i}{n},v_i\right)\middle| i\in[0:n]\wedge v_i\in\mathbf{R}\right\}, \tag{4.29}$$

the ordering relation defined by $(i/n,v_i)\leqslant(j/n,v_j)$ if $i\leqslant j$, and the (open) polygon formed by joining successive points. This polygon, which will be called an n-sided Bézier polygon, is associated with the Bernstein polynomial of degree n,

$$B_n(P,t)=\sum_{k=0}^{n} v_k\phi_{k,n}, \tag{4.30}$$

in which the $\phi_{k,n}$ are the functions equation (4.27), and the values v_i of the vertices of the Bézier polygon are the coefficients (in their given order). This polynomial will be called the Bézier curve associated with the Bézier polygon.

In an interactive design process, the objective is not to approximate the Bézier polygon. The Bézier polygon functions, rather, as the medium by which the shape of the associated Bézier curve can be controlled. The fact that this provides a most effective interactive design tool was Bézier's ingenious discovery. This fact stems from the following properties of the Bézier curve:

1. The Bézier curve has the end-points in common with the Bézier polygon. All other vertices of the polygon are in general not on the Bézier curve.

2. The slope of the tangent vectors in the end-points of the Bézier curve equals the slope of the first and the last segments, respectively, of the Bézier polygon.

3. The Bézier curve lies entirely within the convex hull of the extreme points of the Bézier polygon. Furthermore, it mimics the principal features of the polygon rather well.

Figure 4.3 depicts the Bernstein basic functions $\phi_{k,3}(t)$, $k\in[0:3]$. Figure 4.4 shows a Bézier polynomial, its convex hull, and the associated Bézier curve.

In practice, the Bézier design procedure is heuristic. First, the designer sketches the desired curve manually. Subsequently, he specifies the vertices of the Bézier polygon which he hopes will generate a reasonable first approximation to that curve. If he has no other criteria to go by, he may simply use appropriately selected points on the curve as vertices, as illustrated in Figure 4.5(a) [131]. The next step consists of moving the vertices such that the approximation is gradually improved. If necessary, vertices will be added or deleted. In the example of Figure

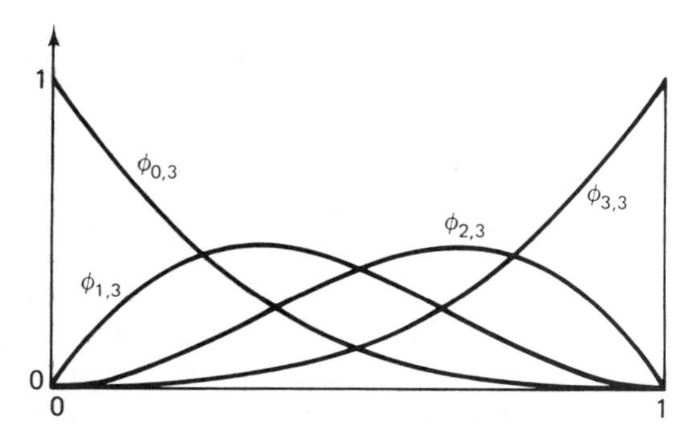

Figure 4.3 Bernstein basic functions $\phi_{k,3}(t)$, $k \in [0:3]$.

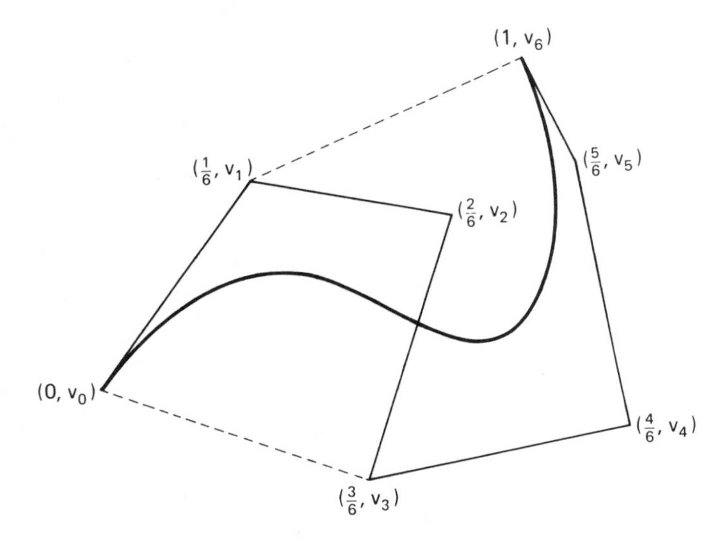

Figure 4.4 Six-sided Bézier polygon, its convex hull (dashed line), and the associated Bézier curve.

4.5(b), the designer first moves the vertices too far away from the curve that shall be approximated [property 2 of the Bézier curve (see page 136) should have told him that]. Finally, a better placement of the polygon vertices is found [Fig. 4.5(c)]. Performed interactively on the CRT screen, this heuristic design method is very fast, as the display provides an immediate response to the user's actions. Of course, the accuracy of approximation is limited by the relatively coarse resolution of the display (the reader should keep in mind that all curves displayed on the screen are approximated by polygons). Therefore, in the creation of precision artwork, there comes the point where the interactive display must be replaced by a plotter. Figure 4.6 demonstrates that not only open curves but also closed curves may be created.

Bézier [16] already pointed out that his design method is not confined to the use of Bernstein polynomials or to polynomials in general. The most promising

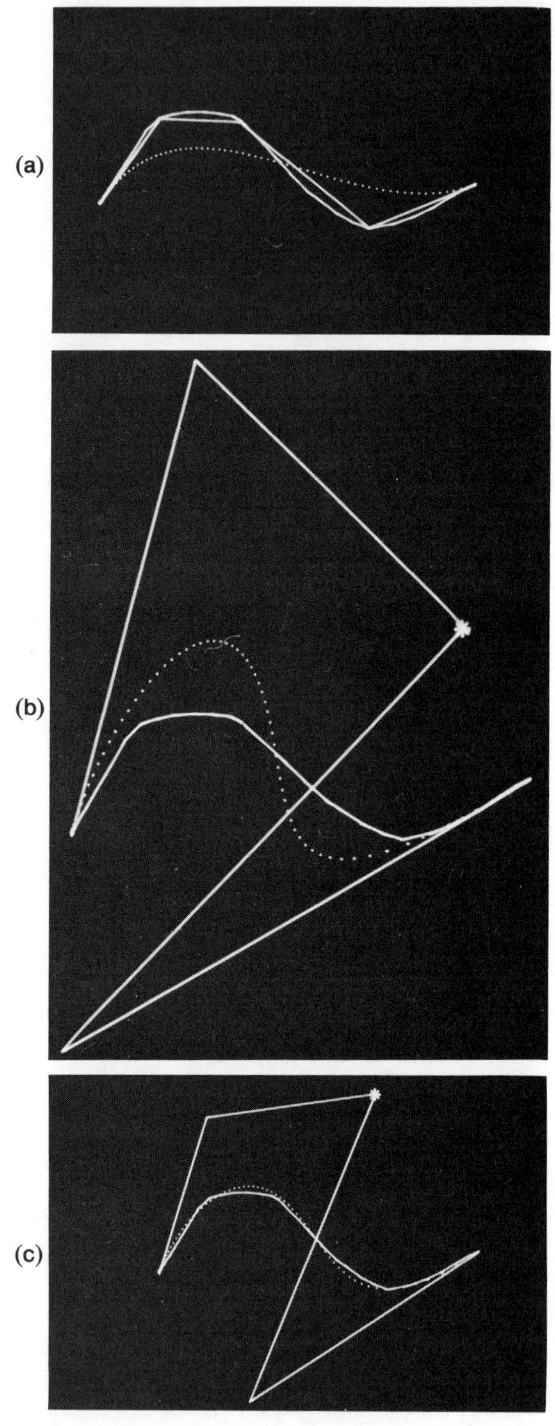

Figure 4.5 Example of the stepwise, interactive approximation of a curve. Solid line: curve to be approximated; dotted line: Bézier curve.

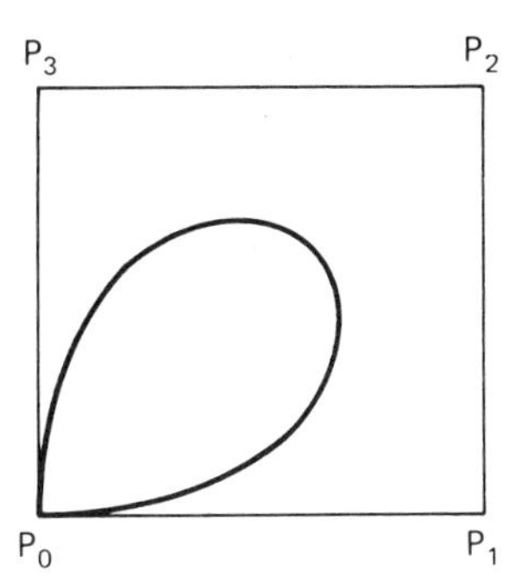

Figure 4.6 Example for the creation of a closed Bézier curve.

choice of basis functions seems to be B-splines [61], [109]. Compared with Bernstein polynomials, B-splines offer the following additional advantages:

1. The B-spline approximation produces a closer fit to the Bézier polygon than the Bernstein–Bézier approximation.

2. Because of the property of B-splines to be of the smallest possible support, a B-spline approximation offers the desirable locality, a feature that the Bernstein approximation lacks entirely.

3. In the Bernstein–Bézier approximation, the degree of the Bézier curve increases proportionally with the number of sides of the Bézier polygon. In a B-spline approximation, both parameters are independent and can, thus, be arbitrarily chosen.

Figure 4.7 depicts a Bernstein–Bézier curve (solid line) in comparison with a

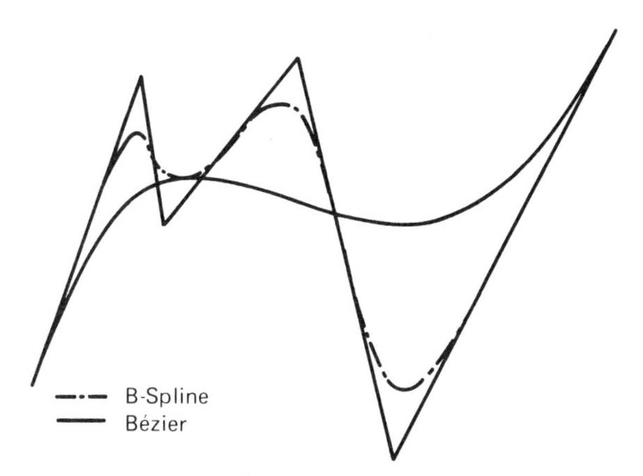

Figure 4.7 Bernstein–Bézier and B-spline curve of degree 3, both produced by the same five-sided Bézier polygon.

Figure 4.8 Demonstration of the possibilities of B-spline approximations.

B-spline curve (dash-dotted line), both curves produced by the same five-sided polygon. Figure 4.8 is another interesting demonstration of the possibilities of B-spline approximations. Both examples are taken from [131].

4.5 GENERAL PRINCIPLES OF SURFACE CONSTRUCTION

Sometimes the shape of a surface can be described analytically, e.g., in the form of cylinders, spheres, revolving conics, etc. However, in most cases, such an analytical description is not given, and the designer must resort to a constructive approach, creating surfaces from simpler data, such as curves or points. Basically, there are three ways in which surfaces can be constructed from such data [62], which will be discussed briefly in this section. Our discussion is based on the fundamental paper of Gordon [62] and the enlightening survey presented by Forrest [41].

In the preceding sections we discussed a variety of methods by which a curve can be constructed by interpolating or approximating a given set of scalar values specifying points or derivatives. A curve is represented (in the general case of a curve in 3-space) by a vector-valued function

$$p(t)=[x(t) \quad y(t) \quad z(t)]$$

(parametric representation in ordinary coordinates). Generally, the curve-generating algorithms may be represented by an operator ϕ_t applied to the vector-valued function $P(t)$ representing the data [41].

$$Q(t)=\phi_t P(t).$$

In all the methods that we have discussed, only a finite number of discrete points of $P(t)$ (or of its derivatives) were used.

Analogously, a point on a surface in 3-space may be described in parametric form by the function

$$p(s,t)=[x(s,t) \quad y(s,t) \quad z(s,t)],$$

and the surface-generating procedure may be symbolically denoted by

$$Q(s,t)=\phi_{s,t} P(s,t), \tag{4.31}$$

where $P(s,t)$ represents the data from which $Q(s,t)$ will be constructed.

It is one of the advantages of the parametric representation of a surface that the user has complete control as to the domain of a surface-constructing operation simply by an appropriate choice of the parametrization. By appropriately specifying subsets of a given domain $[s_{\min}, s_{\max}]\times[t_{\min}, t_{\max}]$, he/she can easily define certain sections of a surface, a feature that is very helpful whenever a surface is to be composed of a number of surface elements or *patches*. This freedom of parametrization can be used to choose $[0,1]\times[0,1]$ as the domain of a surface interpolating or approximating operation. Any other domain can be normalized accordingly.

As it would be too complicated to deal with the bivariate operator $\phi_{s,t}$, ways must be found to approximate $\phi_{s,t}$ by a composition of two univariate operators, ϕ_s and ϕ_t. The most widely used approach to this problem is to form the *tensor product* (alias Cartesian product or cross product) of the two univariate operators, leading to the surface approximation

$$Q(s,t)=\phi_s \cdot \phi_t P(s,t). \tag{4.32}$$

The effect of this operation is that ϕ_t operates on the data $P(s,t_j)$ while ϕ_s operates simultaneously on the data $P(s_i, t)$. The operators ϕ_s and ϕ_t commute, i.e., $\phi_s \cdot \phi_t = \phi_t \cdot \phi_s$, provided that $P(s,t)$ is continuous [62]. If only a finite number of discrete values $P(s_i, t_j)$ of the function $P(s,t)$ are used, we have

$$Q(s,t) = \sum_{j=0}^{n} \sum_{i=0}^{m} P(s_i, t_j) \cdot S_{i,m}(s) \cdot T_{j,n}(t), \tag{4.33}$$

where $S_{i,m}(s)$ and $T_{i,n}(t)$ are interpolating or approximating univariate functions, e.g., taken from one of the function classes discussed in the preceding sections. Usually, S and T are of the same type (e.g., Lagrange interpolates), but this is not mandatory. As the interpolating functions must satisfy the conditions

$$S_{i,m}(s_k) = \delta_{ik} \quad \text{and} \quad T_{j,n}(t_k) = \delta_{jk},$$

where the δ_{ik} and δ_{jk} are Kronecker's delta functions, it is evident that the constructed surface $Q(s,t)$ fits through the set of points

$$\{(s_i, t_j) | (i,j) \in [0:m] \times [0:n]\}.$$

If $P(s,t)$ represents a surface to be approximated by $Q(s,t)$, the approximation error vanishes only in these points.

In matrix notation we may write instead of equation (4.33),

$$Q(s,t) = S \times P \times T^T \tag{4.34a}$$

with S and T being vectors of interpolation functions

$$S = [S_{0,m}(s) \quad S_{1,m}(s) \ldots S_{m,m}(s)], \tag{4.34b}$$

$$T = [T_{0,n}(t) \quad T_{1,n}(t) \ldots T_{n,n}(t)], \tag{4.34c}$$

and P being the matrix of constraints

$$P = \begin{bmatrix} P(s_0, t_0) & P(s_0, t_1) & \ldots & P(s_0, t_n) \\ P(s_1, t_0) & P(s_1, t_1) & \ldots & P(s_1, t_n) \\ \vdots & & & \\ P(s_m, t_0) & P(s_m, t_1) & \ldots & P(s_m, t_n) \end{bmatrix}. \tag{4.34d}$$

The domain of the operation $\phi_{s,t} = \phi_s \cdot \phi_t$ is hence a grid of mesh points, and the interpolation is exact only in these points. Figure 4.9 illustrates the Cartesian product approximation.

Instead of approximating a bivariate operator as a product of two sets of univariate operators, one may employ the principle of superposition. A simple

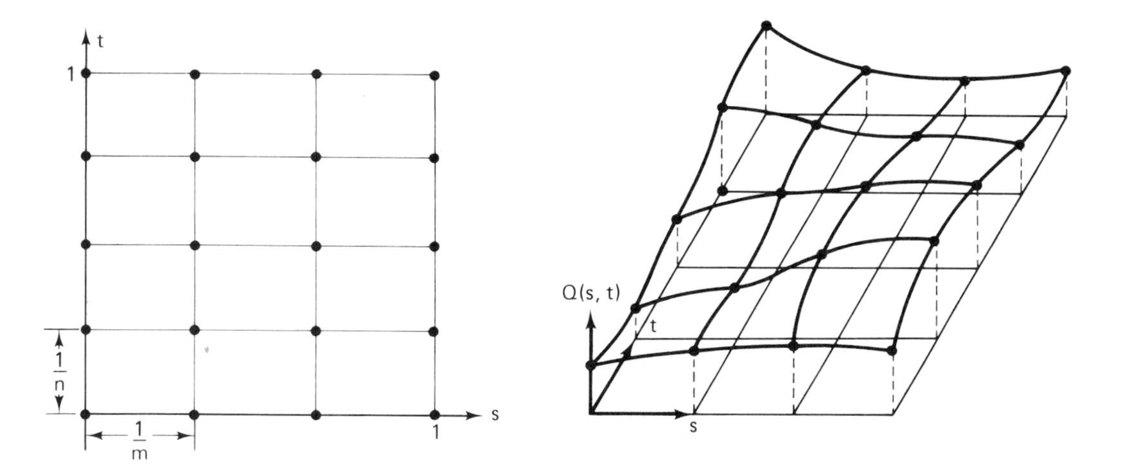

Figure 4.9 Domain of a Cartesian product approximation and a resulting surface $Q(s,t)$. The interpolation is exact only in the mesh points.

addition of the two operators, $\phi_s(s)$ and $\phi_t(t)$, however, would render, in the points of intersection of the two families of interpolates, twice the value of the approximand. To correct this, the following operation must be performed [62]:

$$Q(s,t)=[\theta_s \oplus \phi_t]P(s,t)=[\phi_s+\phi_t-\phi_s \cdot \phi_t]P(s,t). \tag{4.35}$$

If ϕ denotes the set of idempotent linear transformations from a linear space onto a subspace, it can be shown that the algebraic structure $(\phi,\{\cdot,\oplus\})$ is a distributive lattice [62]. If, furthermore, an identity and a complement are defined, the structure consequently becomes a boolean algebra. Gordon [62] proved that the approximation $\phi_{s,t}=\phi_s \cdot \phi_t$ is the minimal approximation, as it interpolates in a minimal number of points—the mesh points—whereas the approximation $\phi_{s,t}=\phi_s \oplus \phi_t$ is the maximum approximation, as it interpolates in a maximum number of points, namely, all points on the mesh lines $\phi_s P(s,t)$ and $\phi_t P(s,t)$. The latter type of approximation is called the *boolean sum approximation* (alias extended Coons or transfinite approximation). The curves along which the interpolation is exact are called *boundary functions* and *blending functions*, respectively. Each of the operators, ϕ_s and ϕ_t, interpolates univariate functions and weights these functions by appropriate blending functions.

The third possibility of surface approximation is simply to use only one of the two operators, thus approximating a surface by one of the two operations

$$Q(s,t)=\phi_t P(s,t_j) \quad \text{or} \quad Q(s,t)=\phi_s P(s_i,t). \tag{4.36}$$

This technique is called *lofting*. Lofting is often used in applications where the surface to be constructed primarily stretches in one direction (e.g., an aircraft fuselage). Figure 4.10 illustrates the lofting technique.

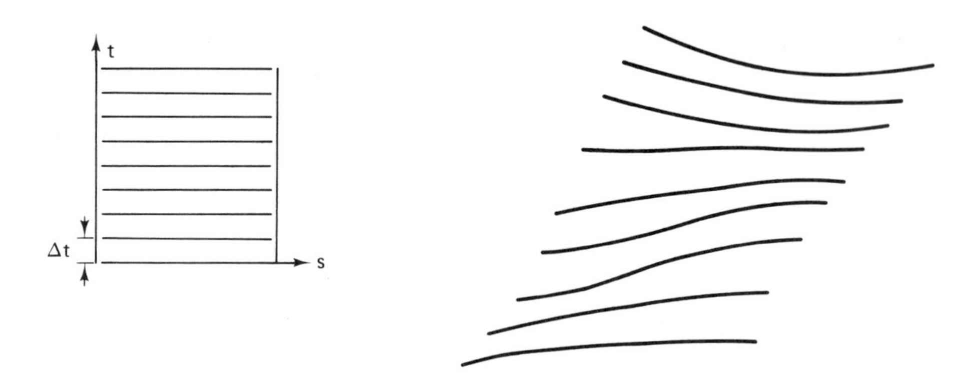

Figure 4.10 Lofting.

It can also be shown [109] that a unique *cubic spline* approximation exists which solves the interpolation problem. Given a bivariate function $F(s,t)$ and a piecewise bicubic polynomial $Q(s,t) \in C^2$ such that

$$
\begin{aligned}
Q(s_i,t_j) &= F(s_i,t_j), && \forall i,j \in [0:n], \\
\frac{\partial Q}{\partial s}(s_i,t_j) &= \frac{\partial F}{\partial s}(s_i,t_j), && \forall j \in [0:n]; \quad i=0,n \\
\frac{\partial Q}{\partial t}(s_i,t_j) &= \frac{\partial F}{\partial t}(s_i,t_j), && \forall i \in [0:n]; \quad j=0,n \\
\frac{\partial^2 Q}{\partial s\,\partial t}(s_i,t_j) &= \frac{\partial^2 F}{\partial s\,\partial t}(s_i,t_j), && i,j=0,m.
\end{aligned}
\tag{4.37}
$$

This approximation is given by

$$Q(s,t) = \sum_{i=-1}^{n+1} \sum_{j=-1}^{n+1} c_{ij}\beta_i(s)\beta_j(t), \tag{4.38}$$

where $\beta_i(s)$ and $\beta_j(t)$ are the cubic B-splines as discussed in Section 4.3. The coefficients c_{ij} can be found by solving a system of linear equations analogously to the two-dimensional case.

Instead of interpolating a given function $F(s,t)$ at a given set of points on this surface, one can alternatively construct approximates on the basis of specifying constraints only at the boundaries and construct the interior of the surface as a "blend" of these boundary conditions. This leads to Coons' notion of surface patches.

An example for the boolean sum approximation is the widely used surface patch technique introduced by Coons [20]. In its simplest form, a Coons surface patch is determined by four boundary curves,

$$P(s,0),\ P(s,1),\ P(0,t),\ P(1,t),$$

and a linear interpolation between these curves (a "linear blend")

$$Q(s,t)=P(s,0)(1-t)+P(s,1)t+P(0,t)(1-s)+P(1,t)s$$
$$-P(0,0)(1-s)(1-t)-P(0,1)(1-s)t-P(1,0)s(1-t)-P(1,1)st. \quad (4.39)$$

We recognize that the two lofting operations, $\phi_t P(s,j)$ and $\phi_s P(i,t)$, are given as the special cases

$$\phi_t P(s,j)=P(s,0)(1-t)+P(s,1)t \quad \text{and} \quad \phi_s P(i,t)=P(0,t)(1-s)+P(1,t)s.$$

More generally, the linear blending functions may be replaced by polynomials of a higher degree. Let $(1-s)$ be replaced in general by $I_{00}(s)$, $(1-t)$ by $I_{00}(t)$, s by $I_{01}(s)$, and t by $I_{01}(t)$. In this notation, the first subscript indicates the position on the unit square of the boundary function being weighted by the blending function, and the second subscript indicates that the blending function is weighted by a univariate function which represents a boundary derivative constraint of the order indicated by the subscript (see Section 4.2). Thus, we have now

$$Q(s,t)=\sum_{i=0}^{1} P(i,t)I_{i0}(s)+\sum_{j=0}^{1} P(s,j)I_{j0}(t)-\sum_{i=0}^{1}\sum_{j=0}^{1} P(i,j)I_{i0}(s)I_{j0}(t). \quad (4.40)$$

These simple Coons surfaces can only be joined with positional continuity. For continuity in the first derivative, which is essential in most applications, the definition of a surface patch must be modified such that, in addition to the boundary curves, boundary derivatives (normal derivative) can also be specified: i.e., in addition of the four curves $P(s,j)$ and $P(i,t)$, the user should be able to specify the cross-boundary slopes $P_t(s,j)$ and $P_s(i,t)$, $i, j\in[0:1]$, with $P_s(s,t)=\partial P/\partial s$ and $P_t(s,t)=\partial P/\partial t$. Therefore, we now apply the operators

$$\phi_t P(s,j)=P(s,0)I_{00}(t)+P(s,1)I_{10}(t)+P_t(s,0)I_{01}(t)+P_t(s,1)I_{11}(t)$$
$$\phi_s P(i,t)=P(0,t)I_{00}(s)+P(1,t)I_{10}(s)+P_s(0,t)I_{01}(s)+P_s(1,t)I_{11}(s). \quad (4.41)$$

The boolean sum is then, with $P_{st}(s,t)=\partial^2 P/\partial s\,\partial t$,

$$Q(s,t)=\sum_{i=0}^{1}\left[P(i,t)I_{i0}(s)+P_s(i,t)I_{i1}(s)\right]+\sum_{j=0}^{1}\left[P(s,j)I_{j0}(t)+P_t(s,j)I_{j1}(t)\right]$$
$$-\sum_{i=0}^{1}\sum_{j=0}^{1}\left[P(i,j)I_{i0}(s)I_{j0}(t)+P_t(i,j)I_{i0}(s)I_{j1}(t)\right.$$
$$\left.+P_s(i,j)I_{i1}(s)I_{j0}(t)+P_{st}(i,j)I_{i1}(s)I_{j1}(t)\right], \quad (4.42)$$

or, in matrix form,

$$Q(s,t)=\begin{bmatrix} I_{00}(s) & I_{10}(s) & I_{01}(s) & I_{11}(s) \end{bmatrix}\times\begin{bmatrix} P(0,t) \\ P(1,t) \\ P_s(0,t) \\ P_s(1,t) \end{bmatrix}$$

$$+\begin{bmatrix} P(s,0) & P(s,1) & P_t(s,0) & P_t(s,1) \end{bmatrix}\times\begin{bmatrix} I_{00}(t) \\ I_{10}(t) \\ I_{01}(t) \\ I_{11}(t) \end{bmatrix}$$

$$-\begin{bmatrix} I_{00}(s) & I_{10}(s) & I_{01}(s) & I_{11}(s) \end{bmatrix}$$

$$\times\begin{bmatrix} P(0,0) & P(0,1) & P_t(0,0) & P_t(0,1) \\ P(1,0) & P(1,1) & P_t(1,0) & P_t(1,1) \\ P_s(0,0) & P_s(0,1) & P_{st}(0,0) & P_{st}(0,1) \\ P_s(1,0) & P_s(1,1) & P_{st}(1,0) & P_{st}(1,1) \end{bmatrix}\times\begin{bmatrix} I_{00}(t) \\ I_{10}(t) \\ I_{01}(t) \\ I_{11}(t) \end{bmatrix}. \quad (4.43)$$

Contrasting with such a boolean sum construction of a surface patch, the Cartesian production construction is considerably simpler:

$$Q(s,t)=\sum_{i=0}^{1}\sum_{j=0}^{1}\left[P(i,j)I_{i0}(s)I_{j0}(t)+P_s(i,j)I_{i1}(s)I_{j0}(t)+P_t(i,j)I_{i0}(s)I_{j1}(t)\right.$$

$$\left.+P_{st}(i,j)I_{i1}(s)I_{j1}(t)\right]. \quad (4.44)$$

If we choose as blending functions the Hermite polynomials discussed in Section 4.2,

$$\begin{aligned} I_{00}(x)&=2x^3-3x^2+1 \\ I_{10}(x)&=-2x^3+3x^2 \\ I_{01}(x)&=x^3-2x^2+x \\ I_{11}(x)&=x^3-x^2, \end{aligned} \quad (4.45)$$

we obtain the matrix

$$Q(s,t)=[s^3 \quad s^2 \quad s \quad 1]\times M\times P\times M^T\times[t^3 \quad t^2 \quad t \quad 1]^T \quad (4.46a)$$

(see Section 4.2), where P is the tensor

$$\begin{bmatrix} P(0,0) & P(0,1) & P_t(0,0) & P_t(0,1) \\ P(1,0) & P(1,1) & P_t(1,0) & P_t(1,1) \\ P_s(0,0) & P_s(0,1) & P_{st}(0,0) & P_{st}(0,1) \\ P_s(1,0) & P_s(1,1) & P_{st}(1,0) & P_{st}(1,1) \end{bmatrix}. \tag{4.46b}$$

It should be mentioned that the modern semiconductor technology allows the design of hardware surface patch generators which generate an opaque surface by employing an appropriate cubic spline approximation and calculate a very dense grid of mesh lines. By controlling the beam intensity as a function of the z-coordinate, the impression of depth can be created [131] (Fig. 4.11).

Bézier's method for the interactive construction of curves can be easily expanded for the case of free-form surface design. If $F:[0,1]\times[0,1]\rightarrow\mathbf{R}^3$ is a vector-valued function from which the values for the Bézier approximation are taken, and if the Cartesian product approach is applied, we have

$$Q(s,t)=\sum_{i=0}^{m}\sum_{j=0}^{n}F\left(\frac{i}{m},\frac{j}{n}\right)\binom{m}{i}s^i(1-s)^{m-i}\binom{n}{j}t^j(1-t)^{n-j}. \tag{4.47}$$

Many of the properties of the univariate case are easily carried over to the bivariate approximation. For example, $Q(s,t)$ coincides with F in general only in the four corner points of the (s,t) unit square. The points $F\left(\frac{i}{m},\frac{j}{n}\right)$ are the vertices of an mn-faced net of line segments, playing a role totally analogous to the role of the Bézier polygon in curve design. Figure 4.12 illustrates a Bézier surface generation.

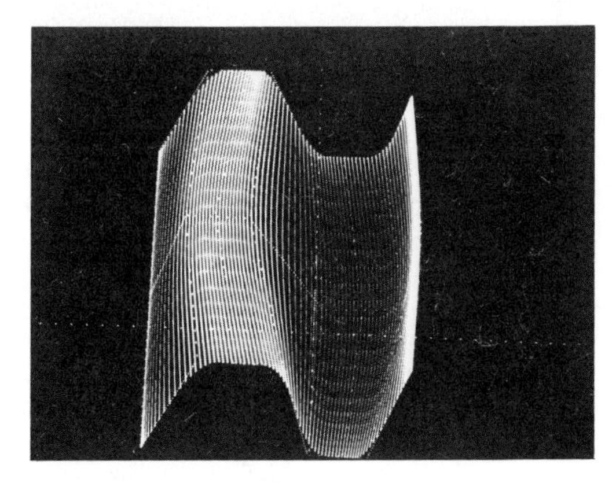

Figure 4.11 "Shaded" surfaces, generated by a dense grid of mesh lines. The shading is accomplished by controlling the beam intensity.

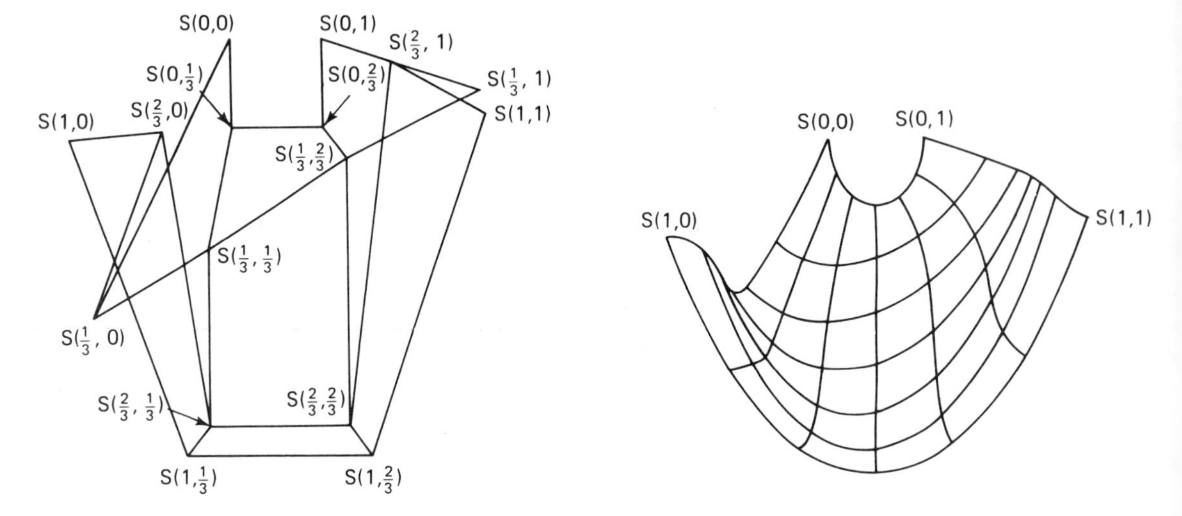

Figure 4.12 Bézier surface generation. The left portion shows the piecewise bilinear surface which is the bivariate equivalent of the Bézier polygon. The right portion shows the associated bicubic surface [131].

For the sake of completeness, we shall also present the approximation formula for the case that the boolean sum is taken instead of the Cartesian product:

$$Q(s,t)=\sum_{i=0}^{m}\binom{m}{i}s^i(1-s)^{m-i}F\left(\frac{1}{m},t\right)+\sum_{j=0}^{n}\binom{n}{j}(1-t)^{n-j}F\left(s,\frac{j}{n}\right)$$
$$-\sum_{i=0}^{m}\sum_{j=0}^{n}\binom{m}{i}s^i(1-s)^{m-i}\binom{n}{j}t^j(1-t)^{n-j}F\left(\frac{i}{m},\frac{j}{n}\right). \qquad (4.48)$$

EXERCISES

1. (a) Plot the curve $x=2\cos t-\cos 2t$, $y=2\sin t-\sin 2t$, $0\leqslant t\leqslant 2\pi$.
   (b) Plot the surface in 3-space $x=-2s+1$, $y=-2t+1$, $z=4(s^2-s-t^2+t)$, over the region $[-1,+1]\times[-1,+1]$. The surface is to be approximated by a grid of curves $[x(s,t_i)\;\; y(s,t_i)\;\; z(s,t_i]$ with $t_i=i\cdot 0.5$ and $[x(s_i,t)\;\; y(s_i,t)\;\; z(s_i,t)]$ with $s_i=i\cdot 0.5$, $i\in[-2:+2]$.

2. A function in 2-space is to be approximated by B-splines of degree $r=3$. Divide the interval $[0,1]$ into five equally spaced subintervals (Section 4.3).
   (a) What are the values of k and q, and what is the subinterval length h?
   (b) How many knots are obtained by this division of $[0,1]$, and where are the knots $t_j=j\cdot h$, $j=0,1,\ldots,q$, located?
   (c) How many functions $\beta_r(qt-i)$ can be defined and for which values of i?

(d) Compute the functions $\beta_r(qt-i)$ for all j and i. What is determined by the result of this computation for a particular j?
(e) How can the obtained result be expressed in matrix form? What kind of matrix will obtained?
(f) Formulate a matrix product that defines the splines $s_r(a_i, t_j)$ for all j. (*Hint*: The result is a vector containing these splines.)

3. Write a program in a language of your choice for the solution of the interpolation problem equation (4.14), using cubic B-splines. The program will be structured into the following subroutines (external procedures). The calling sequence of the subroutines is (we assume the syntax of FORTRAN)

Main program: BSPLINE (M,N,DATA, PNTS)
Calculates $N+1$ uniformly spaced values of the interpolant $s(t)$ from M uniformly spaced knot values as contained in the array DATA. The array PNTS contains the result.

Subroutine 1: CM (M, B)
Generates the coefficient matrix B [see equation (4.25b)] for the specified number of M knots.

Subroutine 2:† INVERT (K, ARG, RES)
Inverts a $K \times K$ matrix ARG, leading to the result RES.

Subroutine 3: BETA (KNOT, VAL)
Calculates the value $\beta_3(\tau)$ for a given τ [see equation 4.15)]. KNOT represents the value of τ, and VAL represents the result.

4. (a) Approximate the function $y=e^t$, $0 \leqslant t \leqslant 1$, using a spacing of $h=0.2$. Perform the calculations either by hand or by using the program of Exercise 3.
(b) Calculate the interpolant at the points $t_i = i \cdot 0.1$, $i \in [0:1]$. Plot the curve. What would you consider a reasonable condition for obtaining a smoother approximation of the given function?

For the convenience of the reader, we list here the matrix B^{-1}:

$$\begin{bmatrix}
-0.4309 & -0.4641 & 1.8565 & -0.4976 & 0.1340 & -0.0383 & 0.0096 & -0.0006 \\
0.1155 & 1.7321 & -0.9282 & 0.2488 & -0.0670 & 0.0191 & -0.0048 & 0.0003 \\
-0.0309 & -0.4641 & 1.8565 & -0.4976 & 0.1340 & -0.0383 & 0.0096 & -0.0006 \\
0.0083 & 0.1244 & -0.4976 & 1.7416 & -0.4689 & 0.1340 & -0.0335 & 0.0022 \\
-0.0022 & -0.0335 & 0.1340 & -0.4689 & 1.7416 & -0.4976 & 0.1244 & -0.0083 \\
0.0006 & 0.0096 & -0.0383 & 0.1340 & -0.4976 & 1.8565 & -0.4641 & 0.0309 \\
-0.0003 & -0.0048 & 0.0191 & -0.0670 & 0.2488 & -0.9282 & 1.7321 & -0.1155 \\
0.0006 & 0.0096 & -0.0383 & 0.1340 & -0.4976 & 1.8565 & -0.4641 & 0.4309
\end{bmatrix}$$

5. Use the tensor product approximation according to equations (4.46) to generate a u-v mesh line representation over the region $[-1,+1]\times[-1,+1]$ of a surface. The surface to be approximated is described in parametric form by the vector of

†It is fair to assume that the reader will find a subroutine for matrix inversion in the public library of the computer system he/she works with. If not, see, for example, G. E. Forsythe and C. B. Moler, *Computer Solution of Linear Algebraic Systems* (Prentice-Hall, Englewood Cliffs, N.J., 1967).

homogeneous coordinates $P(u,v)=[x(u,v) \quad y(u,v) \quad z(u,v) \quad w(u,v)]$.

(a) $P(u,v)=[(-2u+1) \quad (-2u^2+2u) \quad (2u^2-2u+1) \quad (2u^2-2u+1]$

(b) $P(u,v)=[(-2u+1] \quad (-2v+1) \quad 4(u^2-u-v^2+v) \quad 1]$

(c) $P(u,v)=[2(2u-1)(-v^2+v) \quad 4(-u^2+u)(-v^2+v) \quad (2u^2-2u+1)(2v-1)$
$(2u^2-2u+1)(2v^2-v+1)]$

(*Hint*: The $4\times4\times4$ tensor P in equation (4.46a), which is written in equation (4.46b) as a 4×4 matrix whose components are vectors, can also be written as a vector $[A_x, A_y, A_z, A_w]$ whose components are 4×4 matrices. Determine first in each case the matrices A_x, A_y, A_z, and A_w.)

Plot the surfaces, e.g., as a grid of 5×5 mesh lines. Do you recognize their type?

5

Rendering of Surfaces and Solids

5.1 THE HIDDEN-SURFACE PROBLEM

The most general and realistic approach to a rendering of solid objects is to display *shaded pictures*. A shaded picture is produced by recording the "shade of gray" for each point of the picture. This is certainly an adequate approach for TV raster displays; in which case the displayed surfaces may not only vary in their shade of gray but also in their color. The generation of shaded pictures on a line-drawing display that does not provide special hardware support for this purpose, however, is prohibitively expensive, in terms of memory capacity as well as in terms of time. Even if enough refresh memory were available, real-time generation of such pictures is not possible without the aid of special hardware.

What is needed, therefore, is a drastic reduction in the number of points to be stored in the refresh memory and processed by the display processor. Several steps may be taken to reduce this number.

The first step of reduction is achieved by displaying only the contours of the faces of an object, thus reducing the set of all points on a surface to the subset of all points on the boundary of the surface. A second reduction step, quite natural for line-drawing displays, in which the primary graphical primitive is the line segment or "vector," is to approximate contours by polygons. Hence, the set of points on a boundary is further reduced to a much smaller subset of the points that are the vertices of the polygon (the remaining points on the contour being added by the hardware). Each vertex is described by its coordinates in three dimensions in an appropriate coordinate system.

To create the illusion that the interior region of a displayed surface is opaque, it must be ensured that the parts of the object(s) which would be hidden from an observer by the opaque surface are not displayed. The problem of determining the visibility of lines or portions of lines is known as the *hidden-line problem*. Actually, it might be better to speak of the "hidden-surface problem," as the term "hidden-line problem" refers to the particular "wire-frame" representation of objects, i.e., the replacement of the faces of an object by its bounding contours, whereas the term "hidden surface" includes as well the rendering of shaded surfaces. The importance of hidden-line suppression for a realistic view is demonstrated by a simple example (Fig. 5.1).

The hidden-surface problem can be facilitated by a third reduction step, which stipulates that all faces be flat. This introduces the additional property for the vertices of the bounding polygon that they all lie on one and the same plane. More strictly, it is stipulated that the objects be not only plane-faced but also convex. In this case, a simple calculation of the *normal* of a face is sufficient for determining whether this face is a (visible) front face or an (invisible) back face of the solid to which it belongs.

At this point a definition of a convex polygon will be given (which the reader may remember from high school geometry). The reader should also remember that any line divides the plane it is contained in into two half-planes.

Definition: *Convex Polygon*

Given a plane polygon. The polygon is convex if a line can be aligned with every edge of the polygon such that *all* vertices of the polygon that are not on this line are in one and the same half-plane (i.e., all vertices are on the same side of the line).

The question of which parts of the objects of a scene are visible and which are not depends on the point from which the scene is viewed. Also, the question of depth will be significant, i.e., the distance from the various faces to the viewpoint. As a three-dimensional scene can only be rendered in the form of a two-dimensional

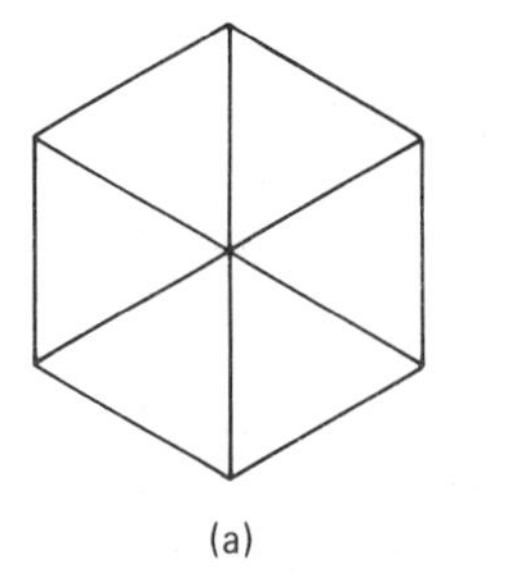
(a)

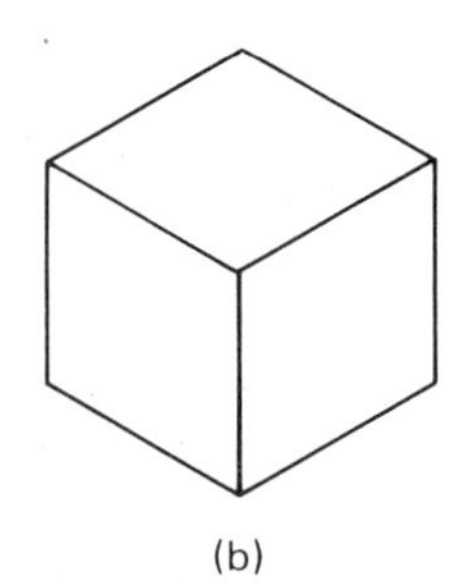
(b)

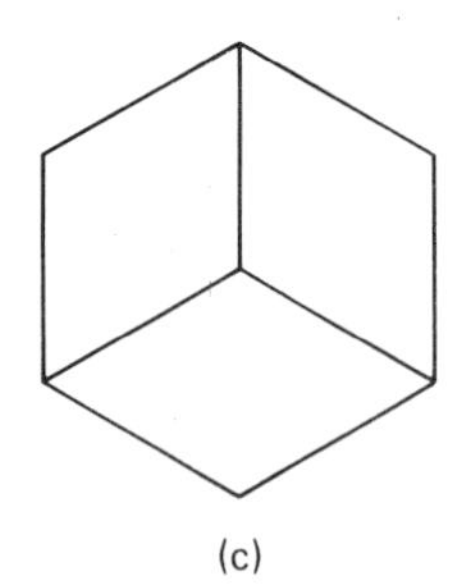
(c)

Figure 5.1 Perspective view of a cube in parallel projection: (A) With hidden lines; (B) three edges are considered to be "back edges" and are eliminated; (C) three other edges are eliminated.

perspective view, a hidden-surface algorithm must include a perspective transformation. If a central projection is applied, a fairly complicated problem arises. Many rays from the observer's eye must be considered, and many points of intersection between these rays with the object faces may have to be calculated. The problem is considerably facilitated if the simple orthographic projection is used. Therefore, most existing hidden-surface algorithms comprise the orthographic projection. If a perspective based on a central projection is wanted, the objects must first be transformed from their original shape into the shape which they would have after application of a central projection. This transformation can be achieved by subjecting the set of vertices of a scene to an appropriate coordinate transform. Such an operation should not be confused with a proper projective transformation, which results in a two-dimensional representation, whereas the shape-transformed object is still three-dimensional. This point is illustrated by an example of a cube (Fig. 5.2).

The adequate form of representing surfaces in a picture file is to have only its contours represented, where a contour is approximated by a (closed) polygon. This holds true regardless of whether the interior of the surface is blank or shaded. Hence, the hidden-line or hidden-surface problem can most generally be characterized as a transformation, mapping a set of objects into a set of "visible parts in 2-space." In the case of shaded objects, the visible part is a visible contour edge of the object surface. An object is defined as a set of coordinates plus a set of relations specifying the object topology. In general, a topology may be diagrammed in the form of a tree, as depicted in Figure 5.3. If we compare this with the picture structure as introduced in Section 3.1, we recognize that surfaces correspond to items and solids correspond to segments.

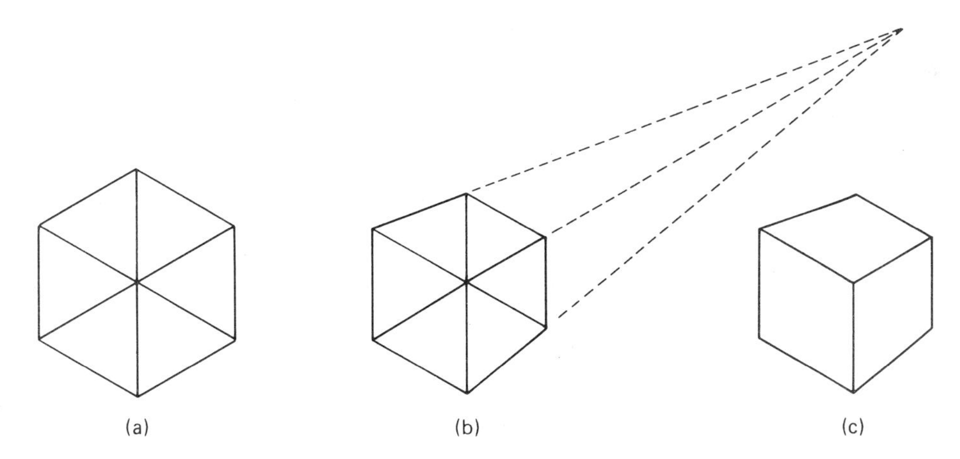

Figure 5.2 (A) Wire-frame representation of a cube in parallel projection; (B) wire-frame representation of the cube after transformation of vertex coordinates to central projection perspective; (C) subsequent orthographical perspective combined with hidden-line suppression yield the view of a central projection without hidden lines.

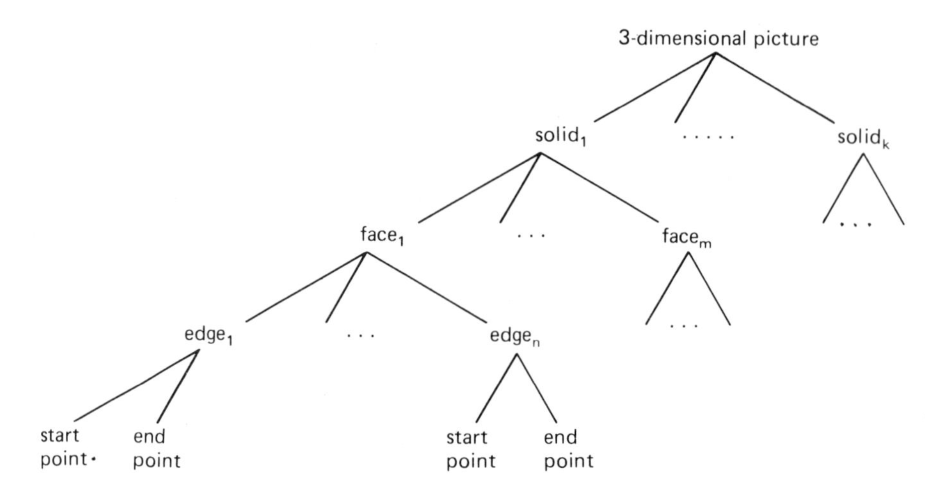

Figure 5.3 Tree-structured object topology.

Such a characterization of the hidden-surface problem implies that after application of the hidden-line algorithm, any structuring of the produced renderings is lost (as the result is simply a set of visible parts). In principle, a new set of relations could be created such that substructures of a picture remain to be individually identifiable. However, this leads to a rather complicated problem that is usually shunned. The rendering of (relatively realistic) views of three-dimensional objects is thus an end in itself; and no attempts are usually made to subject such objects to interactive manipulations.

A hidden-surface algorithm consists of several major steps. Each step provides a particular mapping, and the total algorithm is a composition of such mappings. Consequently, interposed between the domain of a hidden-surface transformation (the set of objects) and its range (the set of visible parts) there may exist a sequence of intermediate representations.

It is now possible to give a formal definition of a hidden-surface algorithm [83].

Definition: *Hidden-Surface Algorithm*

A hidden-surface algorithm is a 5-tuple

$$\text{HSA} = (O, S, I, \varphi, \sigma),$$

where O is a set of objects in 3-space
S is a set of visible segments in 2-space
I is a set of "intermediate representations"
φ is the set of "transition functions" $\{\text{PM}, \text{IS}, \text{CT}, \text{DT}, \text{VT}\}$
σ is a "strategy function".

The meaning of the function names in φ is:

- PM is a function that produces the perspective view ("projective mapping"). Hence, the domain of PM is the 3-space, and its range is the 2-space.
- IS is a function that calculates the point of intersection of two line segments. Hence the domain and range of IS are identical (2-space or 3-space).
- CT is a function that performs a "containment test" in 2-space; i.e., it checks whether or not a point is in a given bounded surface. The result of CT is a boolean variable which is true if the point is contained, and false otherwise.
- DT is a function that performs a "depth test," i.e., it compares two points and finds out which has the greater depth (is farther away from the point of observation).
- VT is a function that performs a "visibility test" for a given surface; i.e., it yields a boolean value which is true if the surface is potentially visible, and false if the surface is totally invisible. The "strategy function" σ specifies the order in which the functions of φ must be applied for obtaining the desired result.

A first classification of hidden-surface algorithms according to the degree of reduction is presented in Figure 5.4. A more detailed taxonomy will be developed after a few representative examples of hidden-surface algorithms have been discussed. But first, we shall take a closer look at the functions of φ.

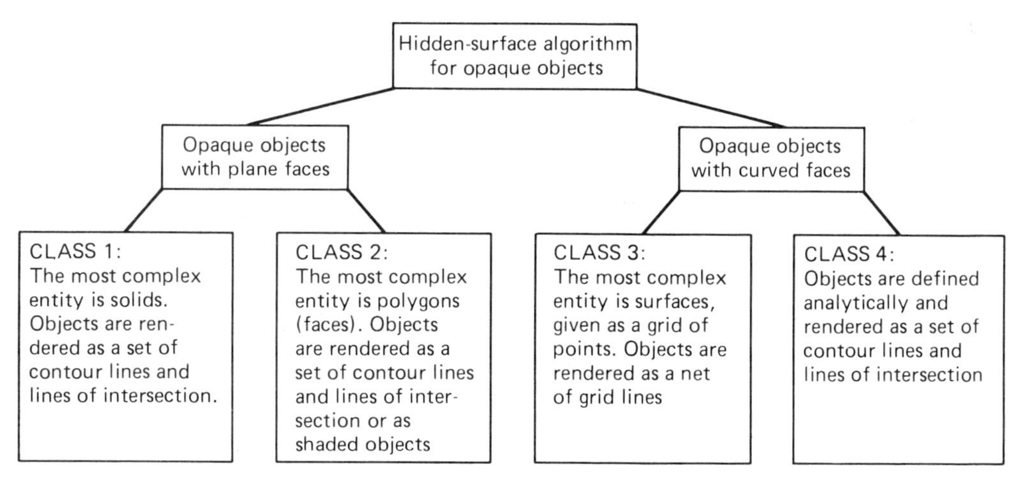

Figure 5.4 Classification of hidden-surface algorithms according to the degree of reduction.

5.2 THE SET OF TRANSITION FUNCTIONS {PM, IS, CT, DT, VT}

The functions of the set $\varphi = \{PM, IS, CT, DT, VT\}$ will now be discussed in more detail. It should be emphasized at this point that for each of the five functions in φ, several variants may exist, depending on the particular hidden-surface algorithm. That means that each of these functions may represent a certain class of mappings [83].

5.2.1 The Projective Mapping PM

The problem of perspective view is elaborately discussed in Chapter 3. As mentioned above, almost all hidden-surface algorithms employ the simple orthographic projection which is almost an identity operation: If (x,y,z) is a point in the object space and (X, Y) is its projection on the image plane, we may simply set $X=x$ and $Y=y$; that is, the x- and y-coordinates are identical in both spaces, and z is dropped.

In Section 5.4 a taxonomy of hidden-surface procedures is presented. There, the reader will notice that one criterion for classifying hidden-surface algorithms is the space in which visibility is determined: object space or image space. This affects the order in which projective transformation and visibility tests are performed. If visibility is tested in the object space, this test must precede the projective transformation. If visibility is tested in the image space, the projective transformation, of course, must come first in order to obtain the image space representation of the objects.

5.2.2 The Function Class IS

IS is defined as a function that yields the point of intersection between two graphic elements provided that the two elements intersect in exactly one point. Graphic elements may be: Two lines, two segments, or a line and a plane. Points of intersection are calculated for various purposes:

1. To find the point on a line segment at which a change of visibility occurs.

2. To find out whether the projection of two polygons on the image plane yields a nonempty intersection.

3. To determine intersections between object elements and a scan line.

4. To function as a subtest within a containment test.

5. To determine intersections between object faces and a line, connecting the point of observation with a special test point in object space.

In Section 3.5, formulas are given for the point of intersection between two lines or a line and a plane, where lines and planes are represented in homogeneous coordinates. The corresponding formulas for ordinary coordinates can easily be obtained from the homogeneous coordinate representation. For the convenience of the reader, we present here the results of such a step.

Intersection of Two Lines: Given two line equations in ordinary coordinates in 2-space,

$$A_1x + B_1y + C_1 = 0$$
$$A_2x + B_2y + C_2 = 0.$$

If

$$\begin{vmatrix} A_1 & B_1 \\ A_2 & B_2 \end{vmatrix} = 0,$$

the two lines are parallel; i.e., there exists no finite point of intersection. If $A_1/A_2 = B_1/B_2 = C_1/C_2$, the two lines are coinciding; i.e., all their points are collinear. Otherwise, we have a point of intersection (x_i, y_i) with the coordinates

$$x_i = \frac{B_1C_2 - B_2C_1}{A_1B_2 - A_2B_1}, \qquad y_i = \frac{C_1A_2 - C_2A_1}{A_1B_2 - A_2B_1}$$

[see equations (3.31)].

Intersection of Two Segments: Given two lines in a 2-space, G and G'. Given the points $P_1, P_2 \in G$, $P_1 = (x_1, y_1)$ and $P_2 = (x_2, y_2)$, and the points $P_1', P_2' \in G'$, $P_1' = (x_1', y_2')$ and $P_2' = (x_2', y_2')$. Let G and G' intersect in the point $I = (x_i, y_i)$. Then the two segments, P_1P_2 and $P_1'P_2'$, intersect in I if

$$\min[\min(x_1, x_2), \min(x_1', x_2')] \leqslant x_i \leqslant \max[\max(x_1, x_2), \max(x_1', x_2')]$$

and

$$\min[\min(y_1, y_2), \min(y_1', y_2')] \leqslant y_i \leqslant \max[\max(y_1, y_2), \max(y_1', y_2')].$$

Intersection of a Plane and a Line: Given:

1. A plane passing through the points $P_i = (x_i, y_i, z_i)$, $P_j = (x_j, y_j, z_j)$, $P_k = (x_k, y_k, z_k)$, P_i, P_j, P_k being not collinear, and given by the equation $Ax + By + Cz + D = 0$.

2. A line passing through the points $P_1 = (x_1, y_1, z_1)$ and $P_2 = (x_2, y_2, z_2)$. The line can be represented by the equation

$$\frac{x - x_1}{\cos\alpha} = \frac{y - y_1}{\cos\beta} = \frac{z - z_1}{\cos\gamma}$$

with

$$\cos\alpha = \frac{x_2 - x_1}{d}, \qquad \cos\beta = \frac{y_2 - y_1}{d}, \qquad \cos\gamma = \frac{z_2 - z_1}{d},$$

$$d = \sqrt{(x_2 - x_1)^2 + (y_2 - y_1)^2 + (z_2 - z_1)^2}\ .$$

Then we have:

1. If $A\cos\alpha + B\cos\beta + C\cos\gamma = 0$, the line is parallel to the plane.

2. If $A\cos\alpha + B\cos\beta + C\cos\gamma = 0$ *and* $Ax_1 + By_1 + Cz_1 = 0$, the line lies in the plane. Otherwise, we have the point of intersection

$$x_i = x_1 - t\cos\alpha$$
$$y_i = y_1 - t\cos\beta$$
$$z_i = z_1 - t\cos\gamma, \quad t = \frac{Ax_1 + By_1 + Cz_1 + D}{A\cos\alpha + B\cos\beta + C\cos\gamma}.$$

Test for an Overlap of Two Closed, Convex Polygons in the Image Plane: Given two closed, convex polygons in the image plane (which may be projections of two faces of three-dimensional objects). Each polygon is specified by a tuple of its vertices; i.e., if F_1 and F_2 denote the polygons, we have $F_1 = (P_{1,1}, P_{1,2}, \ldots, P_{1,m})$ and $F_2 = (P_{2,1}, P_{2,2}, \ldots, P_{2,n})$, with $P_{i,j} = (x_{i,j}, y_{i,j})$. The polygons F_1 and F_2 are disjoint (do not overlap) if the following *minimax test* holds true:

$$\bigwedge_{j \in [1:m]} \bigwedge_{k \in [1:n]} : \begin{array}{lll} \max(x_{1,j}) < \min(x_{2,k}) & \textit{or} & \max(x_{2,k}) < \min(x_{1,j}) \quad \textit{or} \\ \max(y_{1,j}) < \min(y_{2,k}) & \textit{or} & \max(y_{2,k}) < \min(y_{1,j}). \end{array}$$

Usually, the task is not only to find out whether two polygons overlap, but also in this case to find a point in the intersection of the two faces. This may be accomplished by trying to find a point of intersection between the edges of the two polygons, or by testing the vertices of one polygon for containment in the other polygon, and vice versa (see the following section for a containment test). The possibilities of determining distinguished point in the intersection of two polygons are illustrated below.

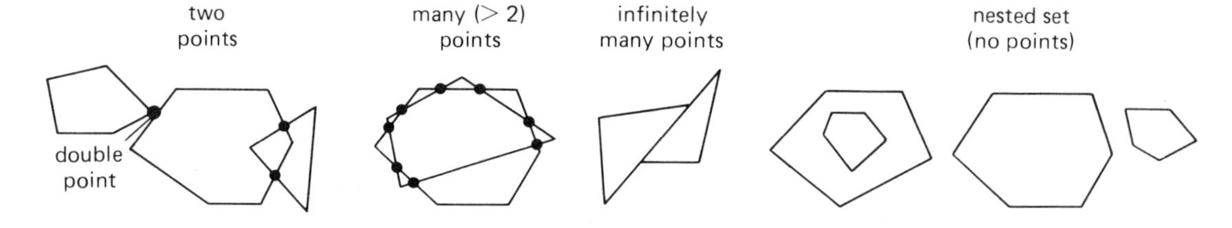

Clearly, there exist at least two points of intersection between two overlapping polygons, or else the polygons form a nested set; i.e., either one of them is fully contained in the other or they are disjoint. Normally, it is more economical to start with the intersection test rather than with the containment test, for the following reasons:

1. The probability for edge intersection is much higher than the probability for full containment.

2. The incidence relation is symmetric, the containment relation is antisymmetric. Therefore, both polygons must be tested for containment, but only one needs to be tested for edge intersection.

3. The containment test (see the following section) includes the calculation of intersections and is, therefore, more complex than the intersection test.

If all possible tests fail to yield a point in the intersection of two polygons, it indicates an empty intersection. However, it is strongly advisable always to determine first whether the two polygons have a chance of intersecting. For such a "prescreening," the minimax test described above is very appropriate. If the probability of an empty intersection is considerably higher than the probability of a nonempty intersection, tremendous savings of execution time may be achieved by prescreening. But even in cases where the probability of an empty intersection is much smaller than the probability of a nonempty intersection, it may still pay to perform the prescreening, the reason being that the simple minimax test requires only a few relational and logical operations, whereas the intersection and containment tests require many elaborate arithmetical calculations. Therefore, their execution time may exceed that of the minimax tests by some orders of magnitude.

5.2.3 The Containment Test CT

CT is an image space test that determines if a point P lies in the interior of a closed polygon F (which is usually the projection of an object face). In elementary geometry, there are two well-known procedures which perform such a test.

Test by Calculating a Sum of Angles: Let $F=(p_1,\ldots,p_n)$ denote a closed polygon in the image plane with the vertices $p_i=(x_i,y_i)$, $i=1,\ldots,n$, and $\underline{p_n}=p_1$. Let p_t be a test point the inclusion of which in F shall be determined. Let p_tp_i denote a segment connecting p_t with p_i, and let p_t be connected in that way with all vertices of F. Let α_i denote the angle between $\overline{p_tp_i}$ and $\overline{p_tp_{i+1}}$. Then p_t is outside of F if $\Sigma_i\alpha_i=0$ and inside of F if $\Sigma_i\alpha_i=2\pi$, as illustrated by the following figure.

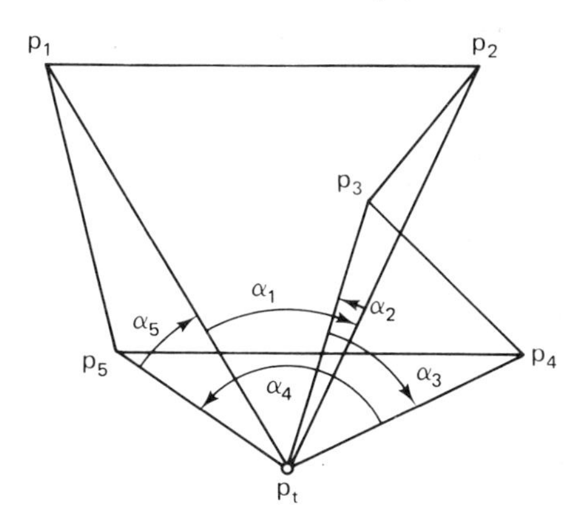

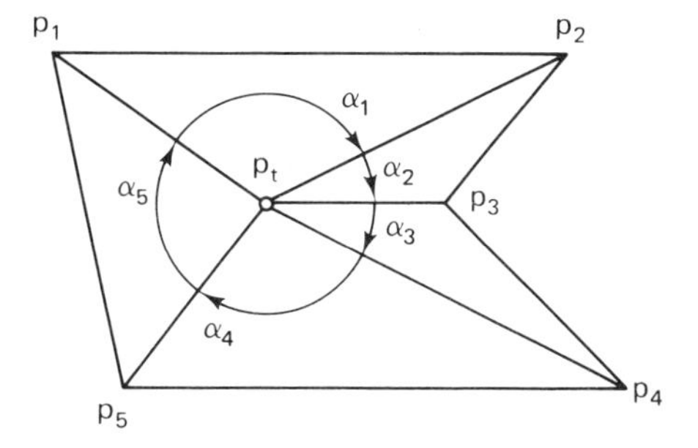

Test by Counting the Number of Intersections: Let R be a ray that starts from the test point p_t and passes somehow through the polygon F but not through one of its vertices. Then p_t is inside the polygon if the number of intersections between F and R is odd, and P is outside F if this number is even, as illustrated by the following figure.

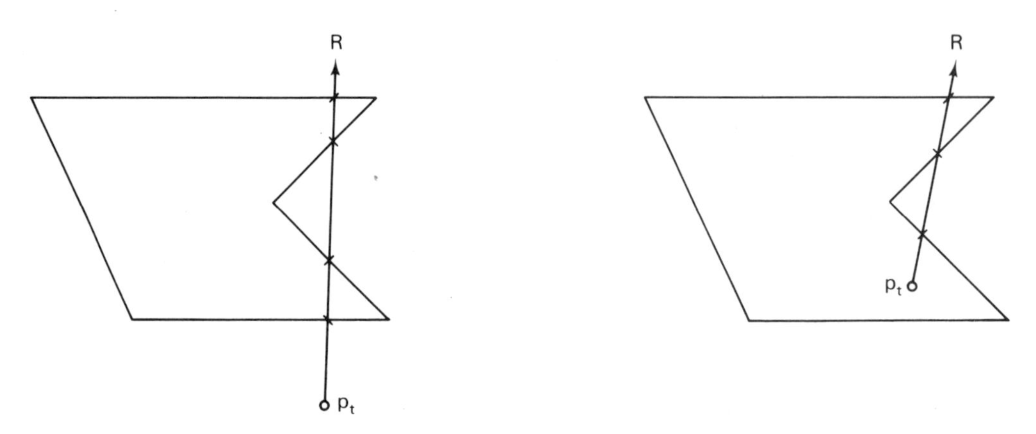

This test may be performed by an algorithm as flow-charted in the following diagram. (See page 161).

5.2.4 The Depth Test DT

The depth test function DT is applied to two overlapping graphical elements and determines which of the elements covers the other, or vice versa. Graphical elements to be compared may be a point and a face or two faces. "Depth" stands for the distance between an element and the image plane or between an element and the viewpoint, respectively, and is thus related to the z-coordinate of the element. Consequently, a depth test usually consists in the comparison of the z-coordinates of the two elements. No hidden-surface algorithm is conceivable that would not include a depth test. Actually, we can find three types of depth tests, which will be labeled DT1, DT2, DT3 and are described below.

Depth Test DT1: DT1 operates in the object space. It compares two graphic elements, a face and a test point, and determines whether or not the test point is hidden by the face. This is accomplished by calculating the point at which the face is pierced by the "line of sight," i.e., the line that passes through the test point and the viewpoint. Subsequently, the distance to the viewpoint is calculated for the test point and the piercing point. Let d_t be the distance of the test point to the viewpoint and let d_p be the distance of the piercing point to the viewpoint. Then the test point is hidden behind the face if $d_t > d_p$ and not hidden otherwise. This method implies that the viewpoint is finite, i.e., the use of a central projection. It is

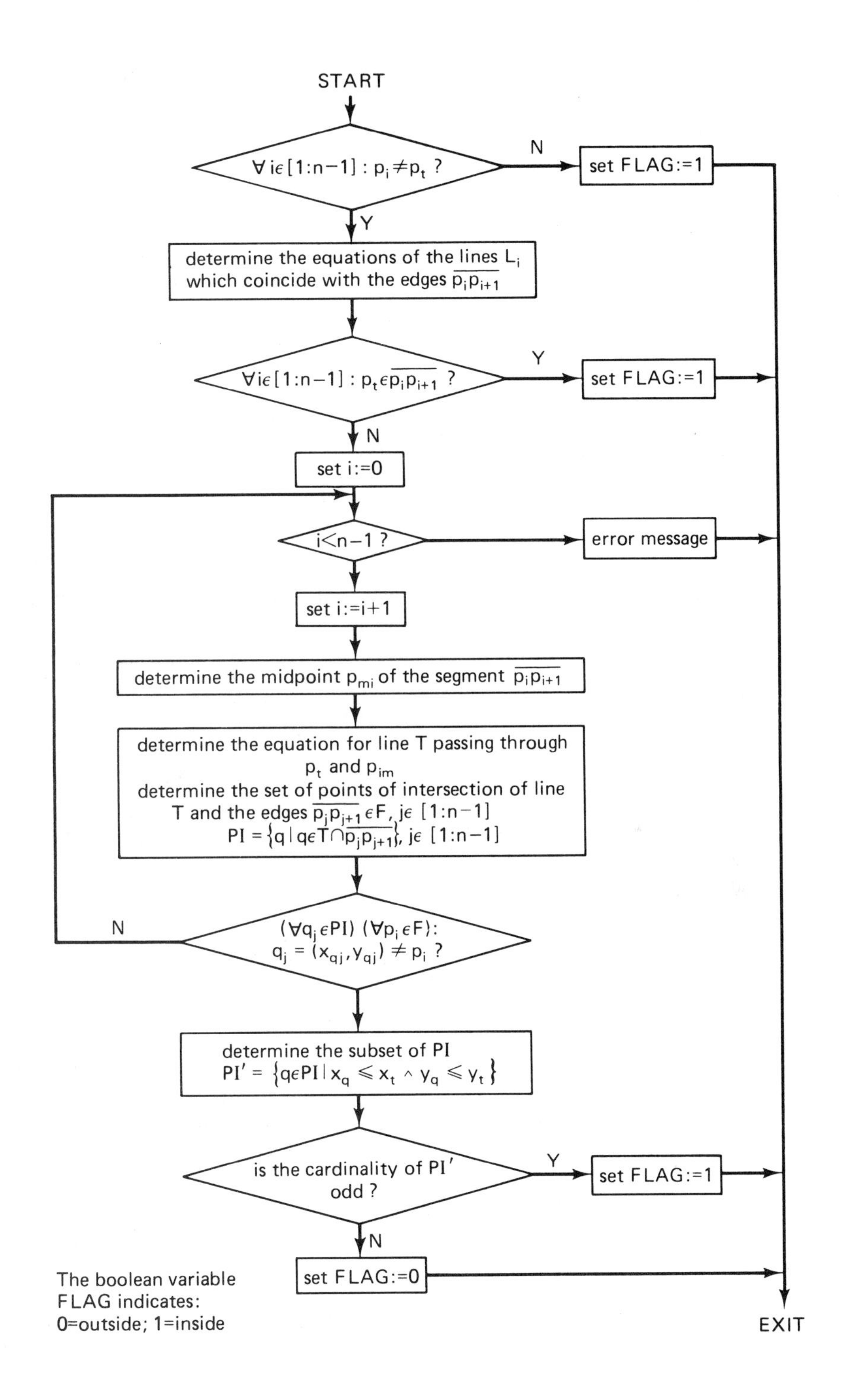
START
$\forall i \epsilon [1:n-1] : p_i \neq p_t$?
N
set FLAG:=1
Y
determine the equations of the lines L_i which coincide with the edges $\overline{p_i p_{i+1}}$
$\forall i \epsilon [1:n-1] : p_t \epsilon \overline{p_i p_{i+1}}$?
Y
set FLAG:=1
N
set i:=0
i<n−1 ?
error message
set i:=i+1
determine the midpoint p_{mi} of the segment $\overline{p_i p_{i+1}}$
determine the equation for line T passing through p_t and p_{im}
determine the set of points of intersection of line T and the edges $\overline{p_j p_{j+1}} \epsilon F$, $j \epsilon [1:n-1]$
$PI = \{q \mid q \epsilon T \cap \overline{p_j p_{j+1}}\}$, $j \epsilon [1:n-1]$
N
$(\forall q_j \epsilon PI)(\forall p_i \epsilon F)$: $q_j = (x_{qj}, y_{qj}) \neq p_i$?
determine the subset of PI
$PI' = \{q \epsilon PI \mid x_q \leqslant x_t \wedge y_q \leqslant y_t\}$
is the cardinality of PI′ odd ?
Y
set FLAG:=1
N
set FLAG:=0
The boolean variable FLAG indicates: 0=outside; 1=inside
EXIT

illustrated by the following picture. DT1 is employed, e.g., by the hidden-surface algorithms of Appel [5] and Weiss [151].

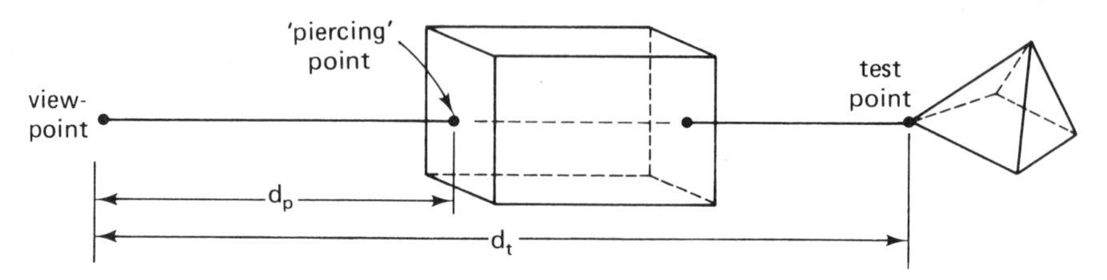

Depth Test DT2 ("Priority Test"): DT2 assumes the orthographic (or parallel) projection and operates partly in the object space and partly in the image space. The graphical elements to be tested are either two faces or a test point and a face. The case that two faces are to be tested will be discussed first. It is illustrated by the following picture. In the example of the picture, two faces, f_1 and f_2, are given in the object space. Let $\mathrm{pr}(f_1)$ and $\mathrm{pr}(f_2)$ denote their respective projections on the image space. Let $p_{1,2}$ denote a particular point in the image space (usually, this is a point in the intersection $\mathrm{pr}(f_1) \cap \mathrm{pr}(f_2)$, e.g., a point of intersection of an edge of $\mathrm{pr}(f_1)$ with an edge of $\mathrm{pr}(f_2)$). The first step in DT2 is to specify such a distinguished point $p_{1,2}$. The second step consists in finding the points $p_1 \in f_1$ and $p_2 \in f_2$, respectively, i.e., the points whose projection on the image space is $p_{1,2}$, by applying the inverse projection pr^{-1}. As the inverse of the projection function is not a function but only a relation, additional information is required in order to determine the points $p_1 = (x_p, y_p, z_1)$ and $p_2 = (x_p, y_p, z_2)$. This additional information is furnished by the equations of the two planes that contain the faces f_1 and f_2, respectively. The coordinates (x_p, y_p) of point $p_{1,2}$ are inserted into the equation of the plane in which f_1 lies and this equation is solved for z_1, as well as into the equation of the plane in which f_2 lies, leading to z_2. The values z_1 and z_2 thus

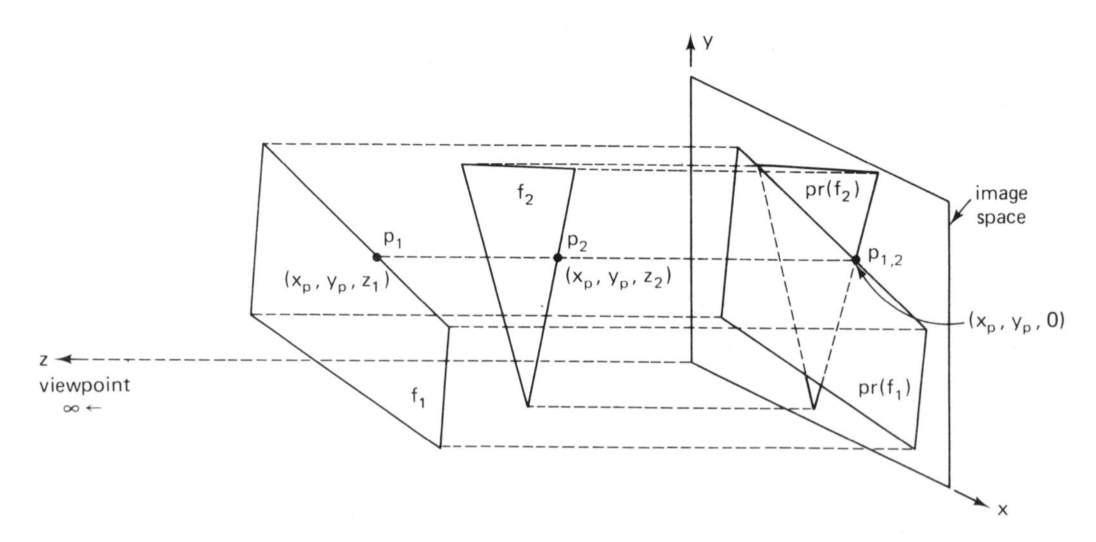

obtained are then compared, leading to the determination of (relative) priority. Hence, we define:

Definition: *Relative Priority*

Let $\mathrm{pr}(f_i)$ and $\mathrm{pr}(f_j)$ be projections of the two surfaces f_i and f_j, respectively. Let $p_{i,j}=(x_p,y_p,0)\in\mathrm{pr}(f_i)\cap\mathrm{pr}(f_j)$, $p_i=(x_i,y_i,z_i)\in f_i$, and $p_j=(x_j,y_j,z_j)\in f_j$ be three points related through the projection as follows:

$$p_i \mapsto p_{i,j} \quad \text{and} \quad p_j \mapsto p_{i,j},$$

implying that $x_i=x_j=x_p$ and $y_i=y_j=y_p$. Furthermore, it is assumed that $z_i \neq z_j$ [in cases where this assumption is not satisfied, another point in $\mathrm{pr}(f_i)\cap\mathrm{pr}(f_j)$ must be found]. Then we say that f_j *has priority over* f_i, denoted

$$f_i \lessdot f_j,$$

if

$$z_i < z_j.$$

Readily, we recognize the relation of relative priority as an irreflexive ordering.

In some hidden-surface algorithms (e.g. the "scan grid" method), a point is given in the first place that is to be tested against a face. In such a *point/surface test*, the crucial task of finding a point in the intersection of the projections of two surfaces is spared. The equation of the plane that contains the surface is constructed, into which the xy-coordinates of the test point are then inserted. Subsequently, the resulting equation is solved for the (only unknown) variable z, and the z-value thus obtained is compared with the z-coordinate of the test point. If the z-value is greater than the z-coordinate of the test point, we may say that the surface has priority over the test point.

The priority test procedure may be complicated by the occurrence of penetrating faces or *cyclic overlaps*, as illustrated below. In this case the result of the test may be $F_1 \lessdot F_2$ or $F_2 \lessdot F_1$, depending on the location of the test point. To avoid such an ambiguity, special steps must be taken: e.g., face F_1 is subdivided along the dashed line into two parts, and the depth test is performed separately for each part.

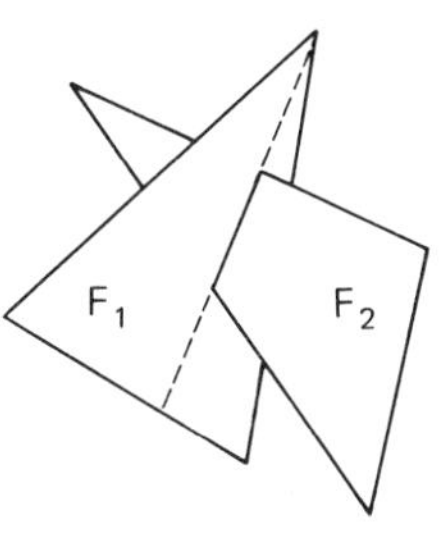

Penetrating faces

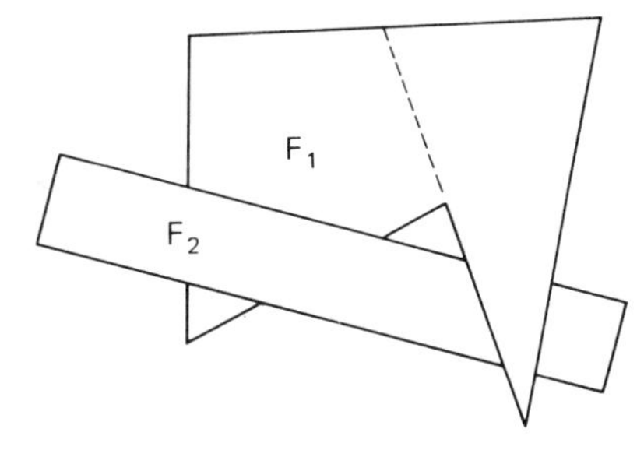

Cyclic overlap

Depth Test DT3: DT3 is the simplest possible depth test. It is applied in scan-line algorithms in which the visibility of surface projections in the image space along a scan line must be tested. If sections of a scan line are formed such that within a section no two edges of projected surfaces intersect with the scan line or no two surfaces penetrate each other in the object space (such a section is called a "sample span," see Section 5.3.4), then visibility can be determined simply by comparing the z-coordinates of the segments in the sample span. Precisely, the segment is visible whose smallest z-coordinate is greater than or equal to the maximal value of the z-coordinates of all other segments in the sample span. A visible segment forms a strip of a visible surface projection.

5.2.5 The Visibility Test VT

The visibility test function VT can only be applied to solids. It determines whether a face of the solid is a "front face," and thus potentially visible, or a "back face," and thus invisible. Needless to say, this test can only be performed in the object space on structured objects. If there are several objects which may mutually hide each other, VT can only state the total invisibility of a face (which is hidden by the volume of the solid it belongs to) but not the total visibility, as a front face may yet be hidden by another object.

In order to perform the visibility test, a *surface normal* N is first defined as an outward-pointing vector, normal to the surface (the normal equation can be directly derived from the plane equation of the surface). Next, a line of sight L is introduced as the line passing through the point of observation and the start point of the surface normal, and the angle between N and L is calculated. Then the surface is potentially visible if this angle is less than or equal to $\pi/2$, and it is invisible if the angle is greater than $\pi/2$. As the "critical angle" is $\pi/2$, it suffices to calculate the scalar product (dot product) $L \cdot N = |L| \cdot |N| \cos\psi_{LN}$ and test its sign, since $\cos\psi_{LN}$ performs a sign change at $\pi/2$. As mentioned above, visibility is completely determined that way if there is only one convex solid. If this condition is not valid, the test can only be used for eliminating invisible faces in a first step. Additional tests must then be applied to the remaining potentially visible faces in order to detect mutual hidings of faces belonging to different objects, or to handle concave objects, or both. Details will be discussed in the description of Appel's algorithm.

Invisibility determinations can be made relatively quickly because only one surface need be found which hides another. Visibility determination, on the other hand, is a relatively slow process, because all possible hiding surfaces must be examined before an element can be considered absolutely visible. Therefore, the ratio of visible surface elements to total surface elements has a significant effect on the total execution time needed to solve a hidden-line or surface problem.

5.3 DESCRIPTION OF FOUR HIDDEN-SURFACE ALGORITHMS

5.3.1 General Remarks

In the following, we describe four hidden-surface algorithms which represent the greater variety of possible approaches to the hidden-line and hidden-surface problem. Of course, more than these four hidden-surface algorithms are in existence. Sutherland et al. [134] describe and classify 10 hidden-surface algorithms, and even their comprehensive survey is not exhaustive. Since there is not sufficient space to describe all these algorithms, we restrict our discussion to the four that we consider particularly representative and interesting. These are:

1. Appel's *quantitative invisibility method.*
2. Encarnacao's *priority method.*
3. Watkins' *scan-line method.*
4. Encarnacao's *scan-grid method.*

Appel's method is representative of the class of object-space algorithms which deal with polyhedrons (solids formed by plane faces) as the graphic entities of highest complexity (see Fig. 5.4). In this domain, Appel's method not only is one of the most efficient hidden-line algorithms, but it is also one of the first ever developed, and thus one of the pioneering endeavors in this area. Furthermore, it is the only scheme that allows a central projection from object space to image space.

Encarnacao's priority method was the first of that type reported in a generally accessible publication [34].† Presently, priority methods probably are the most frequently used hidden-surface algorithms. They operate in the image space.

Watkins developed the probably most widely used scan-line algorithm [149]. It is particularly suitable as a basis for a hardware solution to the hidden-surface problem. Scan-line algorithms operate in the image space.

Encarnacao's scan-grid method differs from almost all other hidden-line algorithms inasmuch as it can be applied to arbitrary curved surfaces. The price for the increased generality is a "brute force" approach that requires considerably more execution time than other hidden-line algorithms which work only on plane-faced objects. The scan-grid algorithm operates in the image space and employs simple point/surface tests.

†Sutherland et al. [134] credit R. A. Schumacker with the invention of the "list-priority" method, based on the fact that such a method was suggested in a study already written in September 1969 [121]. However, the first reference to that study, which was not available to the public and thus not known, dates back no earlier than 1973. This explains why, for example, Newell [106] published a similar method in 1972 without knowledge of Schumacker's work. Encarnacao developed his hidden-line schemes as part of his Ph.D. thesis (published in 1971). He also coined the now commonly used term "priority method" [68]—to the chagrin of his thesis advisor (who happens to be this writer), as this is a rather imprecise label for a process that could more precisely be denoted by the term "ordering method."

Our selection of hidden-surface algorithms shall not imply that the methods presented here are the only "good" and efficient algorithms. Other algorithms may be equally efficient and noteworthy, such as the algorithm developed by Newell et al.

5.3.2 Appel's Method of Quantitative Invisibility

The Appel algorithm operates on solids formed by plane faces that are bound by polygons, i.e., on polyhedrons. The vertices of the bounding polygons must be ordered counterclockwise. The visibility of an object is tested with respect to a finite viewpoint; i.e., central projection is applied. Visibility is determined by testing the edges of the faces for visibility.

In a first step of the algorithm, all edges are eliminated that are hidden by the volume of the body to which they belong. To this end, the visibility test as described in Section 5.2.5 is employed, determining for each face of a solid whether it is totally invisible or potentially visible. Edges bounding a potentially visible face are called *material edges*, and the material edges shared by a potentially visible and an invisible face are called *contour edges*.

In a second step the visibility of all potentially visible edges must be determined. Therefore, Appel considers the *quantitative invisibility* of the points (or of short segments) on a potentially visible edge. The inefficiency of simple point tests for visibility is avoided by the notion that, for a given viewpoint, the quantitative invisibility of a material edge can change only where it passes behind a contour edge or comes out from behind a contour edge—or in terms of the image space projection: the quantitative invisibility of the projection of a material edge can change only where it crosses the projection of some contour edge. At such an intersection, the quantitative invisibility is incremented by 1 if the material edge passes behind a contour edge, and it is decremented by one if a material edge comes out from behind a contour edge. Hence, by determining the quantitative invisibility along a material edge, this edge is subdivided into segments of different quantitative invisibility, the difference in quantitative invisibility between two neighboring segments being exactly 1.

Figure 5.5(A) illustrates the assignment of quantitative invisibility to a test line and its variations whenever this line crosses the contour lines of two (mutually covering) surfaces. Such a test line is called a *material line*, as it passes through the "material" of the solid, as well as through the viewpoint. Figure 5.5(B) illustrates how quantitative invisibility changes as a material line passes behind a solid. Notice that quantitative invisibility can change if a hidden contour line is crossed which forms a concave corner.

The determination of quantitative invisibility of a material edge will be discussed in more detail. First, the quantitative invisibility of the starting point of the edge is evaluated. Therefore, the material line passing through the starting points and the viewpoint O is specified, and the intersections of this line with all given faces are calculated. Of the points of intersection thus obtained, only those

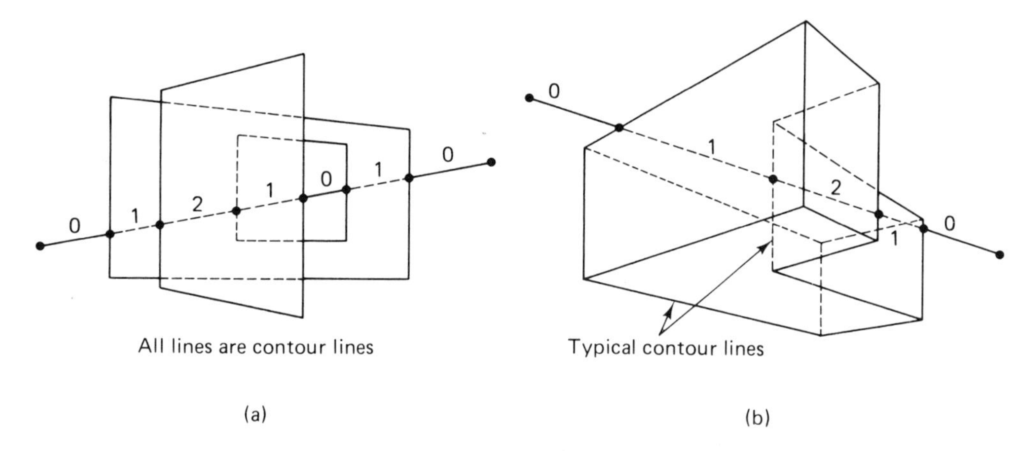

Figure 5.5 Two examples for the assignment of quantitative invisibility indices.

need be further considered which lie on the segment of the material line connecting the viewpoint and the starting point (see the depth test DT 1). For each of these points of intersection, the corresponding face is found. The resulting number of faces is the quantitative invisibility of the starting point.

Figure 5.6 illustrates this procedure. The material line OS has three points of intersection, I_1, I_2, I_3, with the planes that contain the three faces F_1,F_2, and F_3. I_1 lies inside F_1, I_2 lies outside F_2, and I_3 need not be considered, as it does not lie on the segment $\overline{OS}$. Thus, the quantitative invisibility of point S is 1.

Appel's approach to intersecting material edges and contour edges is very interesting. A triangle is formed by the viewpoint and the end points of the material edge. A contour edge will change the quantitative invisibility of the material edge only if it pierces this triangle, i.e., if its point of intersection D with the plane that contains the triangle lies inside the triangle. If such an intersection

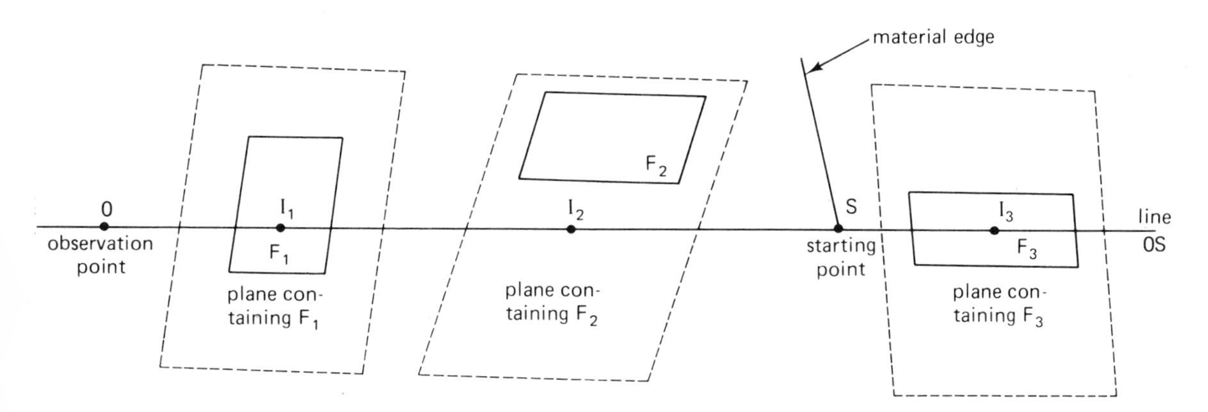

Figure 5.6 Example for quantitative invisibility of a test point.

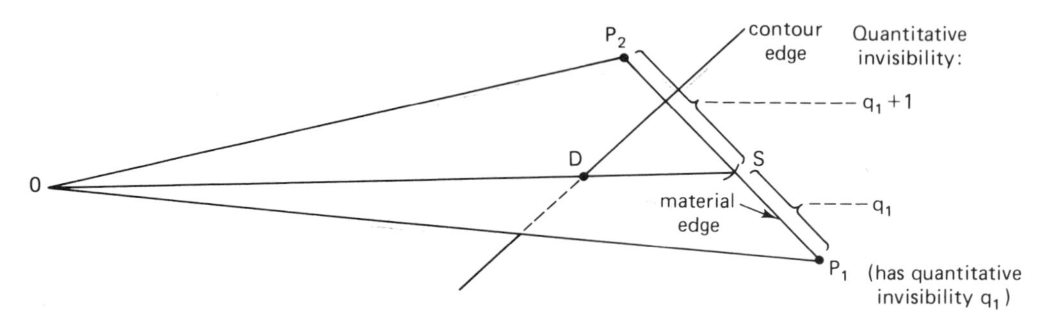

Figure 5.7 Appel's method of computing the quantitative invisibility of a material edge.

exists, the quantitative invisibility is incremented by 1 if the sign of the cross-product of contour edge vector and material edge is positive, and it is decremented by 1 if the sign is negative. The point at which this change occurs is determined by the point of intersection of the line OD and the material edge (Fig. 5.7). That is, if in the example of Figure 5.7 q_1 denotes the quantitative invisibility at the starting point P_1 of the material edge, the whole segment $\overline{P_1S}$ has this quantitative invisibility, whereas the quantitative invisibility for the segment $\overline{SP_2}$ is q_1+1.

In the formal description of a hidden-surface algorithm by a 5-tuple, as given in Section 5.1, we listed a "strategy function" σ which defines for a given algorithm the order in which the functions of the function set φ must be applied. Hence, the strategy function represents the organizational part of a hidden-surface procedure. By a verbal description of the procedure as given above for Appel's algorithm, the strategy function is implicitly specified (Fig. 5.8). As all hidden-surface algorithms are iterative computational schemes (they may even be recursive, e.g., Warnock's algorithm), the strategy function cannot be represented by a mapping in the usual sense. Therefore, we resort to a graphical representation in the form of a directed graph [83], in which the arcs indicate the order of application of the functions of $\varphi=\{\text{PM},\text{IS},\text{CT},\text{DT},\text{VT}\}$. Because of the iterative nature of the process, such a graph will be cyclic, where the cycles represent DO-loops in a software implementation of the algorithm.

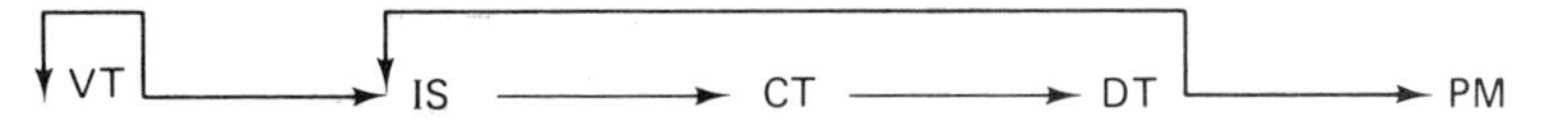

Figure 5.8 Strategy diagram of Appel's algorithm.

5.3.3 Encarnacao's Priority Method

The priority method as originally developed by Encarnacao, Grosskopf, et al., works on polyhedrons. It is applicable not only to convex but also to concave objects, i.e., objects with concave faces, holes, gaps, etc. This is accomplished by a preprocessing step in which a "triangularization" is performed, i.e., a decomposition of all faces into triangles. By such a decomposition, a number of additional edges are introduced which will be tagged as auxiliary edges. Hence, the proper hidden-line algorithm operates on a set of triangles, i.e., on a set of plane, convex faces of the simplest possible type. The triangles are specified by their vertices and thus implicitly by their edges. The object simplification induced by triangularization helps to minimize execution time and storage requirement, with the additional bonus that concave structures can be handled as well. The proper hidden-surface algorithm consists of three major steps: (1) presorting, (2) assignment of priority, and (3) determination of visibility.

Presorting: Presorting eliminates all triangles that are perpendicular to the xy-plane from the list of triangles to be subjected to the subsequent priority test, as such a perpendicular face cannot cover another face. A face is perpendicular to the image plane if the projections of its vertices are collinear.

Assignment of Priority: Given a spatial object decomposed into a set of triangular faces $S=\{s_1,\ldots,s_M\}$. Let $\mathrm{pr}(s_i)$ be the projection of a surface s_i according to the orthographic projection $\barwedge: S \to T$, $T=\{t_i | t_i=\mathrm{pr}(s_i)\}$. Assume that the triangular faces in the object space, as well as their corresponding projections in the image space, are numbered for identification and specified by the coordinate triples of their three vertices.

Then in the first step of the priority method a set of M ordered pairs

$$(t_i, T_i), \qquad i=1,\ldots,M,$$

is formed, in which $t_i \in T$ is a triangle (the projection of the triangular face s_i) and $T_i \subseteq T$ is the set of triangles which have priority over t_i (see Section 5.2.4):

$$T_i=\{t_j \in T | t_i \leqslant t_j\}.$$

The pairs (t_i, T_i), $i=1,\ldots,M$, represent irreflexive orderings in which t_i is the first component in all pairs and the elements of T_i are the second components. They are obtained by applying depth test DT2. As mentioned in Section 5.2.4, first the minimax test is applied to determine which of the objects t_i have a chance of intersecting with other objects t_j. "Isolated" objects t_i' are singled out, as they occur only in pairs $(t_i', \varnothing)$ ($\varnothing$ denoting the empty set). For the remaining $L \leqslant M$ objects, the priority must be determined, requiring $L(L-1)/2$ iterations of the depth test DT2 (which, of course, is fine-tuned to the handling of triangular faces). The test points in the intersection of two triangles required for the determination of priority are obtained by first trying to find intersections between the edges of the two triangles and, if that fails, testing for containment. Cyclic overlap cannot occur because of the triangularization. However, the triangular faces may penetrate each other. This case requires a special treatment. Penetration can occur only in the case where the projections of two faces have a nonempty intersection. Its occurrence can be detected by performing the priority test for *all* points of intersection between the edges of the two projections. If the priority test consistently yields the same result for all these points, there is no penetration.

Determination of Visibility: In each pair (t_i, T_i), T_i is the set of triangles which cover, to some extent, triangle t_i. If T_i is empty, then t_i is fully visible (with the exception of auxiliary edges). For a nonempty set T_i, the extent of coverage of t_i by the triangles in T_i must be determined. This requires the visibility of the edges of t_i which are not auxiliary edges to be determined. Therefore, it is sufficient to describe how the visibility of an edge can be investigated.

Let $e_{i,k} \subset t_i$, $k=1,2,3$, be an edge of t_i. The visibility of $e_{i,k}$ with respect to a triangle $t_j > t_i$ can be obtained by determining the points of intersection which $e_{i,k}$ may have with t_j and by testing whether the end points of $e_{i,k}$ lie inside or outside t_j. A segment of $e_{i,k}$ that intersects t_j is invisible; a segment of $e_{i,k}$ that does not intersect with t_j is potentially visible. Figure 5.9 diagrams the possible cases.

Figure 5.10 illustrates the four cases. It can easily be seen how the visible and the invisible segments (in cases C and D) are determined. Whenever an edge

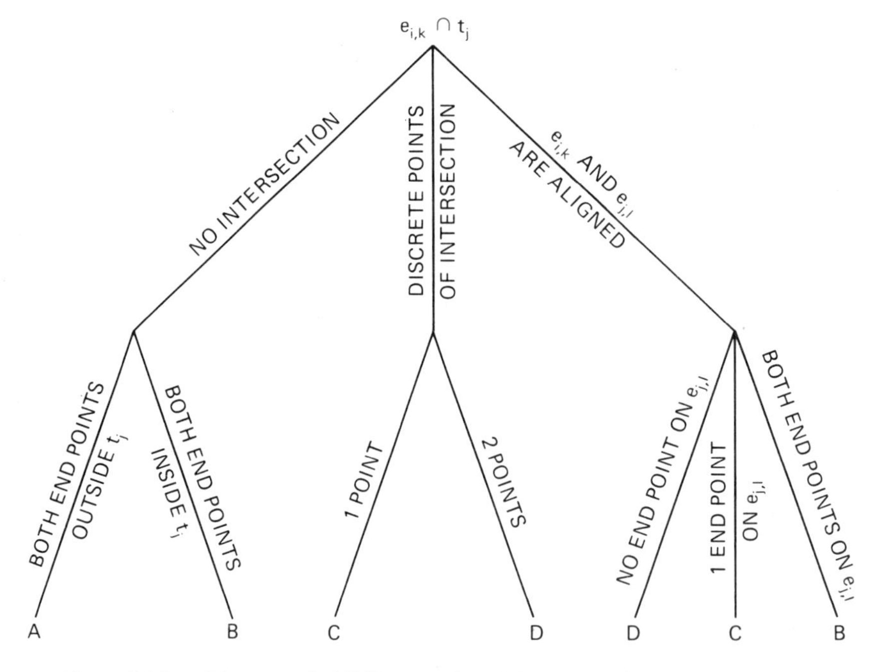

Figure 5.9 Possible cases of visibility ($e_{j,l}$, $l=1,2,3$, are the edges of t_j): case A: $e_{i,k}$ is not covered by t_j; case B: $e_{i,k}$ is totally invisible; case C: $e_{i,k}$ consists of one visible and one invisible segment; case D: $e_{i,k}$ consists of two visible and one invisible segment.

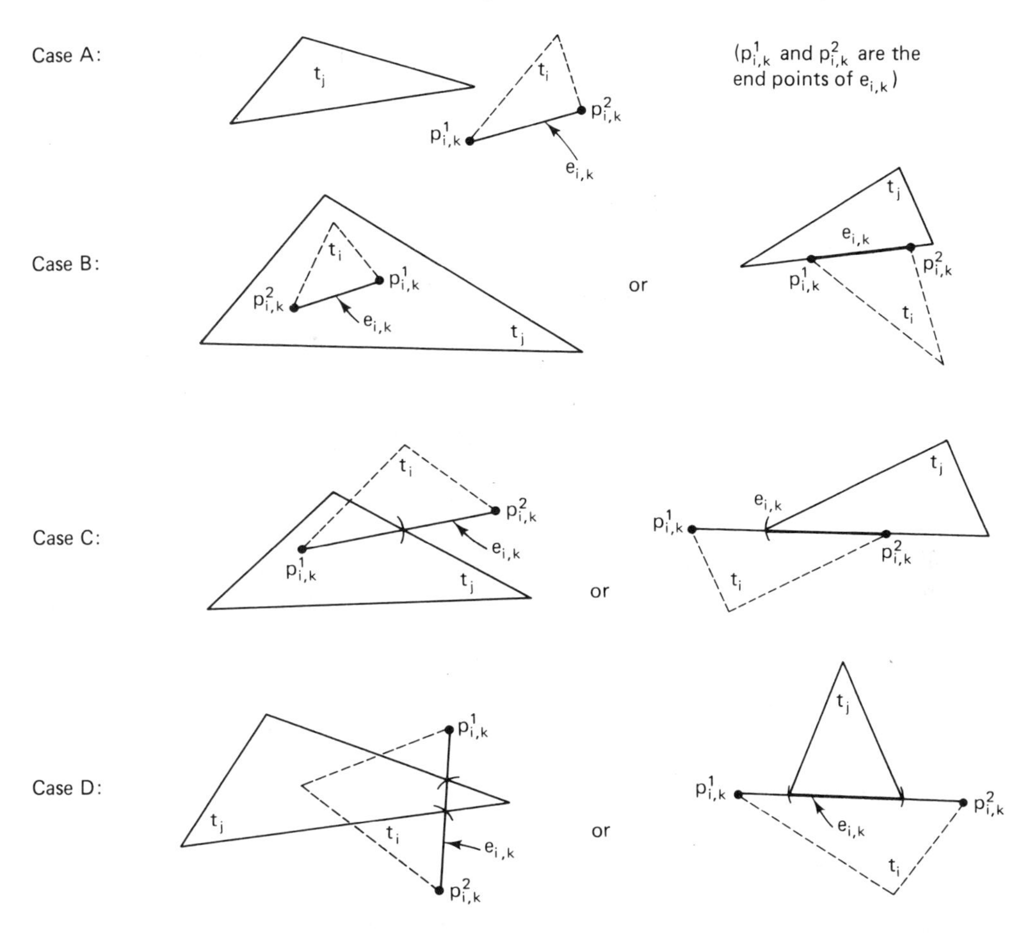

Figure 5.10 Illustration of possible cases of coverage of an edge $e_{i,k} \in t_i$ by a triangle t_j.

$e_{i,k} \subset t_i$ is totally covered by one of the triangles $t_j \in T_i$, it need not be considered further. Whenever the entire triangle t_i is recognized as being totally covered, the processing of the pair (t_i, T_i) can be terminated. As mentioned above, edges marked as auxiliary edges are excluded from the visibility test, as they never are rendered in the first place. However, they are needed to determine the visibility of other edges.

The result of the visibility tests is not triangles (i.e., closed polygons) but individual edges. Hence, whereas the procedure starts from a list of triangular faces, each represented by three coordinate triples, it ends with a list of visible line segments, each represented by two coordinate pairs. The strategy diagram of this method is shown in Figure 5.11. The hidden lines of the object depicted in Figure 5.12 were eliminated with the priority method [34].

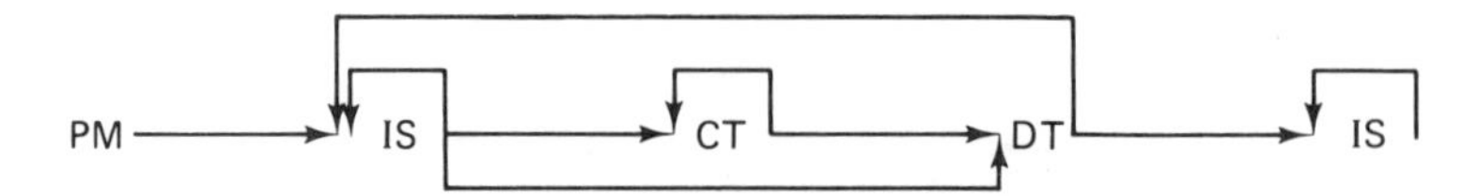

Figure 5.11 Strategy diagram of Encarnacao's priority method.

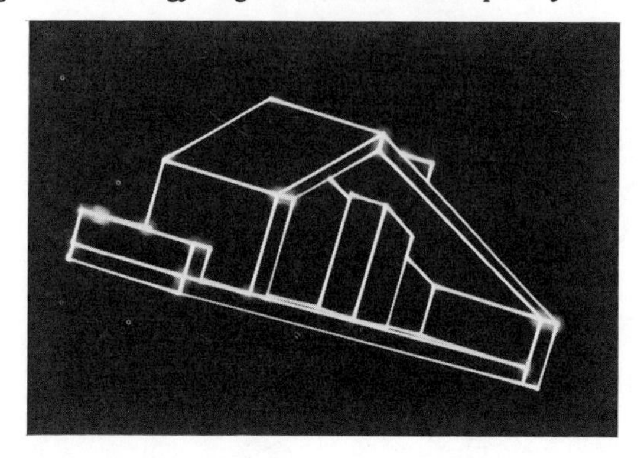

Figure 5.12 Object whose hidden lines were eliminated with the priority method.

5.3.4 Watkins' Scan-Line Algorithm

Like all scan-line algorithms, Watkin's algorithm lends itself toward a hardware realization. It operates in image space on the basis of a raster of scan lines superimposed to the image space representation, which is given as a collection of closed polygons that are orthographic projections of the (plane) faces of three-dimensional objects. Arbitrarily, the scan lines are assumed to be horizontal, i.e., parallel to the x-axis of the image plane coordinate system. All scan lines are processed in the same fashion, and therefore it is sufficient to describe the processing of a scan line. After processing of all scan lines, the visible graphical elements may be rendered in any desired gray shade and the invisible elements suppressed. The algorithm consists of two parts: (1) determination of "sample spans," and (2) determination of visibility.

Two steps must be performed in the determination of sample spans. First, the segments of a scan line are determined which form the intersection between the scan line and one of the polygons. As a result, each polygon in the representation "owns" a set of segments of a scan line—which, of course, may be empty. This is illustrated by the example of Figure 5.13, in which a scan line (m) is depicted together with three polygons, F_1, F_2, and F_3. Polygon F_1 owns the two segments $s_{1,1}$ and $s_{1,2}$ ($s_{i,j}$ = segment j of polygon F_i, formed by the intersection of F_i with a given scan line). Polygon F_2 owns segment $s_{2,1}$, and polygon F_3 owns segment $s_{3,1}$. Note that the segments of F_2 and F_3 overlap in the xy-plane. The segment end points are obtained as points of intersection between the polygons and the scan line.

Isolated, that is, nonoverlapping, segments are rendered visible as they lie on

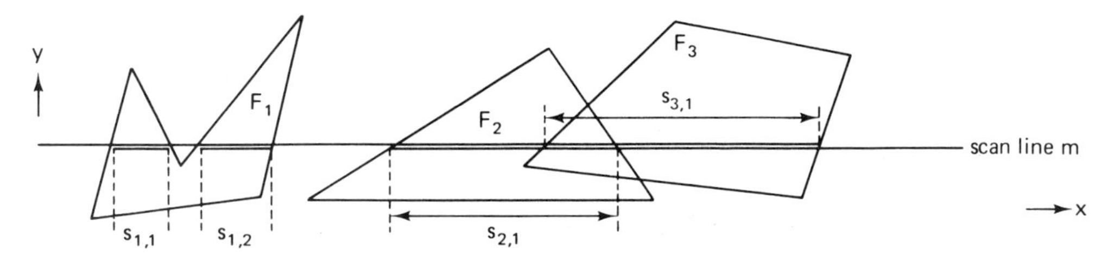

Figure 5.13 Association of scan-line segments and polygons.

the visible part of the image plane projection of a face. In the case of overlapping segments, however, *sample spans* must be formed prior to the determination of visibility. A sample span is a section of a scan line on which no change in visibility can occur. If it could be ruled out that faces in the object space may penetrate each other, the starting points of sample spans simply would be the points at which an overlapping of segments begins or ends. However, in the case of penetrating object faces, the extent of a sample span cannot be determined solely from the image plane projections of the faces, but their xz-plane projection must be considered as well. This is illustrated by Figure 5.14, in which three overlapping image plane projections (of the faces F_u, F_v, and F_w) are depicted, leading to the overlapping segments $s_{u,1}$, $s_{v,1}$, and $s_{w,1}$ on scan line m [Fig. 5.14(a)]. Only a look at the xz-plane projections of the faces reveals fully their relative spatial positions, for which two examples [(b) and (c)] are given. In Figure 5.14(b), there is no intersection of the xz-plane projections and, hence no penetration of the corresponding faces. Conversely, such intersections are found in the example of Figure 5.14(c), indicating that F_1 and F_3 penetrate each other and F_2 penetrates F_1 as well as F_3. Hence, whereas only three sample spans need be distinguished in the case of Figure 5.14(b), six sample spans must be considered in the case of Figure 5.14(c).

Hence, we may define a sample span as a section of a scan line selected such that the following two conditions are satisfied:

1. The number of segments that contain the sample span is constant and >1;

and

2. The xz-plane projections of the faces corresponding with these segments do not intersect (i.e., any such intersection indicates the beginning of a new sample span).

A sample span must be tested for visibility if:

1. The number of segments on the actual scan line differs from the number of segments on the preceding scan line; *or*

2. If the association of segment end points with the edges of projected faces is interchanged when proceeding from the preceding to the actual scan line (in this case, two edges intersect between the two scan lines).

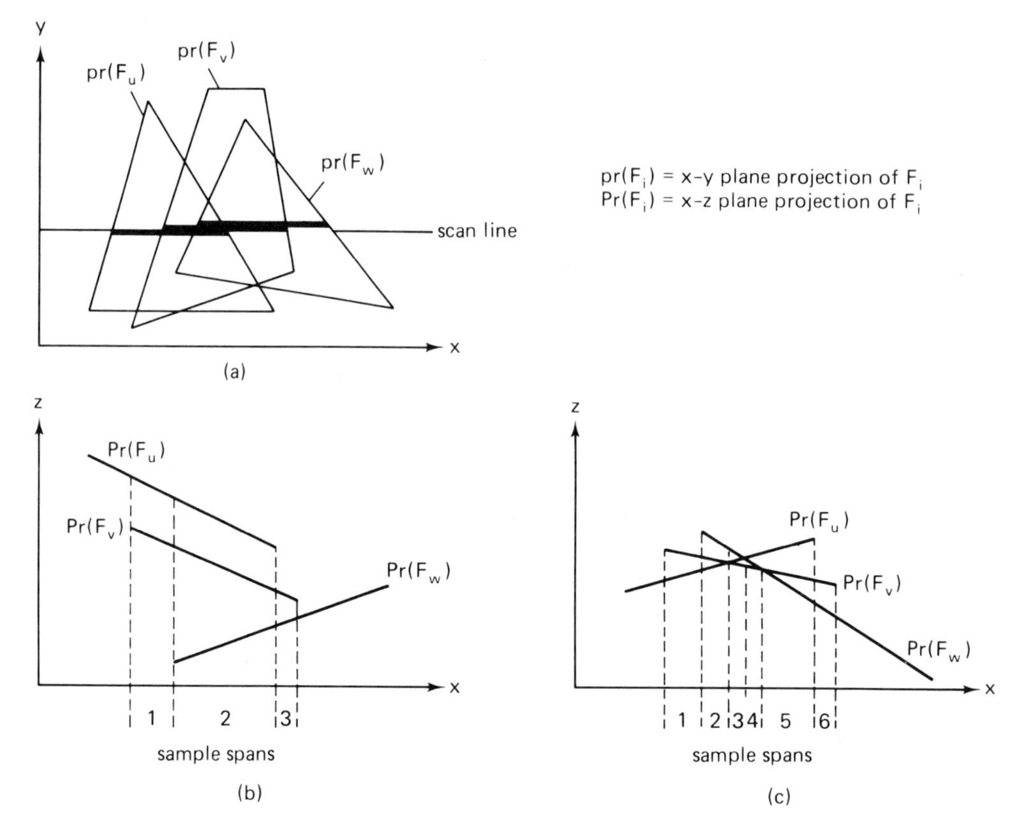

Figure 5.14 Selection of sample spans: (A) image plane projection of three faces; (B) without penetration; (C) with penetration.

The question of which segments in such a sample span covers the other is easily answered by applying depth test DT3 (see Section 5.2.4). The resulting strategy diagram is depicted in Figure 5.15. This procedure is ideally suited for a rendering of gray-shade pictures. In this case, the visible segments are not rendered with a uniform brightness, but with a gray level that is derived from the z-value of the corresponding face. High-quality gray-shade pictures may be the result, as illustrated by Figure 5.16.

Furthermore, the process of forming the segments of a scan line, which requires the calculation of the intersections between the scan line and the edges of the projected surfaces, can be greatly economized if the objects have been previously sorted according to their maximal (or minimal) y-coordinate (depending on whether the scan is performed top down or bottom up). As soon as an object is

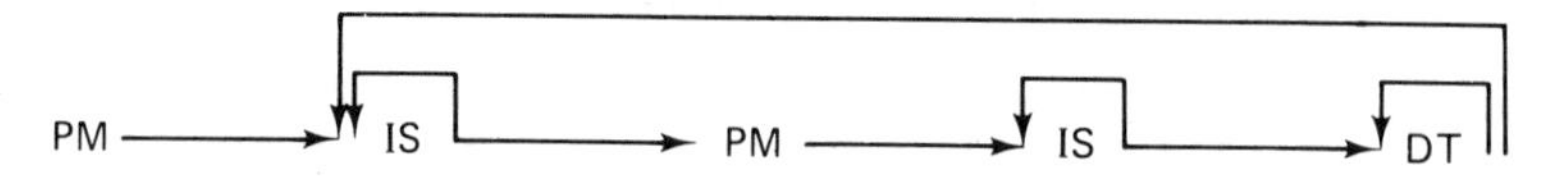

Figure 5.15 Strategy diagram of Watkins' scan-line algorithm.

Figure 5.16 Example of a gray-shade picture. (Courtesy of Evans & Sutherland Computer Corp.)

encountered whose maximal (minimal) y-coordinate is smaller (greater) than the y-value of a scan line, this object, as well as the remainder of the list, can be excluded from the process of calculating points of intersection.

5.3.5 Encarnacao's Scan-Grid Method

This algorithm can be applied to arbitrary curved surfaces which are rendered as a grid of u-v lines (see Chapter 4). The curved surface patches are specified only by their four vertices, and straight lines are used to define their edges. The curved surface of the patch is approximated by four planes formed by inserting diagonals between opposing vertices of the patch. The three-dimensional surfaces defined in the object space are first mapped onto the image plane by the orthographic projection.

The name of the algorithm is derived from the fact that a rectangular two-dimensional raster of grid lines, a "scan grid," is superimposed on the image plane projection of the surfaces (Fig. 5.17). A list of surface patches is constructed for each area of the scan grid. This list reveals which surface patches potentially overlap each of the separate scan-grid areas. This presorting of patches into scan-grid areas permits the visibility of a test point to be determined by considering only those surface patches which lie within the same scan-grid area as the test point. If the scan grid is too coarse, many surface patches will be included in each scan-grid area and the visibility tests will be much more time-consuming. If too fine a scan grid is used, either too much memory will be consumed for the area lists or more time will be spent in the initial presorting than will be saved during the visibility tests. For a typical image with 100–400 total patches, an 11×11 scan grid (121 areas) has been found to be a reasonable size.

The minimax test is used to determine which scan-grid areas will be potentially overlapped by a specific surface patch. In the example given (Fig. 5.18), surface P is determined to potentially overlap scan grid areas a, b, c, and d,

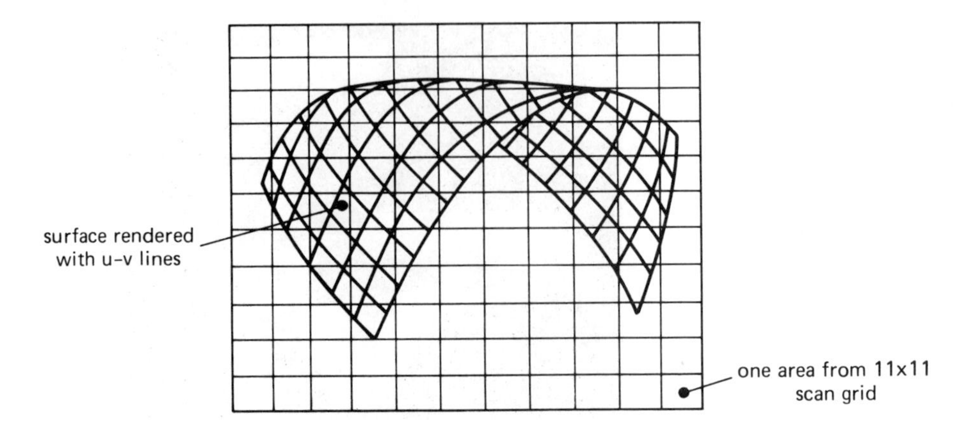

Figure 5.17 Scan grid.

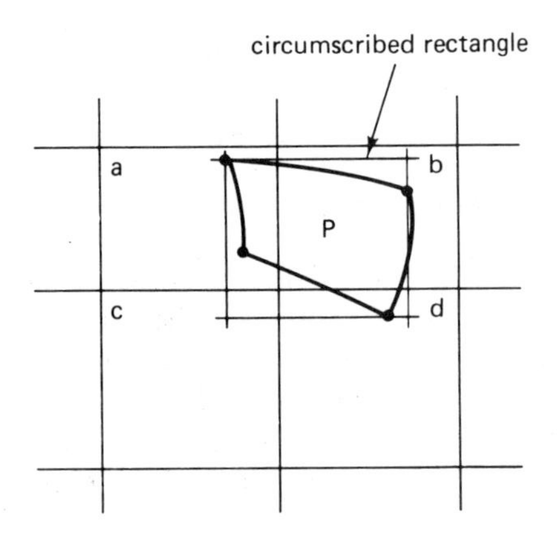

Figure 5.18 Minimax example.

because its circumscribed rectangle as defined by the minimax procedure overlaps scan-grid areas *a*, *b*, *c*, and *d*, even though area *c* is not really overlapped by the patch. The speed and simplicity of the minimax test has been found to more than compensate for the extra containment tests that will occur between test points in area *c* and surface patch *P*.

Once the presorting is completed and the scan-grid lists of surface patches are formed, the main visibility algorithm is executed. All *u* and *v* lines are tested for visibility in the following manner. One *u* or *v* line segment (an edge of a surface patch) is broken into a string of test points evenly spaced between the vertices of the surface patch. Each test point is tested for visibility against the surface patches that lie within the same scan-grid area as the test point. Either none of the surface patches cover the test point, in which case it is visible, or one patch is found to cover the test point, in which case it is invisible. If all test points along the surface patch edge are visible, the entire edge is drawn. If some points are found to be

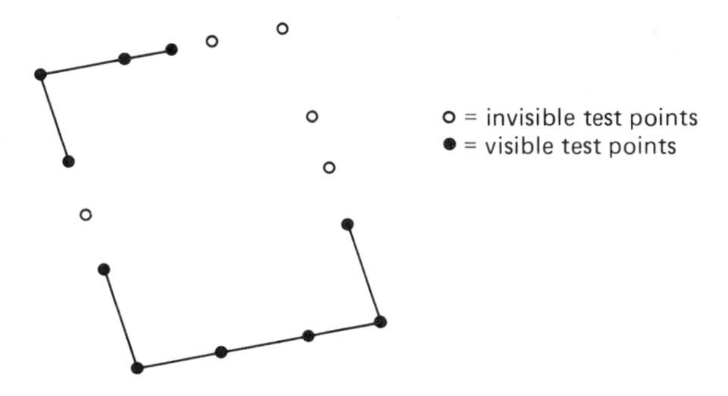

Figure 5.19 Visible portions of an edge.

invisible, only those sections of the patch edge are drawn which lie between visible test points (Fig. 5.19).

Several steps are involved in testing the visibility of each test point. A surface patch is fetched from the appropriate scan-grid list and the patch is immediately discarded if it is the patch from which the current test point was created. Otherwise, a containment test is performed to determine if the patch could possibly cover the test point. Once containment of the test point is found, a depth test must be performed to determine whether the patch hides the test point. Since the surface patches are curved, a single diagonal could be used to break the patch into two triangles which define the two planes that could be used in performing a depth test. This single diagonal introduces a bias based on which of the two possible diagonals is used. This bias is avoided in the algorithm by using both diagonals in the following way. Four triangles are formed using first one diagonal and then the other (Fig. 5.20). The test point must lie in exactly two of these four triangles. Two containment tests are performed to identify the two containing triangles and then two depth tests are performed to determine visibility. Only in the case where both triangles are found to cover the test point is the test point considered invisible. If only one triangle covers the test point, the planar approximation error to the curved surface is greater than the distance between the test point and the two

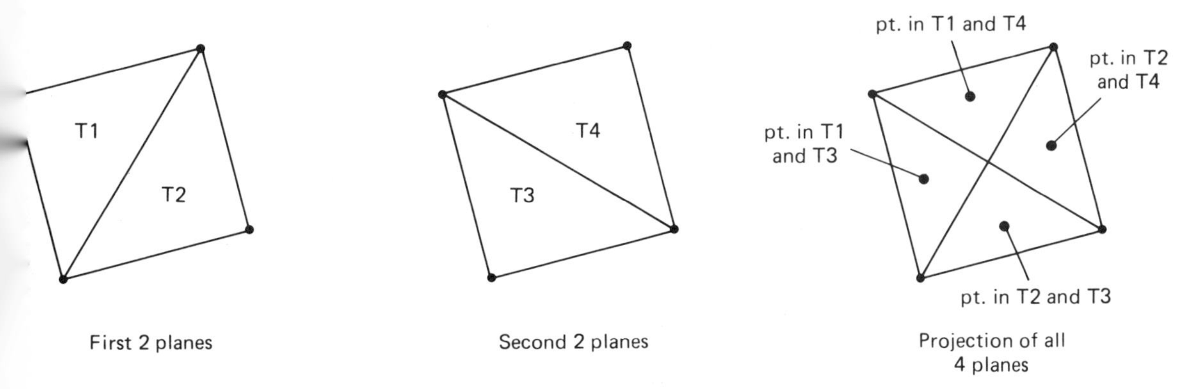

Figure 5.20 Planar renderings of the patch.

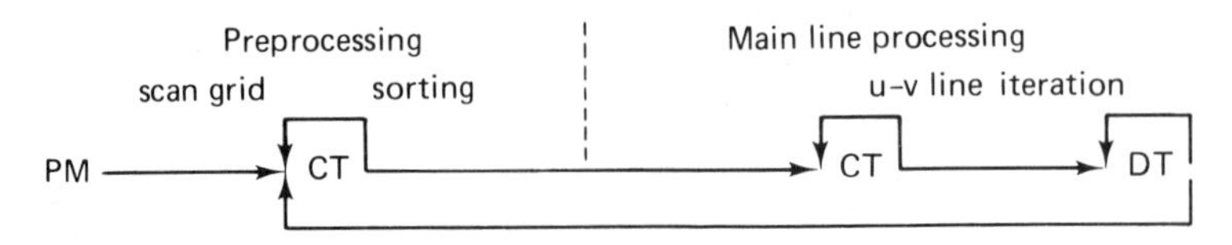

Figure 5.21 Strategy function of the scan-grid algorithm.

planes, and the visibility decision becomes arbitrary. Figure 5.21 depicts the strategy diagram.

In the original algorithm, the list of patches within a scan grid area was searched from the beginning for each succeeding test point along a u or v line. This is a good strategy for testing a series of visible test points, since all patches in the list must be tested to determine that a test point is visible. However, it is an inefficient strategy for a series of invisible test points, because on the average one half of the patch list is searched, only to find that a series of test points is invisible. Savitt modified the original algorithm so that when a patch is encountered that covers a test point, the patch is moved to the top of the patch list because of the high probability that the same patch would hide several adjacent test points along the u or v line. This simple change resulted in an execution speed about one order of magnitude faster than the original algorithm [117].

As originally described, an 11×11 scan grid is a reasonable subdivision for the initial presorting and list building. However, a particular scan grid area may contain an extremely large number of surface patches, and it may thus be advantageous to divide the area into smaller areas. This enhancement works as follows. A scan-grid area with more than eight patches in its area list is subdivided into four subareas, and a temporary list of involved patches is formed for the particular subarea that contains the test point in question (Fig. 5.22). If there are still more than eight patches in this new subarea list, then another subdivision is performed and a new list is created. This process is repeated if there are more than 16 patches in the $\frac{1}{8}$ area list. The subdivision decision threshold of 4, 8, and 16

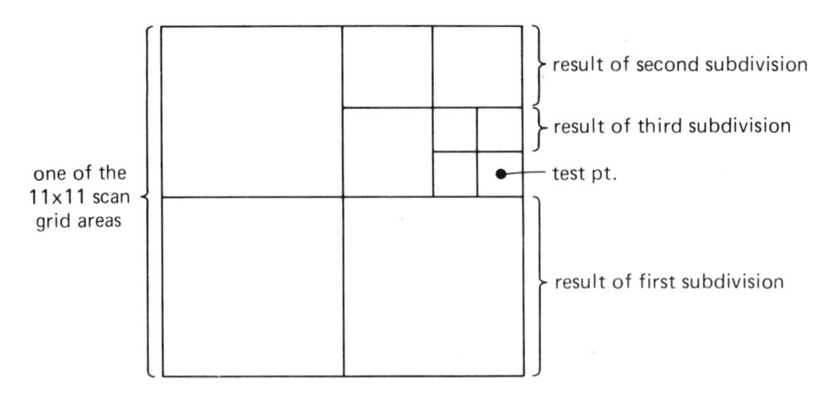

Figure 5.22 Further subdivision of a scan-grid area.

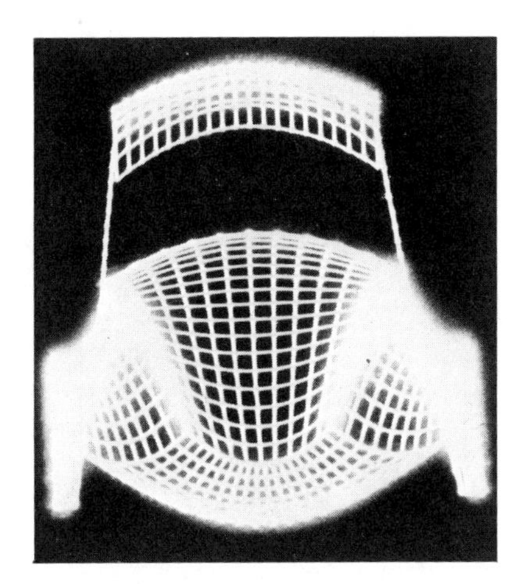

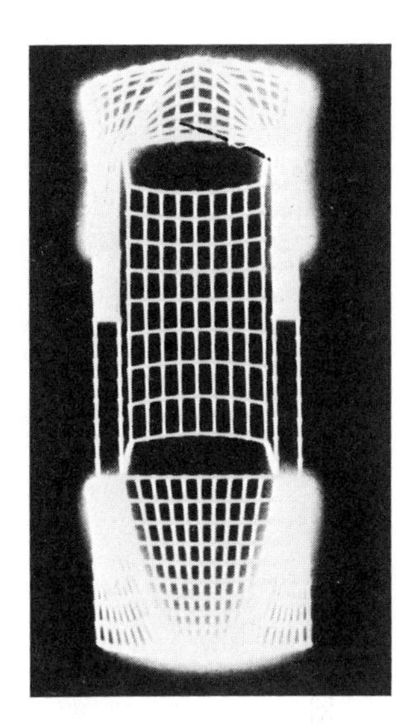

Figure 5.23 Object generated by the scan-grid method and viewed under different angles of rotation.

were found to be the optimal thresholds for one particular computer implementation. The optimal thresholds are determined primarily by the relative speed of the minimax procedure compared to the containment procedure. The general strategy of this local subdivision is to perform more minimax tests, which are fast, in order to reduce the number of the more laborious containment tests. An object generated with this method is depicted in Figure 5.23.

5.4 COMPARISON

As mentioned before, many more hidden-surface algorithms than the four described here are known (see the Bibliography and [134]). Sutherland et al. [134] describe and compare the following methods—named after their inventors—as representatives of the classes of methods applied to plane-faced objects (see Fig. 5.4).

- Class 1: Appel [5], Loutrel [90], Galimberti and Montanari [46], Roberts [112].
- Class 2: Schumacker et al. [121], Newell et al. [106], Warnock [148], Romney [114], Bouknight [10], Watkins [149].

The only representative of class 3 which we can name is Encarnacao's scan-grid method [34, 29, 117]. Class 4 representatives are the methods of Weiss [151] and

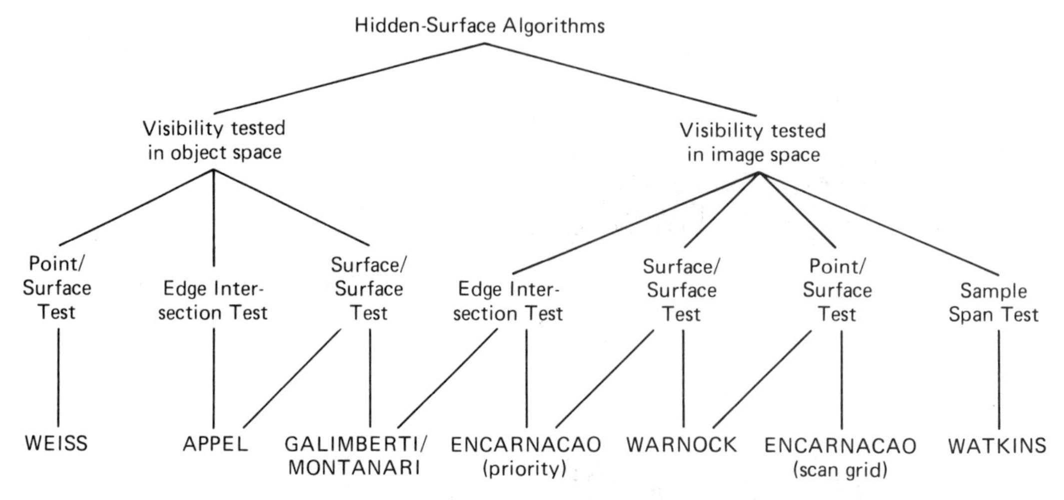

Figure 5.24 Classification of hidden-surface algorithms according to the visibility tests performed (from [83]).

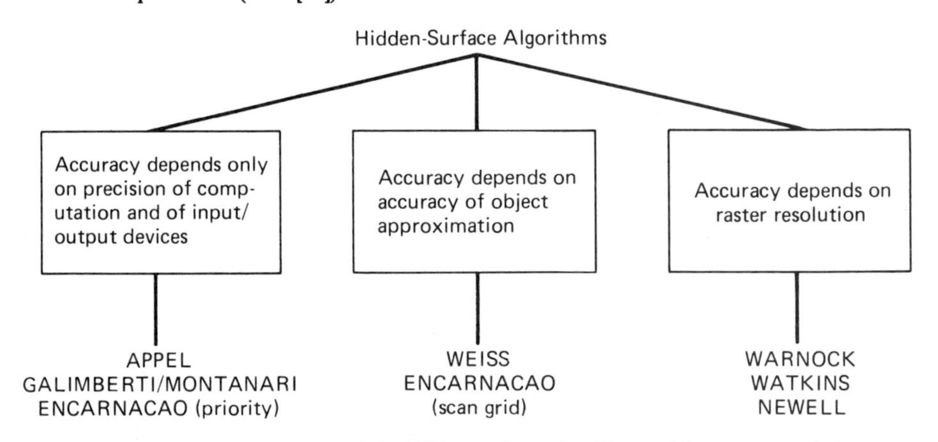

Figure 5.25 Comparison of eight hidden-surface algorithms with respect to their resolution (from [83]).

Mahl [92]. Also to be mentioned are the methods of Kubert [87], The Mathematical Applications Group, Inc. [93], Rougelot and Shoemaker [116], Gouraud [63], and Cutmull [23].†

Other criteria than the degree of reduction [Fig. 5.4] may be used for classifying hidden-surface algorithms, one being space in which visibility is determined—object space or image space—and the nature of the tests which are employed (Fig. 5.24). Another criterion is the resolution obtained (Fig. 5.25).

It would be of great interest to compare existing hidden-surface algorithms with respect to their efficiency. However, a general performance evaluation is not

†A readily applicable hidden-line plotting program is available as ACM algorithm No. 420, *CACM, 15,* 2, Feb. 1972.

possible, as the performance of an algorithm depends not only on the complexity of the scene but also on the "nature" of the objects to be rendered. It is at least difficult, if not impossible, to establish a general measure for the complexity of objects, let alone a measure for their nature. Any complexity measure based on a count of the number of solids, faces, edges, or other graphic elements in a scene is not really significant if it is not supplemented by a measure for the degree of coverage between these elements. Moreover, the efficiency of a hidden-surface algorithm depends very much on the type of objects to be rendered. If the objects of a scene are restricted to a well-defined class, it may be possible to select the best-suited algorithm or even fine-tune this algorithm for the given task. Hence, a comparison can only be given on the basis of certain test pictures; and the results obtained are as significant as the selected test pictures are representative for the class of objects to be rendered.

In order to point out some general tendencies, we present here some results of a performance evaluation of several algorithms performed by Klos [83]. This evaluation is based on a test pattern that consists of a stack of slabs as shown in Figure 5.26. The complexity of the picture here can readily be measured in the number of faces, and the degree of coverage can be measured by the ratio of the number of visible faces to the total number of faces. Of course, absolute figures listing the execution time for the compared algorithms depend very much on the computer system on which the algorithms are executed. The system on which the evaluation was performed is a medium-sized, single-processor time-sharing system (comparable with the IBM 360/67). The hidden-line programs had to compete with many other time-sharing users. Therefore, the absolute time figures may only

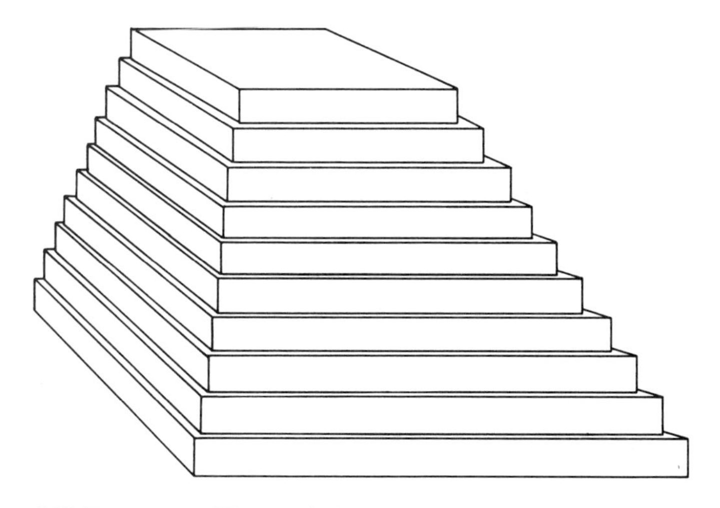

Figure 5.26 Test pattern. The complexity of the scene is measured by the number of faces. The degree of coverage is measured by

$$\delta = \frac{\text{number of visible faces}}{\text{total number of faces}}.$$

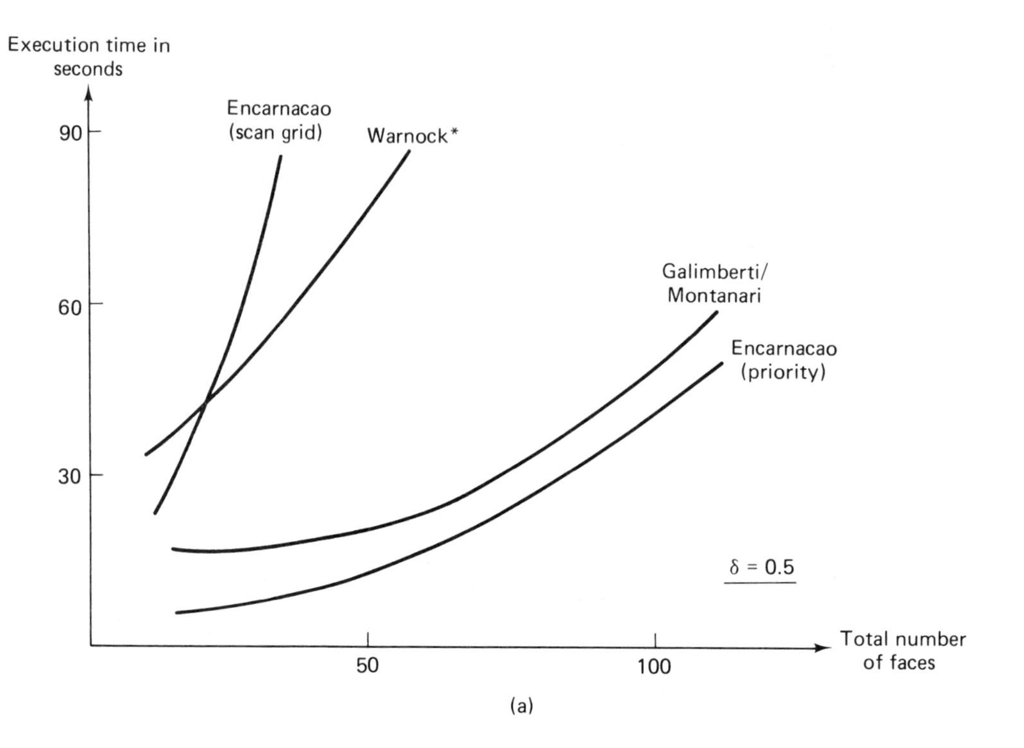

(a)

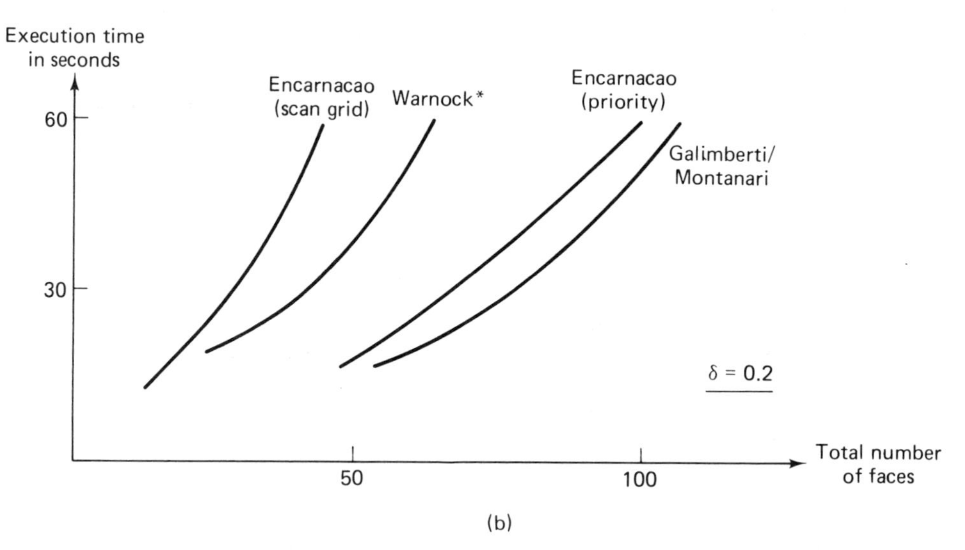

(b)

Figure 5.27 Execution time of the test pattern (Fig. 5.24) as a function of the number of faces, with

$$\delta = \frac{\text{number of visible faces}}{\text{total number of faces}}$$

as the parameter (from [83]). (The asterisk indicates that the image space is subdivided into nine subspaces.)

indicate the order of magnitude of the required execution time (10 seconds), and they provide a relative comparison of the algorithms—with respect to the particular test picture. Some results are diagrammed in Figure 5.27. We recognize that for the given case a certain "efficiency order" can be established, which changes slightly with a change of the picture parameters (the degree of coverage). It shall be emphasized once more that such a comparison is only as significant as the test pattern is representative for the class of objects the algorithm has to deal with. For example, the scan-grid method comes out as least efficient whenever a test pattern is used that consists of a group of convex polyhedra. On the other hand, this method excels all other algorithms it is compared to in its ability to be directly applicable to $u-v$ grid representations of functions of 2 variables. On the other hand, the performance figures of the "front runner" in our example, the priority method, do not include the additional overhead caused by the required preprocessing step in which all objects are first triangularized.

More generally interesting than the absolute execution time of the algorithms may be a statistics of the contributions which the various function classes in $\varphi = \{PM, IS, CT, DT, VT\}$ render to the execution time. For the example given in Figure 5.26, such a statistics is diagrammed in Figure 5.28. Except for the Galimberti/Montanari algorithm, a common feature is that the various test func-

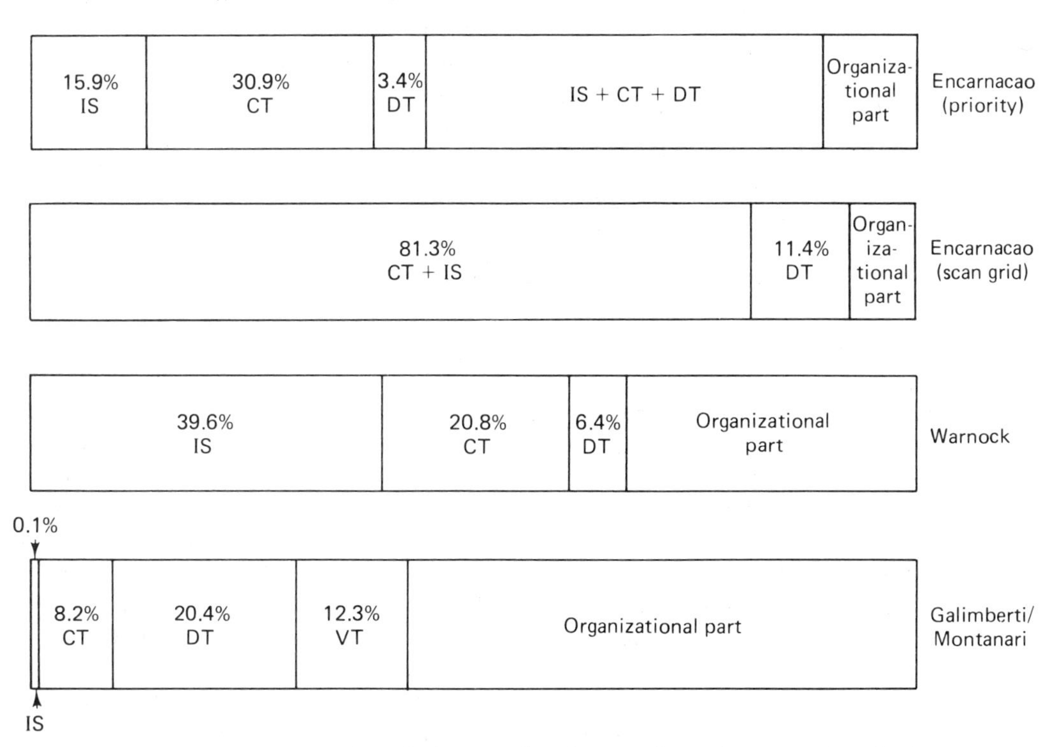

Figure 5.28 Partition of the execution time on the various operations in four hidden-surface algorithms (from [83]).

tions together are far more time-consuming than the organizational parts of the program. The deviation in the case of the Galimberti/Montanari algorithm can be explained by the fact that the surface/surface tests employed in that algorithm produce a maximum of information. The price for this advantage is the limited generality idiosyncratic for that algorithm. As a result, we may endorse the priority algorithms (Schumacker, Encarnacao, Newell) as possibly the best trade-off between performance and generality. Moreover, a good graphical programming system might also contain the scan-grid algorithm as a proper tool for the display of spatial functions [34].

5.5 SORTING

Sutherland et al. [134] characterize the hidden-surface problem primarily as a sorting problem. In fact, the result of a hidden-surface algorithm is a sorting of graphical objects, according to their visibility, into two major classes: the class of visible objects and the class of invisible objects. Most hidden-surface schemes make use of sorting operations in the various steps of the algorithm. In Appel's method, for example, faces are initially sorted into totally invisible and potentially visible faces. Eventually, the edges of the potentially visible faces are subdivided into segments of different quantitative invisibility, a process that may be considered as a sorting of these segments. In Encarnacao's priority method, the triangular faces are sorted first into two classes, one containing all faces that are totally disjoint from all the others, and a second class containing the remaining faces. The faces in the second class are then sorted into the sets T_i according to an irreflexive ordering called "priority." The last step in the procedure, the determination of coverage and visibility within the pairs (t_i, T_i) (see Section 5.3.3), is not a sorting process since the triangles are not ordered. Such a process could be accelerated if it were possible to presort the faces in each set T_i according to the probability of covering the face t_i. In fact, as the edges $e_{i,k} \subset t_i$, $k \in [1:3]$ are tested individually, a certain speed increase may be obtained by moving a triangle t_j that covers $e_{i,1}$ to the top of the list T_i, as there is a higher-than-the-average probability that t_j will also cover $e_{i,2}$ and $e_{i,3}$ (at least one of the end points of $e_{i,2}$ and $e_{i,3}$ is in t_j).

The sorting key is in most sorts one of the tests for containment or visibility. As the statistics illustrated in Figure 5.28 indicate, these operations account for the major part of the total execution time. Hence, although sorting is a major operation in most hidden-surface algorithms, the efficiency of the employed sorting scheme itself has not much effect on the overall execution time. The time consumed by the proper sorting procedure is included in the organizational overhead, which, in return, is only a small fraction of the total execution time. Considerable savings in time can, rather, be obtained by reducing the amount of time-consuming arithmetical operations as required by the calculation of intersections, the containment tests, and the visibility tests, e.g., by performing presorting operations based on simple logical or relational functions.

EXERCISES

1. Consider the figures given here and answer the following questions.

Question	Figure A	Figure B
Is the figure a polygon?		
Is the figure a closed polygon?		
Is the figure a plane polygon?		
Is the figure a convex polygon?		

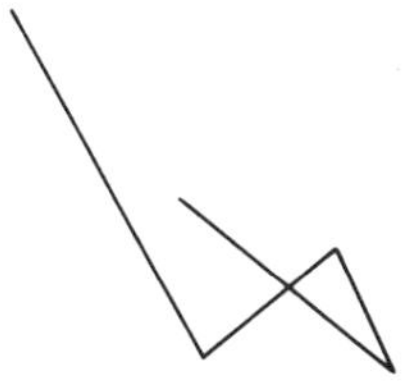

Figure A 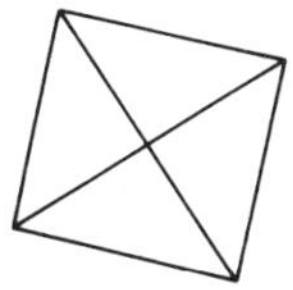 Figure B

2. (a) Name procedures and determine their input and output parameters as well as their calling sequence (in a programming language of your choice) for the following tasks.

(1) To determine whether two segments intersect and indicate the finding through a boolean variable. The two line segments will each be specified by two points. In case there is an intersection, the point of intersection will be returned.

(2) To determine the intersection of a plane and a line (the "piercing point"). The line will be specified by two points, and the plane will be specified by three noncollinear points. A flag variable will indicate whether the line is parallel to the plane, contained in the plane, or has a single point of intersection. In the latter case, this point will be returned.

(3) To determine whether two closed, convex polygons overlap. The polygons will be specified by an array of vertex coordinates. The result of the test will be indicated by a boolean variable.

(b) Flow-chart the three procedures.

3. A common method for converting concave structures into convex structures is to decompose the concave structures into triangles by introducing additional edges (called "auxiliary sets").

(a) Assume that the concave structure consists of a set of closed polygons. Can these polygons be triangularized under all circumstances?

(b) Develop an algorithm for the triangularization of a given polygon and flow-chart an appropriate program.

4. (a) Enumerate the lists needed for the priority algorithm (Section 5.3.3). Identify the information items to be stored in these lists and make assumptions about the list dimensions. Suggest a concept for the memory representation of the lists.

(b) Do you know of any sorting or presorting schemes that might increase the efficiency of the algorithm? If yes, describe them in detail.

(c) Flow-chart a program that implements the priority algorithm and that operates on the basis of the list organization devised above.

5. (a) Enumerate the lists needed for the scan-line algorithm outlined in Section 5.3.4. Identify the information items to be stored in these lists. Suggest a concept for their memory representation.

(b) Flow-chart a program that implements the Watkins algorithm and that operates on the basis of the list organization suggested above.

(c) In the description of the Watkins algorithm we assumed graphic objects which consist of a set of polygon-bounded plane faces of a constant gray shade. However, one often wants to generate curved surfaces with continuously varying gray shading. If such a surface is approximated by a net of connected plane faces of constant gray shade, "pseudo-edges" may appear at the border lines because of the abrupt change in gray shade. Suggest a scheme that eliminates the pseudo-edges by obtaining a continuous gray-level function. How much does this complicate the algorithm?

(d) In the description of the Watkins algorithm given in Section 5.3.4, it is assumed that only the visible sample spans are represented in the refresh memory; i.e., the visibility test is part of the sample span analysis. Modify the algorithm such that the sample spans of *all* (overlapping) faces, visible or not, are stored in the refresh memory (e.g., in the order as given by the scan raster). Furthermore, the gray level of the sample span end points are part of the data associated with the end points and stored in the refresh memory. Hence, it is possible to determine the visibility of a sample span on the readout of the data (e.g., by special hardware). Discuss the advantages and disadvantages of such an approach.

LANGUAGES AND THEIR INTERPRETERS

6

Interaction Handling

6.1 INTERACTIVE INPUT DEVICES

The basic principle of graphical interaction is that of visual feedback: The system may display certain information on the screen, and the user may respond by activating certain input devices to convey a message to the system. The user's response may be explicitly requested by the system, or it may be a spontaneous action whose consequences, in return, may become instantaneously manifest on the display screen. The user's actions are usually called *attentions*. An attention may be to

1. Point to an object on the screen.
2. Position an object on the screen.
3. Draw figures on the screen or on a tablet.
4. Input commands.
5. Input data.

In addition to displaying information on the screen, actions taken by the system may serve the purpose of controlling a sequence of actions of the user. To this end, the system may temporarily enable or disable certain interactive input devices, or it may selectively provide the user with choices for answering questions or giving commands.

For all of these interactions, a number of instruments have been devised which partly substitute and partly supplement each other. These devices and their capabilities will be discussed briefly.

6.1.1 Lightpen

Pointing: The lightpen is the only instrument that allows the user to point directly to an object on the screen, thus identifying it to the system. However, a lightpen can only be used in connection with a *refreshed-picture display* (as contrasted to a *storage tube display*, see Chapter 7). This is a consequence of the method through which the lightpen "sees" objects on the screen.

In a regenerated-picture display, all graphical primitives are repeatedly and consecutively drawn. Let $[t_0, t_n]$ be the time period of a picture refresh cycle. Let $[t_0, t_n]$ be subdivided into a sequence of time intervals $T_{i+1} = [t_i, t_{i+1}]$, $i \in [0:n-1]$, where n is the total number of graphical primitives displayed on the screen, such that during the interval T_i primitive number i is drawn. Hence, we have a bijective mapping $i \mapsto T_i$. A lightpen produces a signal when the CRT beam traverses its aperture or *viewing area*. This signal can be used to determine the current time interval T_i, and, by inverting the mapping, the index number i identifying the currently drawn primitive is known. Actually, not the current interval T_i is determined but the current state of the display processor (instruction counter or data register content), leading to the same result. A lightpen contains a switch that the user activates either by pushing a button or by pressing the head of the lightpen against the face of the CRT. This switch is needed to control the exact instant when the user wants the lightpen to become active. For example, the user can thus make sure that the lightpen is pointed to the proper spot when it is activated. The lightpen operation can be enabled or disabled by the program individually for each graphic entity.

Positioning: Whereas this kind of operation lends itself directly to pointing, it can only indirectly be used for positioning. Two different schemes may be employed for this purpose. In the first case, the user, after having placed the lightpen into the desired position, presses the lightpen switch, thus causing the system to generate a grid of bright dots on the screen. The system then determines which dot the lightpen is pointed to, and the position indicated by the lightpen is the position of that dot. However, two dots that lie both in the viewing area of the lightpen cannot be distinguished. Therefore, the shortcoming of this simple method is its coarse resolution, which may be acceptable in some applications and not in others. The second, more sophisticated method, known as the "pen-track" scheme, avoids this disadvantage. Here, a special symbol called a *cursor* is displayed, usually in the form of a "tracking cross" or an "aiming circle." The cursor can be picked by the lightpen and, subsequently, made to follow the motions of the lightpen. Thus, the

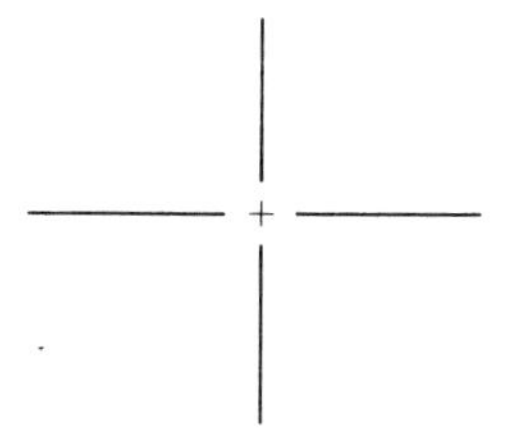

Figure 6.1 Tracking cross.

cursor can be arbitrarily moved around on the screen and positioned at any point of the screen raster. The tracking cross, which may look as depicted in Figure 6.1, usually is attached to the lightpen by pointing to it and pressing the lightpen switch. After the cross has been moved into the desired position, the switch is released and the cross is thus detached. Its center now indicates the coordinates of the point the user wanted to locate. This procedure may be repeated as often as desired. The pen-track procedure is essentially a feedback control process carried out by system software. Such a procedure will be discussed in Section 6.1.6. The pen-tracking procedure of "picking" a tracking cross and causing it to follow any movement of the lightpen can be generalized to pick and move any arbitrary object on the screen. Pen tracking is a very convenient way of composing pictures out of instances of a given set of symbols.

Freehand Sketching: By writing a routine that samples the tracking cross position automatically at a certain rate, an array of coordinate pairs can be obtained which may immediately form the vertices of a polygon (i.e., the obtained points are subsequently connected by straight-line segments). If the sampling points are reasonably close together, the lightpen can be used for freehand drawing of contours or sketches. Such a procedure is also called *inking* [101].

An early application of inking has been to enable the user to put in handwritten characters (of course, this requires an appropriate program for feature extraction and classification in the computer). Lately, inking has obtained more importance as a means of conveying information in medical diagnostic systems with graphical input–output. For example, in display systems for interactive picture processing, where computer-generated graphics can be superimposed onto TV images, the user (e.g., a radiologist) can use the lightpen in connection with an inking procedure for tracing certain edges, contours, or other features in a photographic image (e.g., the X-ray image of a heart, a shading, a lesion, etc.) [52]. Another example for inking usage is AMANDA [99], a medical report generating system. Here, the user is shown sketches of the parts of the human body, and by inking he can indicate to the system the location and extent of a complaint (Fig. 6.2).

INSTRUCTIONS
1 INDICATE IN WHICH SECTION OF YOUR BODY YOUR COMPLAINT LIES WITH THE LIGHT PEN
2 CHECK YOUR ANSWERS AND PICK THE OK COMMAND IF CORRECT

IF YOU MAKE A MISTAKE PICK THE COMMAND AGAIN AND REANSWER

NAME: HERTZ ALLOVER
AGE: 25
SEX: MALE
FEMALE

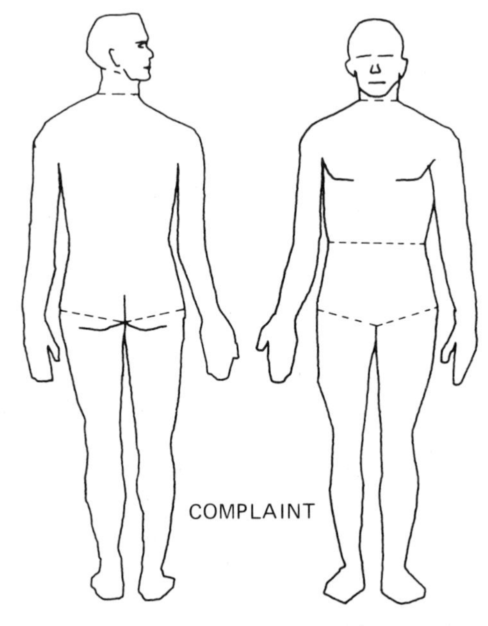

OK

Figure 6.2 Typical "page" of AMANDA, a system for generating medical reports [99]. The system prompts its user to answer questions in a multiple-choice mode using menus, to type in data, and to mark, by inking, in pictures of the human body location and extent of "areas of complaint."

Construction of Figures: For the more precise construction of geometrical figures, freehand inking is, of course, not appropriate. In principle, the construction of figures on a display screen follows very much the same pattern as it is done with paper and pencil. To draw a line on paper, two points are first marked and subsequently connected through a line drawn with a ruler. If a curve is to be drawn, a number of points are first marked and subsequently connected through a line drawn with a French curve. Technically, there are certain differences between the paper and CRT approach. Naturally, in the case of a computer graphics system the ruler or the French curve is replaced by a built-in routine for line generation. If the line is straight, the routine is performed by a hardware vector generator. If the line is curved, the routine is usually performed by software, applying a mathematical interpolation scheme (see, for example, the cubic spline interpolation method discussed in Chapter 4). However, such a mathematically obtained interpolation, calculated by a software routine, can only be displayed on the screen in the form of a polygonal approximation, if the display processor contains a vector generator but no curve generator (and this holds for almost all existing systems). The only way to make the resulting picture sufficiently smooth is to use small vectors.

The specification of end points for a line is a positioning problem. Subsequently, the line-generating routine is invoked by a respective command. Hence, the whole procedure may work as follows. To draw a vector, the user begins by indicating the start point of the vector with the lightpen. As a result, this point will become visible on the screen. The next step is to point to the end point of the vector, which will also become visible. Subsequently, the operator will pick the light button DRAW LINE, and, as a result of this command, the system will generate and display the vector connecting these two points. If an arc of a circle is desired, the command DRAW ARC must be given after three points have been placed. The first point is the center point of the circle, the second is the starting point of the arc, and the third will define the endpoint of the arc.

One modification of the conventional line drawing procedure as described above is a technique called "rubber-banding." Again, the user specifies the start point and end point of a line, but the difference from the conventional procedure is that, as soon as the startpoint has been specified, the system displays the line connecting that point to the current cursor position. Whenever the cursor is moved around on the screen, the line will follow its motions; i.e., it will contract or stretch like a rubber band (thus the name). When the line appears to be in the right position, the user can fix the end point and thus detach the line from the cursor.

Input of Commands: A very popular way of using the lightpen for the input of commands is to display a special character string—usually a key word for the respective command—and to initiate the command execution by picking that string. If the user is given a variety of choices for issuing commands, the respective key words are arranged in a list called a *command menu*. For example, a command menu for picture transformations might be the following:

MOVE LEFT
MOVE RIGHT
MOVE UP
MOVE DOWN
ROTATE LEFT
ROTATE RIGHT
ZOOM UP
ZOOM DOWN

An individual command in such a multiple-choice menu is called a *light button*; for it performs the same function as a control key, albeit it is not activated by pushing a button but rather by pointing to it with the lightpen. If a menu is set up as in our example, each command must be dealt with as an entity with respect to a lightpen pick; i.e., it does not matter which character of the string the lightpen is pointed to. A popular feature is to let the most recently picked light button blink, thus indicating the particular state the system may have been brought into by the command.

Input of Data: There is no direct way of using a lightpen for the input of data. What one can do is to associate, for example, each digit of an integer number with the position of a respective cursor. By picking one of the cursors and translating it in a certain direction, the associated digit is changed. Feedback is provided by displaying the current value of the number [103]. Eventually, a number thus obtained may be put in. In another method, a counter is programmed which, when started, counts up (or down) from a given initial value. The content of the counter is displayed on the screen as an n-digit number (n depending on the resolution one wants to obtain). The rate at which the counter is incremented (or decremented) is chosen so that the user can watch it. The counter may be started by picking a certain digit of the initial number. The counter always counts the units corresponding to the place picked; i.e., in the case of integer numbers, the counter counts ones if the rightmost digit is picked, tens if the next digit is picked, etc. In either method we have some kind of a "digital potentiometer," similar to a thumb wheel, which can be set by the user. Setting may be terminated by releasing the lightpen switch. Although this method looks quite elegant in demonstrations, it generally is rather awkward and time-consuming for real-job applications.

6.1.2 Joystick, Control Ball, and "Mouse"

Positioning: Joystick, control ball, and "mouse" are all devices used for controlling the movements of a cursor on the screen. That is, they are primarily positioning devices. A joystick is a lever than can be moved in two degrees of freedom (front–back and left–right). By activating this lever, the attached cursor can accordingly be steered around the screen in all directions of the compass. The same holds true for the control ball, but in this case the lever is replaced by a ball which can be rotated with the hand in two degrees of freedom. The mouse is a handheld box which has a control ball or two perpendicular wheels protruding at its bottom [35]. By rolling it around on a flat surface, the cursor can be moved in the same way as by turning a control ball or moving a joystick. In any case, the control device provides two input values which become the coordinates of the cursor. Hence, whenever the cursor rests in a certain position, its coordinates are available directly as the control device output.

Pointing: Pointing to an object can only be performed by placing the cursor somewhere on the object. Object identification must then be carried out on the basis of a comparison of the coordinates of the cursor position with the coordinates of the object, for the cursor position is the only available information. Generally, this may be a rather complicated and time-consuming task.

The simplest object, entirely determined by a pair of coordinates, is a dot. However, as it cannot be expected that the user will position the cursor exactly on the dot, the comparison of dot coordinates and cursor coordinates must be performed such that it yields a positive result if both are "sufficiently close." To this end, we define a certain "fuzziness region" around the dot, e.g., as a circle with the point in its center and the *fuzz parameter* as the radius. This leads to a comparison on the base of a distance measurement, an operation that requires the calculation of two squares and one square root prior to the comparison. A simple scheme is to define the fuzziness region as a square with the point in its center. In this case, the absolute values of the difference between the x-coordinates and the y-coordinates, respectively, are separately compared with a fuzz parameter.

In the case of a line, we have two defining points. Furthermore, each end point of a segment may also be an end point of another segment, if the segments form a polygon. The simple scheme of comparing the cursor coordinates with a defining point of the object requires that only one of the two end points be used consistently in the comparison. Actually, many systems operate that way, requiring the user to position the cursor, for example, near the tail of the vector that is to be picked. Such an approach has certain shortcomings. For illustration purposes, consider the following simple figure, in which each end point is an end point of three edges. By which rules shall a line segment be uniquely identified? Moreover, how is the user to know what is head and what is tail?

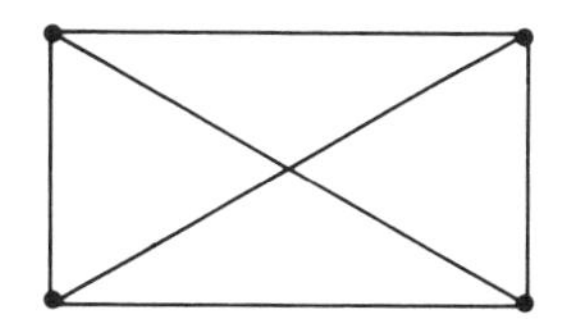

Even in the simpler case of plane polygons, the problem is that the line segments are not displayed as vectors, and thus it is difficult for the user to determine which of the two vertices bounding an edge is its start point and which is its end point. The ordering relation, according to which we speak of vectors instead of lines, is only existent in the program, but is not manifest in the pictorial representation on the screen. These problems can be avoided by measuring the normal distance between the cursor position and a line (Fig. 6.3). The equation for this distance is

$$d = \frac{(y_1 - y_2)x + (x_2 - x_1)y + x_1 y_2 - y_1 x_2}{\sqrt{(y_1 - y_2)^2 + (x_2 - x_1)^2}}.$$

If we have a large number of lines displayed on the screen and, consequently, a large number of such tests, this may become rather time-consuming. Therefore, it is advisable to perform a presorting, sorting out all lines which are not in a certain proximity to the cursor position. Instead of a cursor, we may use "cross hairs", given in the form of two independently controlled lines (one horizontal and one vertical) that extend over the whole screen.

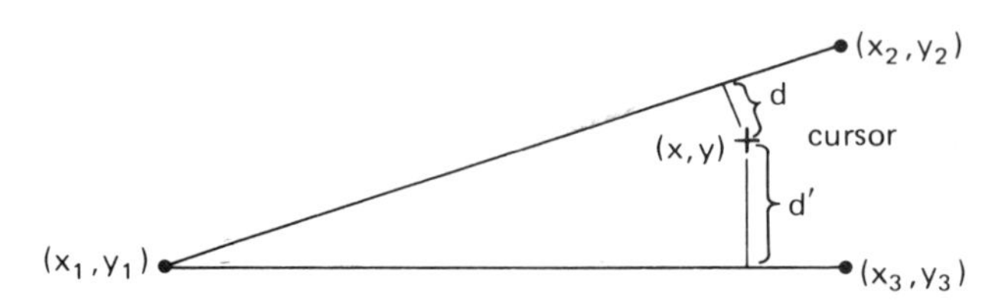

Figure 6.3 Line identification by measuring the normal distance between a line and the cursor position.

Drawing: Only the mouse is suitable for inking, i.e., freehand sketching and drawing of contours, for the mouse has the property, like the lightpen but unlike the joystick or the control ball, that the path on which the hand moves is correlated to the curve generated on the screen. For the construction of objects, however, all three instruments are better suited than a lightpen because of their very nature as positioning instruments.

Input of Commands or Data: For command input, the light button technique as described above can be applied with the minor difference that a light button "pick" now requires a respective positioning of the cursor. Data may be put in by moving a cursor around and displaying its center coordinates, if necessary after a previous scaling. However, such a procedure is even more tedious than the "digital potentiometer" approach discussed in Section 6.1.1.

6.1.3 Tablet

A tablet is a flat surface, usually rectangular, which can be separated from the display. Its use can be: freehand sketching and tracing of graphical features such as edges, contour lines, and maps from paper. Drawing is done with a stylus in the same fashion as one draws with a pencil. This is accomplished by periodically sampling the stylus position and reporting the coordinates to the system. To this end, a number of ingenious methods have been developed. Just to mention a few of the most commonly used schemes, we find methods such as the following [25, 113, 122, 139].

- An electric field is generated on the surface of a semiconductive medium (a special paper) such that the potential picked up by the stylus indicates the stylus location.
- A magnetic field is generated and picked up by an inductance in the stylus such that the wave form and the intensity of the induced voltage is a unique function of the location.
- A spark is periodically generated at the tip of the stylus. Acoustically, the spark causes a sharp sound impulse which disperses in all direction and

which is received by "strip microphones" at two adjacent edges of the tablet. By measuring the propagation time of the sound impulse, a measurement for the distance of the stylus from the two edges is obtained and hence the stylus coordinates.

- A field emitted by the stylus is picked up by an *xy* antenna grid printed underneath the tablet surface. The stylus location is determined by a count of the grid crossings combined with a phase comparison which indicates the stylus position between two grid lines.

The resolution of such a device can be as high as 0.1%, corresponding with a digital accuracy of 10 bits.

A tablet is strictly supplementary to other input devices such as a lightpen, joystick, or control ball, its main application being the tracing of graphical features.

6.1.4 Keyboards

A computer display system usually enables its user to type in alphanumeric text through a keyboard that is either part of the display console or of an attached teletypewriter. Additionally, a number of function or control keys may be provided by which the user can give commands to the system. The function of these control keys must be determined by the application program; i.e., the question of what particular action is initiated if a function key is activated must be specified within that program. Therefore, for different applications the meaning of certain function keys may be different. The manufacturer usually provides "overlays" by which the keys of the control keyboard can be differently labeled according to the application program that is actually run on the system.

6.1.5 Lightpen vs. Cursor Control Devices

Storage tube displays do not permit the use of a lightpen, since in that case a picture is constantly visible on the screen all the time between its creation and its deletion. Therefore, the volatile crossing of the lightpen aperture by a certain line at a certain time, which permits the identification of objects in the case of refreshed pictures, does not occur. Hence, there is no other choice but to use one of the cursor control devices, i.e., joystick, control ball, or mouse. On the other hand, there is no technical obstacle against equipping a regenerated-picture display with such a cursor control device, if one so desires and deems it worth the additional costs.

Our discussion should have shown that a lightpen is somewhat more universal than the cursor control devices. If one has to deal with graphic objects displayed on a CRT screen, it is most natural to point to whatever one wants the computer to work on. Moreover, people who work full-time with computer graphics as a tool for computer-aided design activities emphasize that, under these conditions, there is hardly a substitute for the lightpen [3]. Therefore, a regenerated-picture display should not come without a lightpen, and other input devices should in this case supplement rather than substitute. In the casual use of display systems, however, it is more a matter of custom or taste which instrument one prefers. Interactive computer graphics can be performed with a lightpen as well as with cursor control devices, and it is hard to understand why the lightpen has sometimes been fiercely attacked.

Another point of discussion has often been the question of whether one should provide a variety of input media or whether it might not be more efficient to use only one instrument for each and every action. A "single input device system" somehow facilitates the software that is required for servicing the interactive input devices. However, only two input media are flexible enough for being able to serve all purposes: the alphanumeric keyboard and the lightpen.

In principle, one can design a command language (see Section 10.7) such that mnemonic commands are provided for all conceivable actions of the user. However, such a system is rather inefficient, especially in the case of pointing. Such a language approach rules out freehand drawing or inking. The only advantage of such an approach, its uniformity, does not offset these shortcomings. On the other hand, the lightpen, combined with a light button command input, provides a very versatile and efficient system, and the only weak point is the input of parameter values. In our opinion, a "good" interactive display system, equipped for real-world jobs, should comprise at least the following three input devices:

1. *Lightpen* (or *joystick* in the case of a storage tube display).
2. *Alphanumeric keyboard.*
3. *Function keyboard.*

The use of these instruments will depend on the scope and nature of the application program. For smaller, highly interactive application programs, one may employ the lightpen for all actions except the input of data, for which the keyboard will be used. As a typical example, Figure 6.2 depicts a display obtained in the course of a medical report generating terminal session. Here, the lightpen is used for inking and multiple-choice answering by light button picks. Data are typed in on the alphanumeric keyboard, the result of this becoming immediately visible. For larger computer-aided design programs, especially those which are used for routine jobs, it may be more efficient to employ the function keyboard for command inputs.

6.1.6 A Pen Track Procedure [73]

The execution of a lightpen tracking routine depends, in its details, on the hardware properties of the particular system. As a representative example, we shall briefly discuss such a routine which operates with a cursor in the form of an "aiming" circle with a tracking cross in it. The field of view of the lightpen has approximately twice the diameter of the aiming circle.

In the beginning, the tracking cross is put at a point whose coordinates have been specified as input parameters of the subroutine. If the lightpen switch is pressed, tracking is enabled and accomplished in the following way:

1. A string of dots is put up to the left (west) of the target point. The dots are displayed sequentially, which means, first dot No. 1, then dot No. 2, etc. The dots are visible to the lightpen, and the X-coordinate of the first dot detected within the field of view (dot No. 4 in Fig. 6.4) is stored.

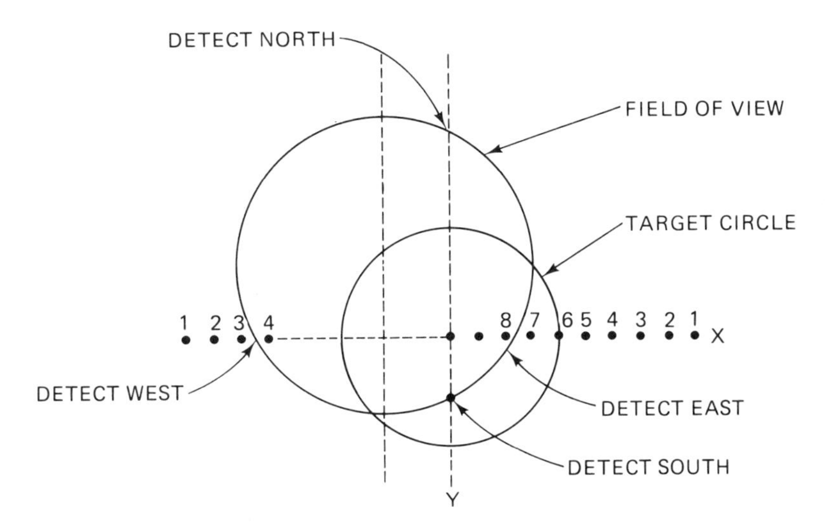

Figure 6.4 Relationship between the tracking pattern and the light pen field of view.

2. Next, a similar east pattern is displayed. The distance between dot No. 1 west and dot No. 1 east is initially twice an assumed "field of view radius" of five raster units.

3. Upon an east dot detection (dot No. 8 in Fig. 6.4), the average of the west and east detected coordinate is taken as a new X-coordinate, and the target is moved accordingly. In the same way, a south–north move is accomplished.

4. If dot No. 1 in the west or south pattern falls within the field of view, the program does not "know" exactly where the field of view is. In that case, the tracking pattern is moved 5 raster units west (south), and the scan is repeated.

5. If dot No. 1 in the east or north pattern falls within the field of view, the tracking pattern is moved 5 raster units east (north), and the assumed field of view radius is incremented by 1 raster unit. Then the east (north) scan is repeated. This permits the assumed field of view radius to be dynamically adjusted.

To predict the next location of the lightpen, the current and the two previous locations of the tracking pattern are used. These three points may be thought of as defining two vectors which are proportional to the velocity of the lightpen movements over the two intervals bounded by these three points. The respective components of these two velocity vectors are averaged, and the result is used to predict the next location of the tracking pattern. It has to be noted that this scheme works best when the lightpen is moved at a velocity which is approximately constant. If the lightpen operator suddenly starts to accelerate the movement of his/her hand, in a quick jerking motion, a point may be reached when she/he loses the tracking cross (this is a common and unavoidable feature of any pen track procedure).

6.2 DEVICE INDEPENDENCE

The modern trend in graphic programming calls for device-independent software. That is, a software package should be applicable on plotters as well as on storage tube displays or high-performance refresh displays. Likewise, input devices should be characterized by their function rather than by technical idiosyncrasies. As functions of input devices, we listed in the preceding section the actions of pointing, positioning, input of constants (numbers or character strings), and input of commands. Corresponding with these functions, Foley and Wallace [38] define the following "virtual" input devices:†

- *Selector* as the pointing device.
- *Locator* as the positioning device.
- *Valuator* as the device for the input of constants.
- *Button* as the device for the input of commands.

The application programmer may now only be concerned with these virtual devices; i.e., he may use service routines for the selector, the locator, the valuator, or the button, regardless of whether selection is performed with a lightpen or by placing a cursor, whether location is performed through a cursor positioning device or by pen track, whether button is a control key or a light button, etc. Of course, such an approach necessitates a software interface between the service routines for the virtual devices and the physical devices, but such an interface may be supplied as part of the programming system which the user need not be concerned with. In the GRIP language for graphic programming introduced in Chapter 10, the notion of device-independent service routines is pursued further.

†The recommendations of the CORE Definition Subgroup of ACM/SIGGRAPH follow along the same line. The only difference is that a distinction is made between the input of numerical data (VALUATOR) and characters (TEXT), besides a different naming (the selector is called PICK and the button is called CHOICE).

6.3 ATTENTION HANDLING

6.3.1 The Role of Interrupts in Display Systems

Generally, the instructions of a program are executed in their given sequence unless a branching instruction occurs. At this point, control is transferred from the current instruction sequence to another one, beginning with the point to which the jump will lead. If this latter instruction sequence is an external procedure (subroutine), control will be handed back at the end of this sequence to the point in the invoking program following the branch instruction. Let A be the instruction sequence currently being executed and B be the instruction sequence to which control shall be transferred. Then, we usually have the following three modes of control transfer [150]:

Normal sequencing:	A keeps control
Branching:	A hands control over to B
Subroutine call:	A lends control to B.

The common feature in all three modes is that a transfer of control is caused by the execution of an instruction. This is not so in a fourth mode of control transfer that is triggered by the occurrence of a special hardware signal, called an *interrupt*. Here, we have the characterization:

interrupt mode: control is taken away from A and given to B.

The program module B is in this case called an *interrupt handling routine*.

Interrupts are instrumental in the input/output (I/O) organization of modern machines; i.e., an I/O request leads to an interrupt of the running program and the transfer of control to an appropriate service routine. Interrupts may also occur if the hardware detects certain irregular program conditions (e.g., memory parity errors or overflows). For the sake of completeness, it will be mentioned that an alternative to an interrupt system is given by a technique termed *polling*. In this case the computer has "sense" lines in lieu of interrupt lines. A special machine instruction causes a SENSE AND JUMP; i.e., a jump is executed only if a certain sense line is being activated by an external device. Polling denotes the process where the program periodically examines the sense lines. When a sense line is found active, indicating that an external device has a message for the computer, the program prescribes what action to take. Polling has been used in the past when savings of even a trifle of hardware were important. With the decrease in hardware cost, priority interrupt systems have become a standard part of the hardware, even in small-scale minicomputers. In a priority interrupt system, the problem of concurrent interrupts, i.e., interrupts that simultaneously request to be served, is resolved on the basis of assigning different priorities to the interrupt input lines. Whenever concurrent interrupts occur, the one with the higher priority is served first.

In computer graphics systems we may find interrupts that help to organize the communication between the host computer and the display terminal. In this respect, the display terminal is just one of several peripheral devices of the host. The service routines for such interrupts are part of the operating system routines with which the application programmer or the end user is not concerned. However, we may also find interrupts of a special quality inasmuch as they have an effect on the display program. These are the interrupts generated by interactive devices such as selectors, locators, valuators, buttons, or special switches. The purpose of such interrupts is to notify the computer that an input is ready to be fetched. The input may be: an object identifier in the case of a selector attention, coordinate values in the case of a locator attention, a key identifier in the case of a button attention, or numerical data in the case of a valuator attention. Another purpose of such interrupts may be to notify the computer about the occurrence of a certain event, such as the activation of the lightpen switch or of other special buttons. Hence, the general characterization of this second group of interrupts is that they signal user's attentions to the application program. The use of interrupts as contrasted to polling offers the advantage of freeing the host from the overhead of repetitive sense instruction execution.

Interrupts are also used in the display processor, e.g., for signaling certain events, such as a frame-time clock inputs, the activation of an input device control switch or of the HALT button, an "edge violation" by the beam (the beam is crossing the screen boundaries). Interrupts of that type are discussed in Chapter 7.

6.3.2 Attention Queue and Task Scheduler

If we rule out polling in favor of the more efficient interrupt scheme, we have two alternatives left for attention handling: (1) to respond immediately to an attention, or (2) to store attentions first in a queue. The second approach offers the advantage of protecting the recipient of an interrupt message from having to receive it at an instant when it is not yet ready for it. Furthermore, it is a good rule to spare the user the trouble of having to write interrupt-handling routines, thus leaving that task to the system programmer. This requires that such routines are as general as possible so that the casual user may utilize them in a program as prefabricated building blocks. Consequently, these interrupt-handling routines will not communicate directly with the application program. For example, data put in by an interrupt-handling routine will not be passed on directly to the program as they are received, but will be put into a special form of a shared file, termed an *attention queue*. The recipient of such a message will then examine the queue from time to time and remove a message. The entries in the attention queue may be input data and/or device identifiers, supplied by the alphanumeric keyboard, the function keyboard, a pointing or a positioning routine in the display processor, etc. The program, which examines the attention queue and removes the entry from its head, interprets this entry and decides which process—if any—to invoke in response, may be named a *task scheduler*.

In the case that mnemonic commands typed in on the alphanumeric keyboard have to be processed, the task scheduler must be extended into a *language processor* by adding a *lexical analyzer* that recognizes terminal symbols, interprets the character string between two terminal symbols, and replaces it by a token which is more convenient for the task scheduler to handle.

In simpler systems, the necessity of an attention queue and a task scheduler is avoided by introducing the following stipulations: (1) a user's attention is only permissible if it has first been "prompted" by the application program; and (2) the system, after it has prompted an attention, suspends the execution of the program until the attention has occurred. Hence, concurrent attentions are prohibited, and no provisions need be made for such a case. An attention may be prompted by displaying a text, explaining to the user that it is now his/her turn to select one out of an offered choice of attentions. Of course, one sacrifices in such a case the possibility that, while an attention is being served by the system, the user may generate further attentions.

One may raise the objection that such a scheme need not be based on the use of interrupts. While the system idles in a wait loop until the occurrence of the prompted attention, it might as well perform a polling. Polling is not only a simpler mechanism than the handling of interrupts, but also, from the viewpoint of programming, a "cleaner" method. The advantage of using interrupts even in this simple approach, however, is that it allows for the execution of another program during the wait, e.g., in a foreground/background mode of operation or, more generally, by multiprogramming. Since this is a matter that concerns the operating system of the host computer rather than its user, and since this problem is not specific for the operation of interactive computer graphic systems, we shall not dwell on it.

If the display terminal in a computer graphics system is furnished with local computing power, the attention queue can be maintained in the terminal. That is, attentions are handled by the local processor, and the attention source identifiers and input data are stored in a "message table." From time to time, whenever the host computer is ready to receive such a message, this table is transmitted to the host. Such an organization is almost mandatory in the case that the display terminal is part of a general-purpose time-sharing system. A system that is organized in such a fashion is discussed in the appendix.

Johnson [77] lists the following four principles which can be employed for implementing the task scheduler.

1. Program flow modification (synchronous).
2. Tabular specification.
3. Dynamic procedure declaration (asynchronous).
4. State-diagram approach.

The last method is equivalent to the first one (it differs only in its formal representation, not in its procedural essence). The third method is certainly the most efficient approach; alas, it is not existent in any of the common high-level programming languages, and thus its implementation would require a language extension. The second method is rather inadequate and may give rise to programming pitfalls. This leaves the first method as the scheme that is mainly used.

The first method is based on the use of an IF statement or a sequence of IF statements, respectively. In lieu of a sequence of IFs we prefer the CASE clause as a more efficient construct. A simple example in PL/I is

IF LIGHTPEN *THEN* ⟨program unit-1⟩ (*ELSE* ⟨program unit-2⟩).

LIGHTPEN is here a boolean variable which indicates an activation of the lightpen switch. Other examples for the use of IF are considered in Chapter 10, where it is demonstrated, in particular, how one can make sure that an anticipated attention has really happened when the IF statement occurs, without having to program a polling loop. This technique consists of providing interrupt-sensitive functions such as, in the example of GRIP: PICK; POINTIN; TYPEIN; VALUEIN; KEYIN; and BUTTON. The execution of such a function is delayed (by a WAIT loop) until the attention has occurred, on which event the respective function value is returned. In the normal case that an attention is called for by the program (exceptions are spontaneous terminations of a program by an overflow condition in the system or by the user pressing the HALT button), one of these functions is invoked prior to the IF statement.

Tabular specifications have been used in earlier systems. An example, taken from GSP [75], is to set up a table by a FORTRAN subroutine call

CALL SETUP ((attntype(1),routine(1)),...,(attntype(n),routine(n))).

Such an approach exhibits serious shortcomings. The reason is that the decision of the task scheduler as to what process to invoke next will only in the most simple cases be based solely on the information given by the attention message. Normally, such a decision will also depend on the state of the program at this point. For example, the program may, in the course of its execution, send a request several times to the user for an activation of the lightpen switch. Each time the lightpen switch activation is given a different meaning. As the lightpen switch is one of the attention sources specified in the table, the table would in this case have to be respecified dynamically. At best, such a dynamic respecification of the attention table lacks the program transparency we should strive for; and in the worst case there is the danger that such a process will get out of hand.

The most efficient principle for task scheduling is the dynamic procedure declaration. This could be provided in PL/I by the ON statement, if the language were extended such that the user could arbitrarily specify an ON CONDITION. An even more elegant approach would be the inclusion of event-driven constructs.

In PL/I such a construct is given in the form [76]

ON ⟨on-condition⟩ ⟨on-program-unit⟩

(the SYSTEM option is irrelevant in this case). The special feature of this construct is that the ON statement is executed as it is encountered in statement flow, whereas the on-program-unit is executed only when the associated interrupt occurs. Any names used in the on-unit, however, are in the scope of the block encompassing the ON statement and not in the scope of the block in which the on-unit is executed. If two ON statements, specifying the same condition, are internal to the same invokation of the same block, the effect of the former ON statement is nullified when the latter is encountered. If the second ON statement occurs in a block which is nested in and initiated by the block containing the first ON statement, the action specification of the prior ON statement is temporarily suspended (or "stacked"). The reader will notice that such a mechanism offers the perfect instrument for asynchronous attention handling.

6.3.3 Finite-State Model of an Interactive Program

Task scheduling based on program flow modification is characterized by the fact that a branching is performed at certain points in the program (we may also say at certain states of the program), where the branch destination is determined by a user's attention. This suggests a model in the form of a finite-state automaton, whose input alphabet is the set of possible attentions, whose set of states is the set of branching points, and whose output are the actions that result in the execution of the sequence of statements between two branching points. Thus, we have a finite-state automaton in the form of the *MEALY* (or "transition assigned") machine.

For illustration, let us consider the simple example of a "rubberband" line-drawing procedure given by Newman, who introduced the notion of representing an interactive display program by a finite-state model [104]. In this example, the user's actions consist in a sequence of five attentions:

1. Press lightpen switch to start pen tracking.
2. Move cursor to starting point of a line.
3. Press lightpen switch to fix starting point.
4. Move cursor to mark end point and display line connecting the two points.
5. Press lightpen switch to fix end point and stop tracking.

The resulting automaton has three states: the initial state, from which it is switched by a lightpen switch attention into the state in which the starting point is fixed, from which it is switched by a lightpen switch attention into the state in which the end point is fixed, from which it can be switched back into the initial state. The state diagram is illustrated (Fig. 6.5) in the following customary way: states are indicated by the nodes and transitions are indicated by the edges of a graph. Each transition is associated with a pair (input, output), where input is an attention and output is the response of the program. NOOP stands for the empty response.

Sutherland and Newman [105] developed a procedural language, named DIAL, which consists of an ALGOL 60 subset extended by three statements: DURING, ENTER and ON. *DURING ⟨state identifier⟩ DO* delimits the program unit to be executed during the phase the system is in the specified state (the second delimiter is given by the next DURING statement). *ENTER ⟨state identifier⟩* (which is another way of saying GOTO⟨label⟩) causes the program to enter a new state. *ON⟨event name⟩ DO* is similar to the PL/I WAIT statement. Such a nomenclature is certainly interesting inasmuch as it is directly derived from the finite-state model of an interactive display program. However, it has to be questioned whether this justifies the implementation of a special language dialect (of which there are already too many), as a similar effect can be easily achieved within the framework of existing, widely used languages such as PL/I or APL.

For example, the program represented by the state diagram above could be written in PL/I as follows (LS stands for the event given by a lightpen switch attention,[†] (XS, YS) are the starting-point coordinates, and (XC, YC) the current cursor coordinates):

```
STATE_1: WAIT(LS);
         GOTO STATE_2;
STATE_2: ⟨program unit for cursor positioning by pen tracking⟩;
         WAIT(LS);
         XS=XC; YS=YC;
         GOTO STATE_3;
STATE_3: ⟨program unit for cursor positioning and generation of line⟩;
         WAIT(LS);
         XT=XC; YT=YC;
         GOTO STATE_1;
```

In our opinion, such an approach, which introduces state names as labels, shows the relation between program and finite-state model as clearly as any special language could conceivably do. Therefore, neither from a programming language viewpoint nor from that of the task scheduler mode of operation is there anything idiosyncratic about the state-diagram model. However, the model is of great conceptional value, as it may more clearly reveal the intrinsic mechanism of an interactive display program than the program itself, especially if the program is not structured according to the finite-state model. As a simple example, let us have a

[†]Contrasting to a boolean semaphore, an event is not simply a two-valued variable but a change of the value of such a variable; e.g., in the case of the lightpen switch, it may be the change from "make" to "break."

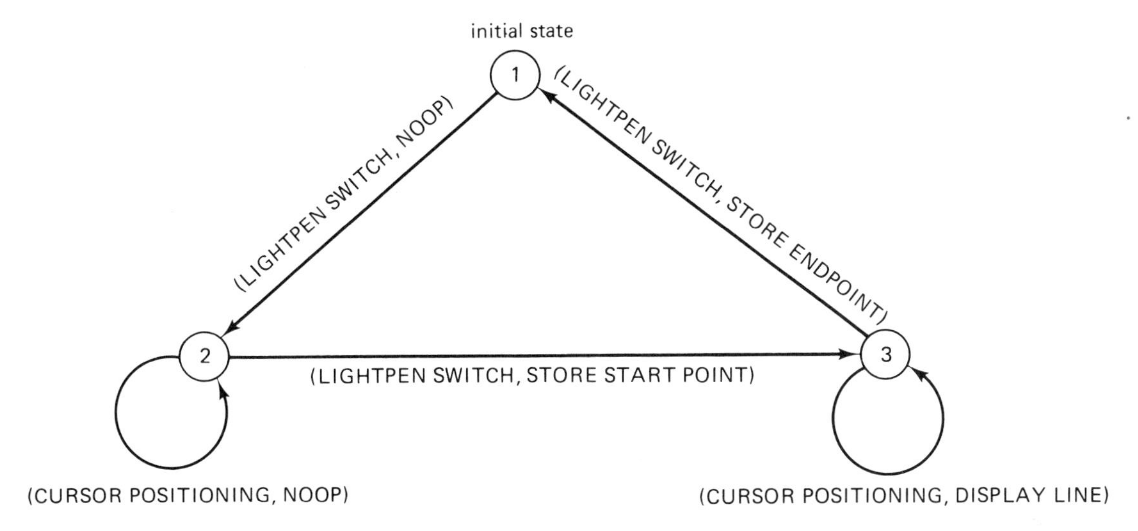

Figure 6.5 State diagram for the rubberband line-drawing procedure.

second look at the rubberband line-drawing procedure discussed above. Here, the state diagram (as well as the PL/I program structured accordingly) reveals at first glance that such a program will not work properly, as it is lacking any provision for exiting from it. The program as it is written would force the poor user to rubberband forever.

EXERCISES

1. Assume that pictures are defined in the standardized screen domain $[0:999]\times[0:999]$. Assume the existence of the following item-generating procedures:

DOTSET: Generates a set of dots;
POLYGON: Generates a polygon;
TEXT: Generates a text string.

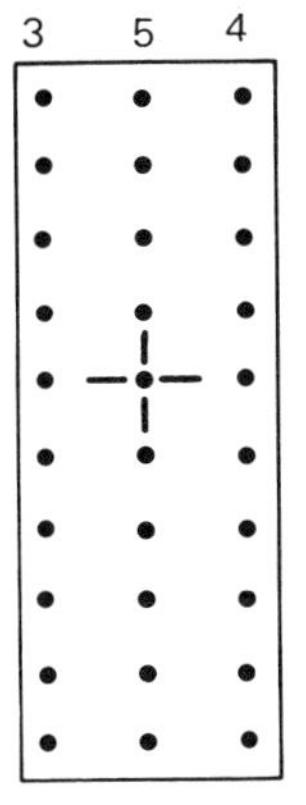

Furthermore, assume the existence of a procedure, PENTRAK, which allows the user to pick a cursor by the lightpen and move it into a desired location. On the release of the lightpen switch, the procedure returns the actual coordinates of the cursor position. For syntactic details, see Section 10.4.

Write a program in a language of your choice for Newman's "light handle" as shown in the accompanying diagram. The light handle, which allows the input of a number $n \in [0:999]$, is to be displayed in the lower right corner of the screen. The spacing between the dots is 20 units in horizontal and in vertical direction, and the size of the box is 50×200 units. Make sure that the cursor (the tracking cross) can rest only on one of the dots. Therefore, when the cursor is detached from the lightpen, it will not stay at its current position but will assume the coordinates of the nearest dot. The input number is displayed on top of the box, as illustrated.

2. Zahn [157] suggests for future high-level programming languages "event-driven" control constructs, e.g., given in the form

> **BEGIN UNTIL ⟨event⟩$_1$ OR...OR ⟨event⟩$_n$:**
> **⟨statement list⟩$_0$;**
> **END;**
> **THEN⟨event⟩$_1$→⟨statement list⟩$_1$;**
> **⋮**
> **⟨event⟩$_n$→⟨statement list⟩$_n$;**

Assume the existence of such a construct in the language of your choice and write a program for the rubberbanding described in Section 6.3.3. Assume that the event consists in the release of the lightpen switch as signaled by a variable LS. Assume the existence of the procedures POLYGON and PENTRACK for the generation of a polygon and for a pen track, respectively.

3. (a) Describe the sequence of steps required for interrupt handling in a computer. For supplementary reading on this matter, see a book on computer organization (e.g., [137]).

(b) Compare control transfer through polling to control transfer through interrupts. Why does polling save hardware? How is it possible to implement attention handling through program flow modification without having to program polling loops? What advantage is offered by such an organization?

4. Write a FORTRAN subroutine for a generalized rubberbanding procedure. The calling sequence shall be CALL RUBBER(ITYPE,IX1,IY1,IX2,IY2). The routine shall provide for its user an exit through the pick of a light button named ACCEPT. ITYPE is an input parameter that indicates the type of graphic object that is to be rubberbanded as follows.

If ITYPE=0, then (IX1,IY1) are input parameters which specify the absolute coordinates of the starting point of a vector. The end point of the vector can be interactively moved with the tracking cross. When the ACCEPT light button is picked, the absolute coordinates of the vector end point are returned in (IX2, IY2).

If ITYPE = 1, then the tracking cross is displayed and then moved with the lightpen to the starting point of a vector. The cross is then moved again to specify the end point of the vector. Each time the lightpen switch is released, the tracking cross moves from one end of the vector to the other so that both ends of the vector can be alternately moved. When the desired position is obtained, the ACCEPT light button is picked and the starting point of the vector is returned in (IX1, IY1) and the endpoint is returned in (IX2, IY2).

If ITYPE = 2, the tracking cross is displayed and moved to the desired vertex of a rectangle. The cross is then moved again to the desired vertex opposite from the first to complete the specification of a rectangle. Each time the lightpen switch is released, the cross moves to the opposite vertex of the rectangle so that the rectangle can be easily moved to any place on the screen. When the desired placement of the rectangle is achieved, the ACCEPT light button is picked and the two opposite vertices are returned in (IX1, IY1) and (IX2, IY2).

If ITYPE = 3, the tracking cross is displayed and moved to the desired center of a circle. The cross is then moved to any desired point on the radius of the circle and the circle will enlarge to follow the tracking cross. Each time the lightpen switch is released, the tracking cross alternates between the circle center and the circle radius. When the desired placement of the circle is achieved, the ACCEPT light button is picked and the circle center is returned in (IX1, IY1) and the tracking cross position on the circumference of the circle is returned in (IX2, IY2).

In all four cases, the rubberbanded vector, rectangle, or circle is erased when the light button ACCEPT is picked.

Assume the existence of the procedures POLYGON, CIRCLE, and SAMPLE. POLYGON and CIRCLE allow the generation of vectors, rectangles, and circles, respectively (for syntactic details, see Section 10.4). SAMPLE is a procedure that generates a tracking cross which can be "attached" to the lightpen by picking it and keeping the lightpen switch pressed. While the cross is attached, its current coordinates are returned as a continuous stream of data (flowing at a fixed rate). The release of the lightpen switch terminates the sampling procedure. What distinguishes SAMPLE from PENTRAK of Exercise 1?

5. Develop strategies for using (a) a lightpen or (b) a joystick (1) as a selector, (2) as a locator, (3) as a valuator, (4) as a button. Flowchart the respective procedures.

6. (a) Describe the steps in the construction of a line with a lightpen (1) using the conventional line-drawing procedure, (2) using rubberbanding.

(b) What kind of display system does not allow the use of a lightpen, and why?

(c) Is a tablet a sufficient input device for all four virtual-input-device functions of an interactive display system? If not, which user attentions are not possible?

(d) Enumerate and discuss alternative lightpen positioning procedures.

7

The Display Processor

In this chapter, the functional and organizational aspects of the display processor are discussed. We distinguish three functional parts of a display processor:

1. The display controller.
2. The display generator.
3. The display console.

For easier understanding, we discuss these three components separately and in reverse order. Our considerations will, in general, refer to the *vector display* (or "line-drawing" display) with directed beam and picture refresh,† i.e., a system that allows for the drawing of arbitrary vectors connecting any two randomly addressable points of a given raster or for the writing of characters at any arbitrary position. Additionally, raster displays will be discussed.

7.1 THE DISPLAY CONSOLE

The typical display console contains the *cathode ray tube* (CRT) with deflection and intensity control circuits, a control keyboard, an alphanumeric keyboard,

†The ACM/SIGGRAPH glossary defines *directed beam* as "the CRT method of tracing the elements of a display image in any sequence given by the computer program, where the beam motion is analogous to the pen movements of a flat bed plotter. This contrasts with the raster scan method, which requires the display elements to be sorted in the order of their appearance."

and a lightpen or other pointing or positioning devices. Other accessories, such as a hard-copy device or a tablet for the input of line drawings, may also be part of the console.

7.1.1 Typical CRT Specifications

The screen size of the CRT typically is in the range of a 16- to 24-inch diagonal in the case of rectangular tubes and a 20-inch diameter in the case of circular tubes. This results in a usable rectangular area ranging from 10×10 inches to 13×16 inches (14×14 inches for the 20-inch-diameter tube). The average spot size (spot diameter) of the CRT beam can be kept down to typically 0.02 to 0.015 inch. With special provisions (special correction circuitry), the spot size may be further reduced to about 0.01 inch. These ratings usually give the mean of a Gaussian distribution which the spot size is assumed to have over the whole screen area.

The resulting *number of distinguishable points* is calculated on the assumption that two points which have a distance from each other of half a spot size can be distinguished. Thus, we obtain the formula

$$\text{number of distinguishable points} = 2 \times \frac{\text{maximal displacement}}{\text{spot diameter}}.$$

With an absolute resolution of 0.5×0.015 inch$=0.0075$ inch and a maximal displacement of 14 inches, this results in approximately 2000 distinguishable points in each coordinate direction. However, such a figure is only valid if the *repeatability* of the display matches the resolution as given by the spot size. Deflection amplifiers may have a certain drift, and deflection coils may have a certain hysteresis. Therefore, when the beam is deflected from one edge of the screen to the opposite one and back to the old position, it may not exactly reach the old position, but there may be a small displacement. The maximum of the displacement is called the *repeatability*. A typical value for the repeatability is 0.02 inch, i.e., the same as the spot size, so that this effect does not further limit the obtainable resolution.

The *addressable number of raster units* of the CRT screen is usually chosen as the power of 2 that comes closest to the number of distinguishable points, for not much could be gained by addressing more points than are distinguishable. Consequently, the typical addressable screen raster is 2048×2048.

The *resolution* obtained cannot be equated with the *precision* of a display. Any dimension specified in the display program will only be met with a certain limited accuracy by the respective representation on the screen. The limiting factor is here the nonlinearity of the deflection circuitry, which is typically in the magnitude of percentage points. However, the limited precision does not matter, as graphics display system are only used for a visualization of graphic patterns and objects, not for the generation of precision graphics or artwork. Whenever a precise "hard copy" of a graphic representation is desired, a plotter may be used.

Vector display CRTs use a phosphor that has a longer persistence than the one used in standard TV picture tubes. The reason is this: whereas the standard TV raster regenerates a picture at a rate of 60 frames per second (fps), a vector display system is designed to generate a flicker-free picture at a lower refresh rate, e.g., at 40 or 30 fps. The trade-off for the longer persistence is a lower brightness.

To avoid interference or beat between the picture refresh rate and the power-line frequency, the picture refresh is often synchronized to the line frequency or an integer fraction of it, leading to a refresh rate of either 60 fps or 30 fps (50 fps or 25 fps in Europe).

7.1.2 Input Devices

The various input devices that a display console may be equipped with were already discussed in Chapter 6.

1. The lightpen as a pointing device.

2. Joystick, control ball, "mouse," and tablet as positioning devices.

3. The control keyboard for command input and the alphanumeric keyboard for the input of data and text strings.

The only device that can provide a certain feedback to the user other than via the display screen is the control keyboard, if the display processor comprises a keyboard lamp register that can be set by the program, resulting in the illumination of certain keys. Thus, the system can, for example, indicate to the user which control functions are enabled at a certain instant.

Figure 7.1 presents a simplified diagram of a lightpen circuit. Usually, the lightpen consists of an aperture, fitted with a lens that is connected by a bundle of fiber optics (called "light pipe") to a photomultiplier tube. The output of the photomultiplier tube is amplified and fed to the input of a Schmitt trigger. Hence, whenever the aperture or *viewing area* of the lightpen is exposed to a light source exceeding a certain brightness threshold, the photomultiplier responds with an impulse and the Schmitt trigger is switched. Presently, lightpens exist which use a simple photodiode (or phototransistor) directly connected to an amplifier. Integrated circuits make it possible to put the amplifier into the lightpen case, which is approximately the size of a ballpoint pen.

The viewing area of a lightpen is relatively wide and not well defined in size and shape. Actually, it depends on the angle under which the lightpen is pointed to the screen, yet it would be unreasonable to require the user to hold it precisely perpendicular to the screen pane. This fact may cause problems when one of some narrowly spaced objects shall be individually picked. Another problem may be parallaxis. It can be mitigated by putting a light source together with some optics into the pen, thus causing the projection of an "aiming circle" on the screen that indicates the viewing area.

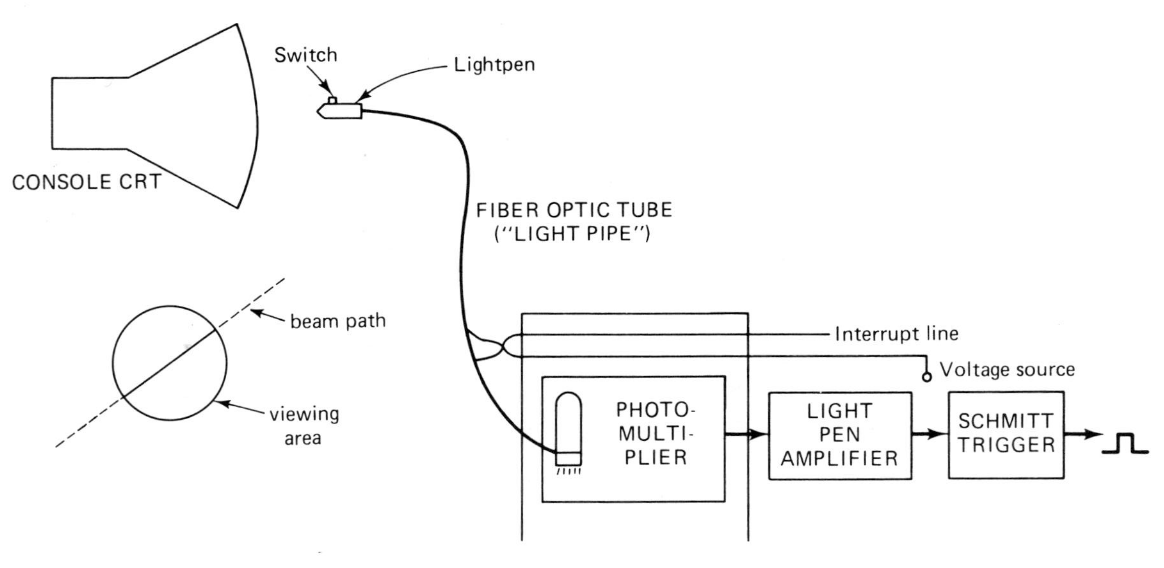

Figure 7.1 Lightpen circuit.

7.2 THE DISPLAY GENERATOR

7.2.1 Position Generator and Dot Generator

The position generator is used to arbitrarily position the CRT beam. Technically, the position generator consists of two digital-to-analog converters (DACs) for horizontal and vertical deflection, whose input range (length of the data word) and precision must correspond with the number of raster units in each coordinate direction (e.g., 11 bits corresponding with 2048 raster units).

The beam displacement from a given location to another one requires the application of step functions to the horizontal and vertical deflection circuits. The driving force in the deflection system, an electromagnetic or electrostatic field, however, can only change with a certain maximal rate. Thus, the total settling time required for a certain movement of the beam depends on the distance the beam has to travel from the old to the new position. Typically, the positioning time range is 1–2 microseconds. Of course, the intensity is switched off during the time the beam moves.

The position generator can be expanded into a dot generator if a provision is added to intensify the beam for a short period of time after it has reached its new position, thus creating the appearance of a dot on the screen.

7.2.2 Character Generator

Four methods of alphanumeric character and symbol generation can be distinguished:

1. The starbust principle.
2. The Lissajous principle.
3. The stroke principle.
4. The dot matrix principle.

The starbust principle and the Lissajous principle were used in the early days of computer graphics systems but are now obsolete. The stroke principle is the one predominantly used in vector displays, whereas the dot matrix principle is practically the only one used in TV raster displays. For the sake of completeness, we will briefly explain all four schemes.

Starbust Character Generator: The CRT beam is caused to write the pattern shown in Figure 7.2. This pattern is composed of 24 strokes and called a *starbust* because of its characteristic appearance (sometimes, a starbust with only 16 strokes has been used). An actual character is generated by brightening some of the strokes and blanking out the others. The reader can easily verify that in this way all uppercase letters of the alphabet and all digits as well as some special characters

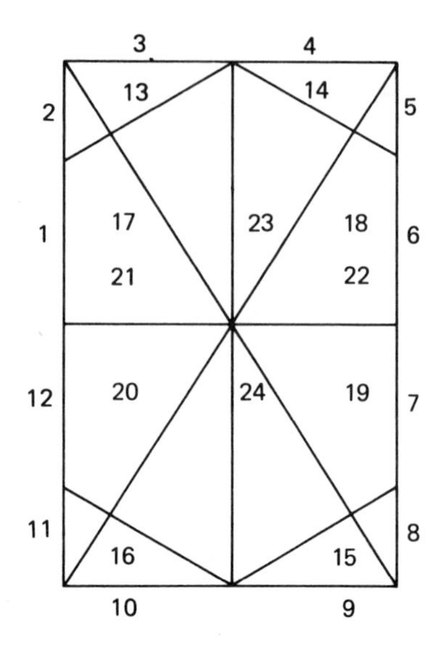

Figure 7.2 Starbust.

can be generated. Each character is coded by a 24-bit word, each bit representing one of the strokes. If a bit is 0, the corresponding stroke is blanked out. If a bit is 1, the corresponding stroke is visible.

The starbust principle has serious disadvantages, and these are the reasons it is no longer used. The first shortcoming is that the quality of the characters is poor. The second is that it is slow, for the entire starbust must be written by the beam (although only part of them become visible), resulting in a relatively long time to write one character. Furthermore, this approach requires either a code conversion

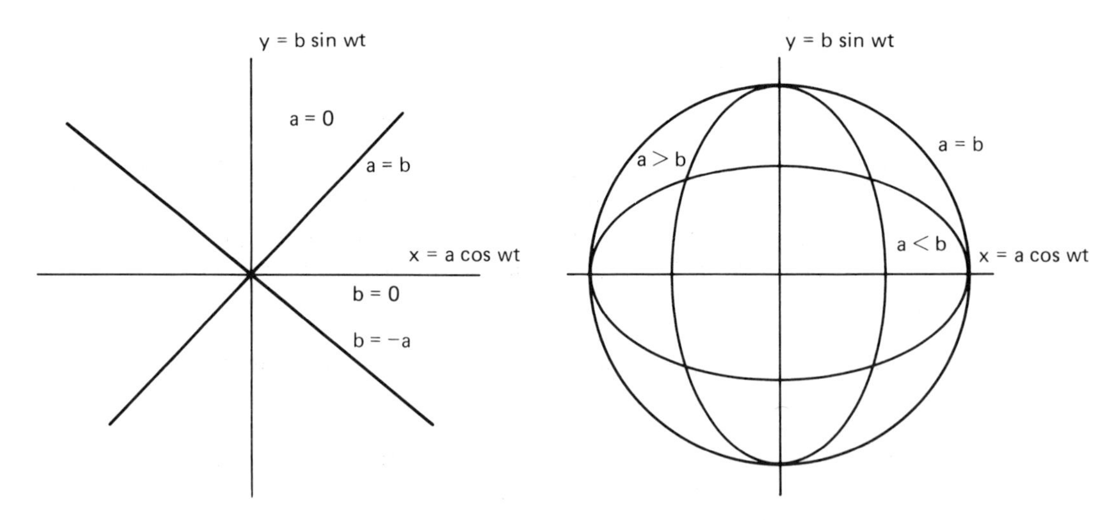

Figure 7.3 Generation of strokes and circles as Lissajous figures.

from the standard character code (e.g., ASCII) into the special 16- or 24-bit code or, as an alternative, the characters are generated as subpictures, in which case the 24-bit code for each character must be stored in the display file.

Lissajous Principle: In the Lissajous scheme, as the name indicates, Lissajous patterns are used for the composition of characters. The various possibilities of generating certain "primitives," such as strokes, arcs, circles, of which the characters can be composed, are shown in Figure 7.3. The Lissajous scheme is particularly apt to generate very smooth lowercase letters, composing them of strokes, half-circles, and circles written in a certain sequence, which are, together with the position of each primitive, determined by a decoding and switching matrix. The disadvantage of this scheme is that its hardware implementation is expensive and relatively slow.

Stroke Principle: Figure 7.4 illustrates the way a stroke character generator works. As in the Lissajous case, the stroke character generator is an analog device that produces separate wave forms for the x and y deflection of the CRT beam, along with an on–off intensity pattern. The hardware expenses of such a device are not

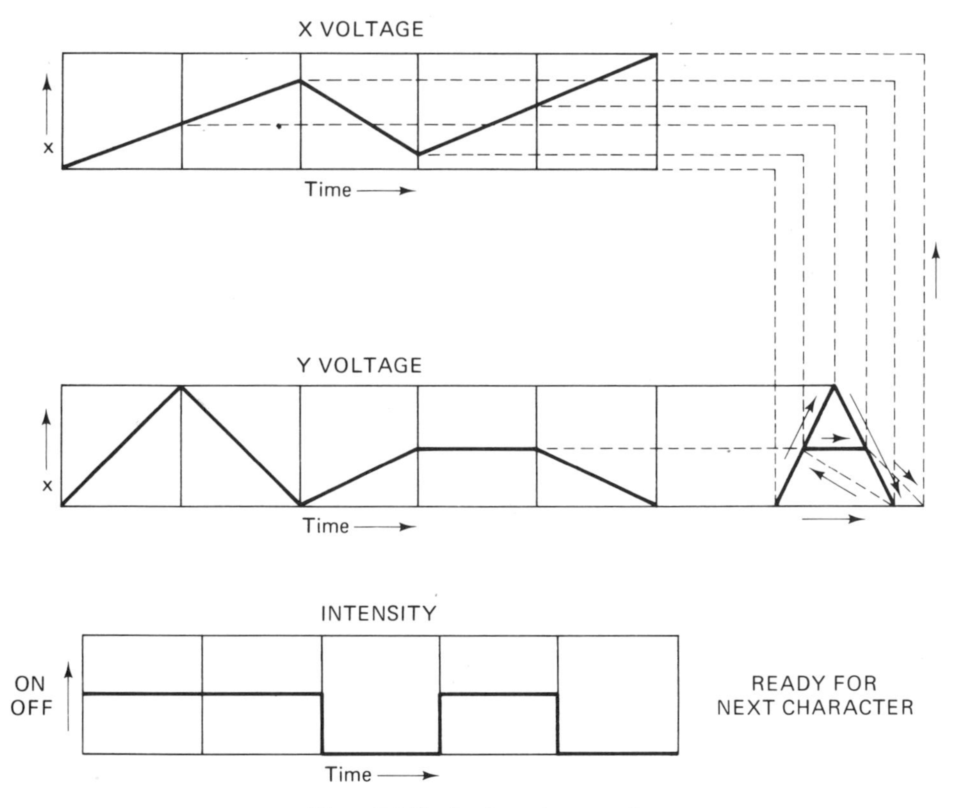

Figure 7.4 Stroke character generator.

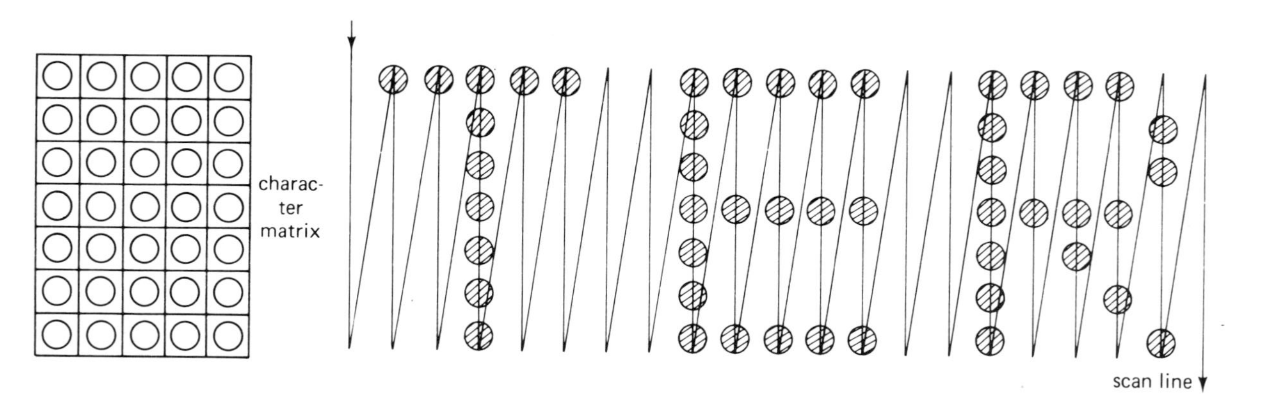

Figure 7.5 Dot-matrix character generation.

low, but are reasonable. However, the expenses are much higher than those of the subsequently discussed dot matrix generator. The speed of operation of the stroke character generator is high and the quality of the characters is very good. The average time for writing a character is typically in the order of 10 microseconds. The character size is arbitrary.

The latter fact explains why this more expensive scheme is widely used in vector displays. Whereas the very inexpensive dot matrix character generator produces characters with a satisfactory quality as long as these characters are small, it degrades if the characters become bigger. In a vector display, however, one wants to have the choice between several character sizes, including fairly big characters for headings.

Dot Matrix Principle: Dot matrix character generators work strictly digitally and are therefore very reliable and inexpensive. They produce symbols of a satisfactory quality as long as the size of the symbols is kept small enough—about the size of teletype or line printer characters. Therefore, dot-matrix character generators are used in alphanumeric displays (terminals which display only text) and in low-cost displays (storage tube or TV raster displays). Figure 7.5 illustrates its operation.

A character is generated by writing a sequence of bright dots. To this end, the CRT beam is deflected by a scan-line raster so that it passes successively through the fields of a character matrix, e.g., the 5×7 matrix shown in Figure 7.5. In each field of the matrix the beam can be either switched off or switched on. A character is defined by a bit pattern of as many bits as the matrix has fields. In our example a $5 \times 7 = 35$-bit word is needed for each character. If a bit is 0, the beam is off; if a bit is 1, the beam is on. The bit pattern for the letters of our example is

Line															
Line 1	1	1	1	1	1	1	1	1	1	1	1	1	1	1	0
2	0	0	1	0	0	1	0	0	0	0	1	0	0	0	1
3	0	0	1	0	0	1	0	0	0	0	1	0	0	0	1
4	0	0	1	0	0	1	1	1	1	1	1	1	1	1	0
5	0	0	1	0	0	1	0	0	0	0	1	0	1	0	0
6	0	0	1	0	0	1	0	0	0	0	1	0	0	1	0
7	0	0	1	0	0	1	1	1	1	1	1	0	0	0	1

For a character "font" of k characters, k such words must be stored, e.g., in a read-only memory (ROM). Figure 7.6 explains the organization of such a memory for horizontal scan. Three input lines are used to determine the actual scan line (out of seven, the eighth line is blank), and $\log_2 k$ lines are needed to code the characters of the font (e.g., for the entire ASCII set of 128 characters, 7 bits are needed). The ROM has five output lines for the 5-bit pattern given for each one of the seven scan lines of a character. Hence, the total capacity of the ROM must be $5 \times 2^{10} = 5K$ $(K = 1024)$. Such a memory capacity can be easily accommodated in one single dual-inline MOS device (presently selling for a few dollars). Dot-matrix generators are available for horizontal as well as for vertical scan.

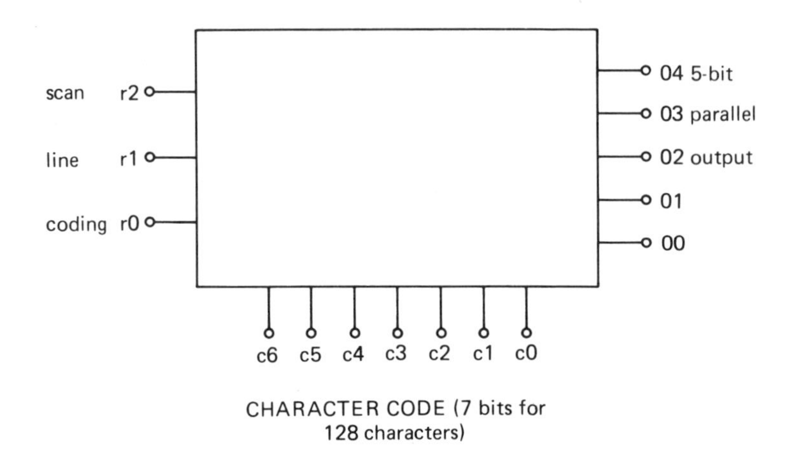

Figure 7.6 Dot-matrix character generator using a ROM.

By slightly slanting the characters as indicated in Figure 7.5, their appearance can be improved. The scan-line generator may be an analog sawtooth generator or, with better results, a digital counter which is combined with a digital-to-analog converter. The counter is alternatingly counting up and being reset. Such a device is more reliable, and it produces a "staircase" function instead of the straight slope of a sawtooth, causing the CRT beam to rest for a while in each dot position. By this measure a higher intensity of the characters can be achieved.

In TV raster displays, the required scan-line raster is already in place. Therefore, character generator devices for horizontal scan are used in this case. A string of characters is now not written characterwise, as depicted in Figure 7.5, but linewise for the whole string. The organization of TV raster displays is discussed in Section 7.5.

7.2.3 Vector Generator

Technically, vector generators are circuits that must convert a step function (as given by the ΔX and ΔY increments between the vector end points) into such wave forms of the x and y deflection voltages that, as a result, a straight-line segment is written connecting these two points.† This can be accomplished by analog as well as by digital networks. In the analog case, analog integrators or filters may be used. In the digital case, integrators may be used as well, but this time in the form of DDA integrators (DDA = digital differential analyzer). Depending on the respective technique, two principally different modes of operation may be distinguished: constant-time and constant-rate vector generation.

†Actually, the parametric expression for a vector with start point (X_s, Y_s) is $x(t) = X_s + \Delta x(t)$ and $y(t) = Y_s + \Delta y(t)$. For the sake of simplicity, in the following considerations we drop the additive constants, X_s and Y_s.

Constant-Time Vector Generation: The time for writing a vector is constant regardless of its length. Thus, the writing speed varies with the vector length, resulting in a varying beam brightness on the screen. This effect must be compensated by an appropriate beam intensity control—the price for this relatively simple scheme. Several presently employed techniques will be briefly discussed.

A1. The RC-Filter Method: Let $\Delta X = X_t - X_s$ and $\Delta Y = Y_t - Y_s$ be two voltages representing the two increments of a vector whose start point is (X_s, Y_s) and whose terminal point is (X_t, Y_t). In the RC-filter method the deflection voltages, $\Delta x(t)$ and $\Delta y(t)$, used for driving the CRT beam from position (X_s, Y_s) to position (X_t, Y_t), are obtained by applying the voltages ΔX and ΔY to two identical RC networks. As ΔX and ΔY are step functions, the RC-filter response is, with $RC = T$,

$$\Delta x(t) = \Delta X(1 - e^{-t/T}); \qquad \Delta y(t) = \Delta Y(1 - e^{-t/T}). \tag{7.1}$$

These two equations form a parametric expression for a straight line. However, it takes the time interval $(0, \infty)$ to draw the entire line. Even if we consider that it is sufficient to reach the terminal point with an accuracy of $\frac{1}{2}$ raster unit, it takes $T \log_e 2k$ seconds if k is the number of raster units by which the beam is to be moved.

The speed at which the beam is moved is a function of time, since we have

$$\frac{d}{dt}\left(\sqrt{\Delta x^2 + \Delta y^2}\right) = \sqrt{\Delta X^2 + \Delta Y^2}\,\frac{1}{T} e^{-t/T}. \tag{7.2}$$

As the beam intensity is reciprocal to its speed, a special intensity control circuit is needed to obtain constant intensity. According to equation (7.2), the length $\sqrt{\Delta X^2 + \Delta Y^2}$ of the vector must thus be calculated, and a third RC circuit with time constant T must be applied to generate the function $e^{-t/T}$ for intensity control.

A2. The Ramp-Function Method: The disadvantage of the simple RC-filter method can be avoided by replacing the exponential functions in the parametric form of a straight line by a linear parametric form

$$\Delta x(t) = \Delta X \cdot r(t); \qquad \Delta y(t) = \Delta Y \cdot r(t), \tag{7.3}$$

with the "ramp function"

$$r(t) = \begin{cases} 0 & \text{for } t \leqslant 0 \\ t/T & \text{for } t \in [0, T] \\ 1 & \text{for } t \geqslant T \end{cases} \tag{7.4}$$

The ramp function could be produced by a filter that responds to a step input with a ramp output. However, such a filter is expensive. It is much less expensive to use a simple analog integrator for generating the ramp function waveform and multiply it by ΔX and ΔY, respectively. The multiplication can be performed by using *multiplying digital-to-analog converters* (MDACs). DACs are required anyway for converting Δx and Δy from its digital form into an analog form.

A3. The Integrator Method: Instead of using an analog integrator to generate a ramp function and then multiply that ramp function by ΔX and ΔY, respectively, ΔX and ΔY may be directly applied as inputs to two analog integrators. Thus we obtain

$$\Delta x(t) = \int_0^T \frac{\Delta X}{T}\, dt = \Delta X \cdot \frac{t}{T} \qquad \text{for} \quad 0 \leqslant t \leqslant T \tag{7.5a}$$

$$y(t) = \int_0^T \frac{\Delta Y}{T}\, dt = \Delta Y \cdot \frac{t}{T} \qquad \text{for} \quad 0 \leqslant t \leqslant T. \tag{7.5b}$$

Constant-Rate Vector Generation: Here, the time required for writing a vector is proportional to its length. The constant writing speed guarantees a constant brightness of the CRT beam. This approach is technically more expensive (except for short vectors) but avoids the necessity for a specific intensity control circuit.

Constant-rate vector generation can be accomplished by using analog integration, but now with the modification that both integrands, ΔX and ΔY, are divided by the norm of the vector to be drawn, i.e., by

$$L = \sqrt{\Delta X^2 + \Delta Y^2}\,. \tag{7.6}$$

Thus, we have

$$\Delta x(t) = c_0 \cdot \int_0^t \frac{\Delta X}{L}\, dt' = c_0 \cdot \frac{\Delta X}{L} \cdot t \tag{7.7a}$$

$$\Delta y(t) = c_0 \cdot \int_0^t \frac{\Delta Y}{L}\, dt' = c_0 \cdot \frac{\Delta Y}{L} \cdot t. \tag{7.7b}$$

The integration is terminated when the terminal values, $\Delta x(t) = \Delta X$ and $\Delta y(t) = \Delta Y$, have been reached. This requires that the integrator outputs be monitored by analog comparators which switch when the signals $(\Delta x - \Delta X)$ and $(\Delta y - \Delta Y)$, respectively, change from negative to positive. The vector write speed is now constant, since we have

$$\frac{d}{dt}(\Delta x^2 + \Delta y^2)^{1/2} = c_0. \tag{7.8}$$

The necessity to divide ΔX and ΔY by L seems to be a serious disadvantage of that scheme. However, this causes no major problems, as the division can be performed by an analog circuit in the following way. In an analog integrator, the factor c_0 in equations (7.10) is

$$c_0 = \frac{1}{R_0 \cdot C}, \tag{7.9}$$

where R_0 and C are the resistor and capacitor used in connection with an operational amplifier to perform the integration. In order to perform the division, we replace the constant resistor R_0 by a variable resistor R such that

$$R = R_0 \cdot L. \tag{7.10}$$

Such a variable (switched) resistor can be easily obtained by employing a network of electronic switches and resistors similar to the one used in a DAC. With (7.9) and (7.10), we have

$$\frac{1}{RC}\int u\,dt = c_0 \int \frac{u}{L}\,dt. \tag{7.11}$$

The use of analog integrators for vector generation raises speed and accuracy problems. In particular, it is high-speed analog switching that limits the obtainable speed and causes distortions in the generated vectors. Moreover, analog circuits exhibit a relatively poor stability over temperature and over time—as compared with digital circuits—and thus cause calibration and adjustment problems. Linear distortions may be compensated to some extent. However, since such a compensation works accurately only for a given length or a given slope of the generated vector, it is the main effect of a compensation to make test patterns look good rather than to really improve the quality of vector generation in general. For all these reasons, digital vector generators tend to replace the analog solutions.

Digital vector generators also employ the symmetrical integrator method. Because of the discrete operation of digital circuits, however, the integral is approximated by a sum ("rectangular" or "Euler" integration). Hence, we have instead of equations (7.7),

$$\Delta x(i \cdot \Delta t) = \sum_{h=0}^{i} \frac{\Delta X}{N} = \frac{\Delta X}{N} \cdot i, \qquad i = 0, 1, \ldots, N \tag{7.12a}$$

$$\Delta y(i \cdot \Delta t) = \sum_{h=0}^{i} \frac{\Delta Y}{N} = \frac{\Delta Y}{N} \cdot i, \qquad i = 0, 1, \ldots, N \tag{7.12b}$$

In order to obtain the constant-rate vector writing mode, N should be equal to the length L of the vector to be written. However, an incorrect N will still result in the correct vector length (after N steps ΔX and ΔY are reached regardless of N). Thus, a deviation of N from the true length L results only in a change of intensity, to which the human eye is rather insensitive. Therefore, L may be rather coarsely approximated by N.

On the other hand, the choice of N must ensure that the steps in $\Delta x(i \cdot \Delta t)$ and $\Delta y(i \cdot \Delta t)$, respectively, are smaller than the beam spot size, for only then will the vector appear as a continuous line. This condition is satisfied if the step size is made smaller than 1 raster unit. Thus, we have the conditions for N,

$$\frac{|\Delta X|}{N} < 1 \quad \text{and} \quad \frac{|\Delta Y|}{N} < 1$$

or

$$N > \max(|\Delta X|, |\Delta Y|). \tag{7.13}$$

An undesirable feature of this scheme is the required division of ΔX and ΔY, respectively, by N. The division can be replaced by a simple shift if N is chosen as a power of 2. Hence, we finally obtain the condition

$$N = 2^{\lceil \log_2 \max(|\Delta X|, |\Delta Y|) \rceil} \tag{7.14}$$

($\lceil x \rceil$ denotes the next higher integer of x). This is also a sufficiently good approximation of the vector length L as far as intensity control is concerned (see the following section).

The circuit for calculating Δx and Δy, respectively, is depicted in Figure 7.7(A). The screen domain is assumed to be $[0:1023] \times [0:1023]$. The accumulator in Figure 7.7(A) must be given twice the wordlength of the coordinate registers (i.e., in our example a length of 20 bits). To understand this fact better, the reader may imagine the case $\Delta X = 1023$ and $\Delta Y = 1$. Here, $N = 1024$ and $\Delta Y / N = 1/1024 = 2^{-10}$. Thus, the total range of Δy (or Δx, respectively) is $2^{-10} \leqslant \Delta x, \Delta y < 2^{10}$, leading to a length of 20 bits. Of course, only the first 10 bits need be used for the display. The spacing of the generated dots by less than 1 raster unit guarantees a connected "bead" of dots; however, such a line may still exhibit a "staircase" behavior, distorting the quality of the generated vector. This effect can be mitigated by allowing more than the minimum of 10 bits for the display of Δx and Δy, respectively. Figure 7.7(B) shows the influence of the number of bits used for display on the vector quality. A very good behavior is obtained if 12 bits are used, and since 12 bits is a standard size of a fast digital-to-analog converter (DAC), it is preferable to use such a wordlength for the display. An additional improvement is obtained by rounding rather than truncating Δx and Δy. This can easily be accomplished by starting the process of accumulation not from zero but from the initial values $\Delta x = \frac{1}{2}$ and $\Delta y = \frac{1}{2}$. Hence we have now the algorithm

$$\Delta x_0 = \tfrac{1}{2} \qquad \text{and} \quad \Delta y_0 = \tfrac{1}{2}$$

$$\Delta x_i = \Delta x_{i-1} + (\Delta X / N) \quad \text{and} \quad \Delta y_i = \Delta y_{i-1} + (\Delta Y / N), \qquad i = 1, \ldots, N$$

$$(\Delta x, \Delta y) \; \textit{displayed} = (\Delta x_i, \Delta y_i) \; \textit{truncated}.$$

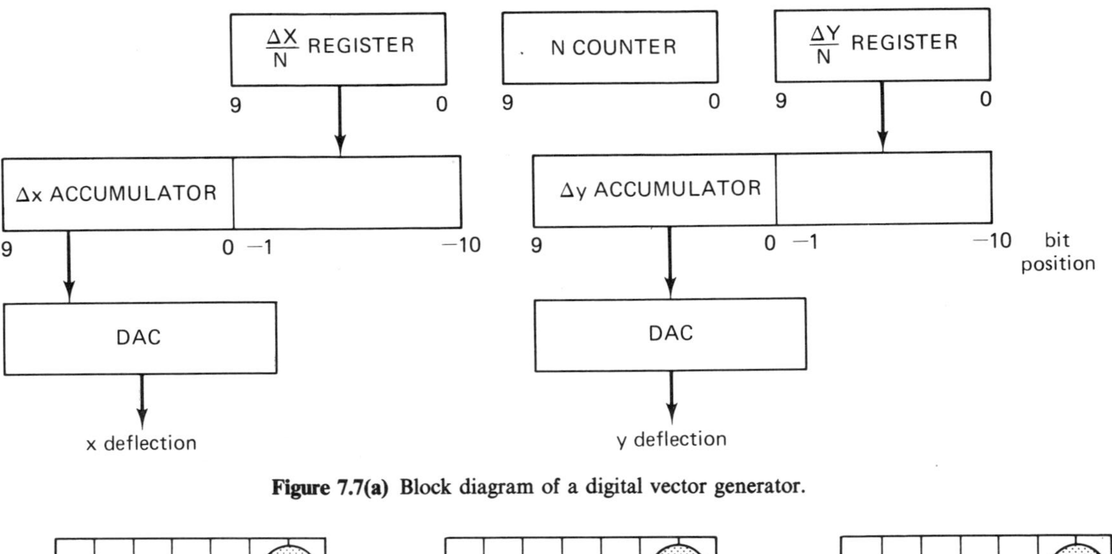

Figure 7.7(a) Block diagram of a digital vector generator.

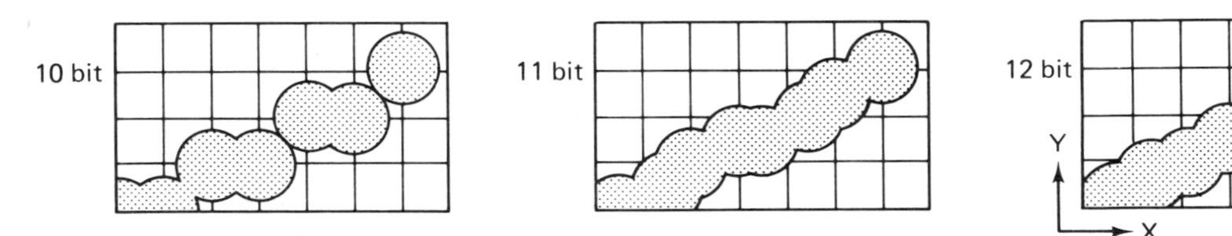

Figure 7.7(b) Influence of the number of bits on the vector quality.

For positive increments, ΔX and ΔY, N can be easily determined by shifting ΔX as well as ΔY (for example, in their respective accumulators) until the first 10 bits have become zero. The remainders, which are exactly $\Delta X/N$ and $\Delta Y/N$, are then transferred into the respective $\Delta X/N$ and $\Delta Y/N$ registers, and subsequently the accumulators are set to the initial value $\frac{1}{2}$. One complication arises if ΔX or ΔY is negative and if negative values are, as usual, represented in the two's complement. In this case, a two's complement representation must first be converted into the absolute value of the number. N itself is loaded into a counter, which is decremented by 1 in each step. The vector-generation process is terminated when the counter content has reached zero.

Intensity Control: Whether we use the constant-time scheme, which requires an intensity control, or the constant-rate scheme outlined above, we must always calculate the norm of the vector according to equation (7.6). This could be rather expensive (if done by hardware) or time-consuming (if done by software). On the other hand, since the intensity control, which compensates for the variable brightness of a line written at a variable speed, need not be very accurate, we may approximate L by an expression simpler to calculate. It is therefore perfectly sufficient to use the expression

$$\max(|\Delta X|, |\Delta Y|) + \tfrac{1}{2}\min(|\Delta X|, |\Delta X|) \tag{7.15}$$

[which we may derive by dividing in equation (7.6) the smaller component by the greater one, developing the resulting function into a power series, and truncating after the linear term]. The operations required for equation (7.15) are: comparison, one-place shift, and addition. Satisfactory results are even obtained if only the term $\max(|\Delta X|, |\Delta Y|)$ is used, requiring only a comparison. Additionally, since we replaced the exact norm by a coarse approximation anyway, it is unnecessary to use all the bits of ΔX or ΔY, respectively. By the same token, we may use a constant intensity as long as the length of the vector is less than a certain minimum in order to save execution time. However, even if an isolated short vector appears at the right intensity, curves composed of such vectors may be noticeably brighter.

Hence, a simple intensity control scheme may work as follows: First, the first m bits (of highest significance) of ΔX and ΔY are checked to verify that they are all zero. If confirmed, the vector is recognized as a short vector to be written with a certain constant intensity. If unconfirmed, it must be determined which component, ΔX or ΔY, is greater in order to provide the normalization for intensity control.

7.2.4 Time Requirements

The display on the screen must be refreshed at least 30 times per second to be flicker-free. That means that all the vectors, dots, and characters of which the representation is composed must be generated in approximately 30 milliseconds. If a picture consists of 1000 vectors, the time for positioning and generating a single vector must be less than 30 microseconds (actually less than that because some time will already be consumed by the memory cycle time and by the display processor). On the other hand, the number of 1000 vectors may easily be exceeded in a complex representation. Therefore, it is extremely important that the display generators work very fast.

Typical values for the speed of operation of the display generator components in a high-performance vector display system are about the following:

- Positioning time: 2–20 μseconds; depending on the magnitude of the displacement (e.g., a good performance is 2 μseconds for <1 inch, 10 μseconds for 1–10 inches).
- Vector-generation time: 2–20 μseconds; 4 μseconds for short vectors (e.g., 5% of full screen), 20 μseconds for the longest vectors.
- Character-writing time: 4–10 μseconds; 4 μseconds for the smallest and 10 μseconds for the biggest size.

We recognize from these figures that a high-performance vector display is capable of writing in maximum about 10,000 primitives (short vectors and small characters) and in minimum about 2000 primitives (long vectors and large characters). However, it must be noted that these figures hold true only if all the vectors are contiguous so that no new positioning is required each time. A realistic estimate for a mix of short and long vectors may be approximately 3000–5000 primitives.

7.3 THE DISPLAY CONTROLLER

7.3.1 Typical Display Processor Instruction Set

We define a display processor program as the sequence of instructions that is required for the drawing of a given scene in a given appearance on the CRT screen. As in the case of a general-purpose computer, these instructions might be classified into memory-referencing and non-memory-referencing instructions. However, it makes more sense to categorize display processor instructions into *file-organization, primitive-generation*, and *communication* commands.

There is a very distinct difference of instruction format between a display processor and a general-purpose processor. One of the basic notions of the von Neumann architecture is that instructions deal with memory contents (which may vary) rather than with data (which are constant). Thus, instructions refer in general to memory cells which represent variables of the program. In a display processor program, however, we have only constants, representing control parameters or point coordinates or character code. In the execution of such a program, no new data are created. Hence, it would be a waste of memory space if the primitive-generating instructions referred to memory addresses under which the data could be found. Rather, it is more economic and efficient to pack these data directly into the instruction word. For that reason, only the instructions that organize the stream of words from the display file memory for continuous picture refresh are memory-referencing instructions.

File-organization instructions provide for the generation of the picture refresh word stream, for interrupts of that stream and jumps to interrupt service routines, and for jumps to and returns from subpictures ("symbols"). Typical instructions of that type may be, e.g.,

SET ADDRESS REGISTER
INTERRUPT COMMAND
JUMP
JUMP AND STORE
JUMP TO SUBPICTURE
JUMP RETURN
SKIP NEXT WORD

The *address register* plays the role of the instruction counter in a general-purpose computer. *Interrupts* may be employed to synchronize the picture refresh with a "frame repetition clock." They are also used to detect *attentions* such as the activation of a control key or of the lightpen switch. Moreover, an interrupt may occur when the lightpen responds to light, when the HALT button is pressed, or when a line displayed on the screen crosses one of the edges of the viewing area of the screen. In addition to these "hardware" interrupts we have "software" interrupts, which may occur whenever an INTERRUPT COMMAND, a JUMP TO SUBPICTURE, or a JUMP RETURN is encountered in the display processor program. Certain bits in the interrupt instruction may function as "template" bits; i.e., the interrupt instruction is executed only if a given template is matched. Another bit may be used to enable or disable the synchronization of picture refresh with the clock. Certain interrupts may be transferred to the host computer, together with a code specifying the interrupt source. At the same time, the display is halted. The interrupt code is actually a memory location at which a service routine for the respective interrupt is stored. After execution of the service routine, the host computer may restart the display. For the host computer, the display system is just one of several peripheral devices which may send interrupts.

The JUMP and SKIP instructions are the same as (unconditioned) jumps or skips in a general-purpose processor. That is, a direct, relative, or indirect addressing scheme may be used in the case of the JUMP instruction, with the additional provision that, like the INTERRUPT instruction, a JUMP or SKIP may (or may not) be synchronized with the frame repetition clock. Conditioned Jumps are not provided for, as a display processor program is branch-free. [Objects displayed on the screen can be switched off either by skipping their respective code segment or (even simpler) by changing their appearance to zero intensity.] The use of these instructions for display file organization and the generation of word streams for picture refresh will be discussed in Chapter 8.

At this point, we are primarily interested in the *primitive-generating* instructions, among which we distinguish three classes:

1. System mode instructions.
2. Entity initialization instructions.
3. Primitive specification instructions.

System Mode Instructions: System mode instructions may set the display processor in either one of two modes:

- Graphic mode (drawing of dots, vectors, arcs, etc.).
- Character mode (writing of character strings).

Depending on the entered mode, the subsequent instructions for the generation of primitives are interpreted in different ways. Thus, the number of instructions which can be distinguished with a given length of the operation code is increased, an important feature if only a short wordlength (e.g., 16 bits) is available. Furthermore, it is possible to attach some other parameters, which are pertinent to the subsequent execution of a whole string of primitive specification instructions, to the ENTER GRAPHIC MODE and ENTER CHARACTER MODE instruction. The ENTER CHARACTER MODE instruction may encompass the specification of the character size (for all characters of a string have the same size). The ENTER GRAPHIC MODE instruction may comprise a coordinate-type code (absolute or incremental). In both cases, a special bit may enable or disable the analog scissoring circuit (see Section 7.3.4). In the absolute mode, the operand contains the actual screen coordinate, X or Y, of a dot or of the end point of a vector. In the incremental mode, the operand contains the distance, ΔX or ΔY, between the current and the new beam position in the drawing of dots or between start point and end point in the drawing of vectors. Usually, a program is written such that all coordinates are either absolute or incremental. Hence, it is certainly a more economic solution to let this parameter be part of the ENTER GRAPHIC MODE instruction rather than to attach it to each individual primitive-generating instruction.

Entity Initialization Instructions: Entity initialization instructions initialize the generation of graphic entities (usually an *item*, i.e., a set of dots or a set of line segments or a string of characters). This includes the specification of the initial position of the CRT beam (the origin of this particular item), of the lightpen pick status (enable/disable), and of the appearance (intensity or color, blink status, and line style) of this entity. A representative set of initialization instructions may be, e.g.,

SET CONTROL REGISTER
POSITION X
POSITION Y

The control register performs the functions of storing and routing the control parameters related to the following control functions:

• Enable/disable lightpen	(a particular bit set to 1 or 0)
• Enable/disable cycle timer for pen-track operations	(a particular bit set to 1 or 0)
• Enable/disable blinking	(a particular bit set to 1 or 0)
• Select beam intensity level or gray level or color	(a combination of k bits for 2^k gray levels or 3 bits for the three prime colors)
• Select line-drawing style	(a combination of 2 bits for four styles)

The various intensity levels may be, for instance: dim/medium/bright/very bright; the line styles may be: solid/dotted/dashed/dashed–dotted. In the POSITION X and POSITION Y instructions, a flag bit may indicate whether the variation of the current content of the respective register will be absolute or relative. In the first case, the current content is replaced by the new one. In the second case, the current content is incremented by the new one.

Primitive Specification Instructions: Primitive specification instructions cause the actual generation of primitives in the preset mode (graphic or character), in the preset position, and in the preset appearance. These instructions consist of an instruction code and of data—character code or coordinate values, respectively. A representative set of primitive specification instructions may be, for example,

SHORT POINT
SHORT VECTOR
SHORT BLANK
LONG POINT
LONG VECTOR
LONG BLANK
LONG DEFERRED
CHARACTER DATA

All these instructions except CHARACTER DATA refer to the graphic mode. The inclusion of "long" and "short" primitives provides for economical use of the available display file memory. In a "short point" specification, the increments in x and y to the preceding point are limited, e.g., to ± 31 raster units, while a "long point" specification means that the increment can be the maximal number of raster units, e.g., ± 1023. In "short vectors" the components of the Euclidean length of the vector are restricted in the same way. In the case of SHORT POINT and SHORT VECTOR, both components, ΔX and ΔY, can be accommodated in a single instruction word, whereas the specification of a long point or vector requires two words. Naturally, the coordinates specified in SHORT POINT or SHORT VECTOR instructions are incremental. On the other hand, picture transformations, except translation, require absolute coordinates (see Chapter 3). Furthermore, in the case of scaling and rotation, the maximal length of a short vector may be exceeded after the transformation. Hence, short vectors are mainly used in subpictures written for the generation of "symbols" of higher complexity. An example is the approximation of a circle by a polygon consisting of a greater number of short vectors.

Communication commands are commands given by the host computer in order to organize a communication between host computer and display processor. Communication commands may be part of the display processor program stored in the display file or they may be entered into the display processor from a separate source, that is, directly from the host computer. (If the display file exists somewhere in the memory of the host computer anyway, this difference is meaningless.) Some examples of communication commands are:

SET LAMP REGISTER
SET ADDRESS REGISTER
READ KEYBOARD REGISTER
READ ADDRESS REGISTER
READ X POSITION REGISTER
READ Y POSITION REGISTER
READ INSTRUCTION REGISTER
READ CONTROL REGISTER
START DISPLAY
STOP DISPLAY

The SET LAMP REGISTER command loads a certain code word into the lamp register (see Fig. 7.8), causing the illumination of the corresponding key of the function keyboard. The READ KEYBOARD REGISTER command reads the content of that register, which is a code word indicating which key has been pressed last. The meaning of the other commands is self-explanatory.

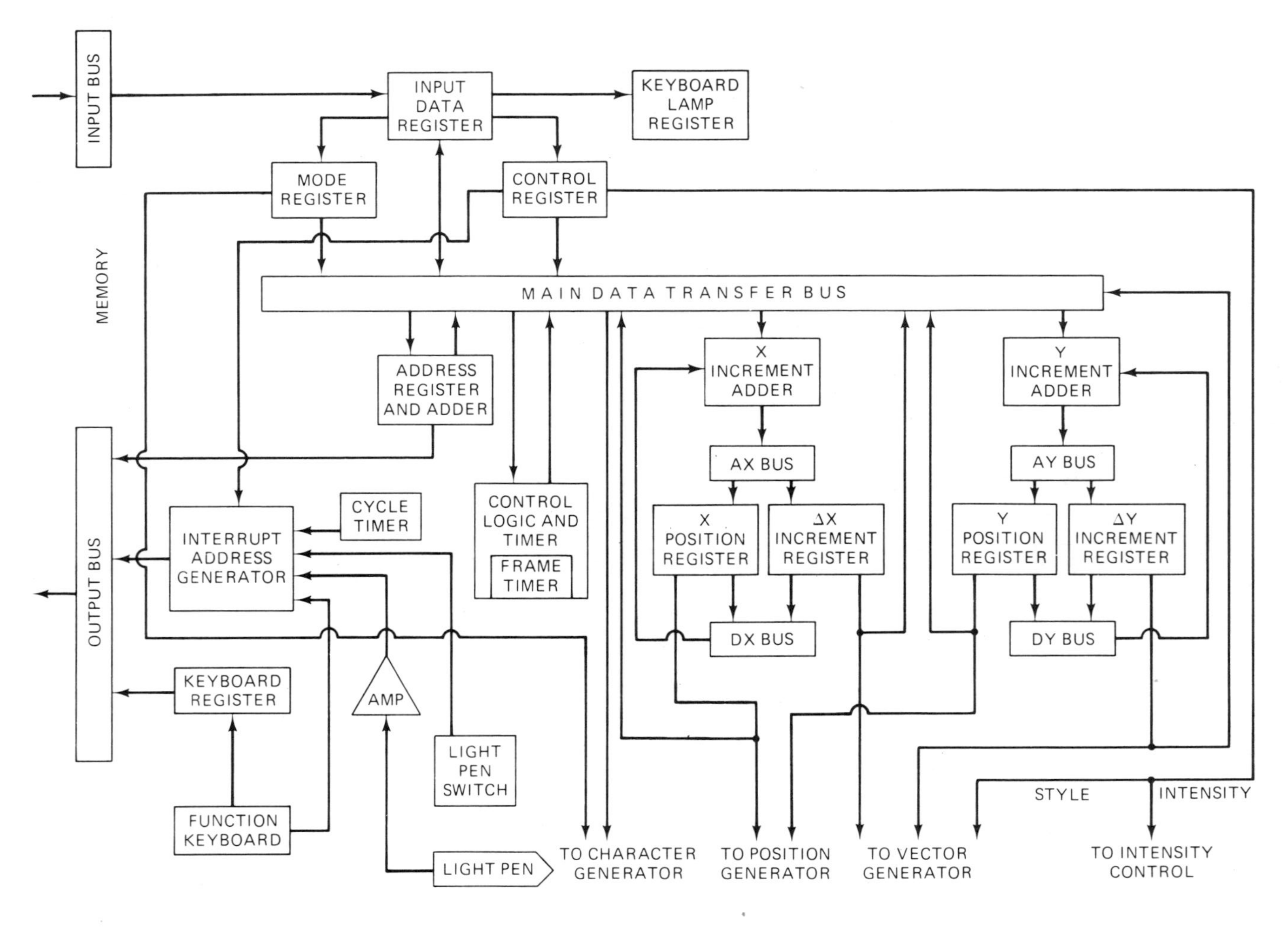

Figure 7.8 Block diagram of a display processor [74].

7.3.2 Instruction Formats

The typical word length of a display processor is 16 bits, thus conforming to the common byte format and matching the standard wordlength of contemporary minicomputers. With few exceptions the 16-bit format is sufficient to accommodate, in a single word, the display processor instructions which we encountered previously. The number of instructions for file organization and primitive generation is about 16 or more. This suggests a 4-bit operation code (OP-CODE), especially if a distinction is made between the GRAPHIC MODE and the CHARACTER MODE as discussed above. In a 16-bit word, this leaves 12 bits for data or addresses. A 12-bit address field allows the addressing of 4K memory cells —a value which may fall short of what is actually needed for the display file of a frame consisting of several thousand primitives. Moreover, no bits would be left to indicate whether an address should be absolute, relative, or indirect. A satisfactory solution to this problem is to let the memory referencing instructions—as we have seen, these are strictly the JUMP instructions—consist of two words, the first for the instruction code and the second for the jump address. This has the additional advantage that the first word provides enough space to specify the particular nature of the jump instruction (JUMP, JUMP AND STORE, SUBPICTURE JUMP, JUMP RETURN), so that only one of the 16 possible combinations of the 4-bit OP-CODE needs to be used (without such a measure a 4-bit OP-CODE might not be sufficient).

If we assume a screen domain given by $[-1023:+1023]\times[-1023:+1023]$, the 12-bit data field can accommodate exactly one component, ΔX or ΔY, plus a tag bit which specifies whether the value is an x- or a y-value. Therefore, for a long point, vector, or blank, two words are required. As a result, the drawing of a vector or of a dot must be deferred until both components have been loaded into the respective registers. This is accomplished by using first a LONG DEFERRED instruction (whose data field supplies ΔX), followed by a LONG VECTOR or LONG POINT instruction (whose data field supplies ΔY).

In the case of SHORT POINT, SHORT VECTOR, and SHORT BLANK instructions, on the other hand, both increments are packed into one single word. A 12-bit data field permits the allocation of 5 bits plus sign for an increment. Thus, the maximal increment in this case is 31 raster units.

The character font of any better display system encompasses the full ASCII set of 128 characters, requiring a 7-bit character coding. Hence, it is not possible to have more than the code word for one character accommodated in a 12-bit data field. As characters occur normally in strings, such a format would be wasteful. Two particular schemes are in use, which utilize the display file memory more economically.

In the first case, character strings are treated similar to subpictures; i.e., a WRITE TEXT instruction contains a reference to another part of the display file memory where the code for the character string (two characters per word) is stored. In the second case, a WRITE TEXT instruction is directly followed by a number

Table 7.1: INSTRUCTION FORMATS

(A) *JUMP Instruction Format*

bit positions	15	14	13	12	11	10	9	8	7	6	5	4	3	2	1	0
1st word									FRM	INI	SPI	REL	DEL			
2nd word	IND															

OP-CODE: JUMP
JUMP AND STORE
JUMP TO SUBPICTURE

Parameters:
- FRM = 1 means that execution is delayed until a frame time clock impulse occurs. Thus, the refresh rate can be synchronized to the frame time clock (60 Hz) or submultiples of it.
- INI = 1 means that an index register is incremented by 1 each time an instruction is executed.
- SPI = 1 means that the instruction is executed only if the index register is negative ("skip on positive index").
- REL = 1 means that the content of the following address word is relative to the content of the address register (REL = 0 means that the content of the address word is the effective address).
- DEL = 1 means that the instruction is immediately executed even if the execution of a preceding primitive-generating instruction by the display generator has not been completed. This provides a look-ahead feature (DEL = 0: no look-ahead).
- IND = 1 means that the following address is indirect.

(B) *JUMP RETURN, SKIP NEXT WORD, NO OPERATION, and INTERRUPT Format*

1	JUMP RETURN	FRM	INT	SPI		
1	SKIP NEXT WORD					
1	NO OPERATION					
1	INTERRUPT	FRM	INT	SPI	other parameters	TEMPLATE

Parameters as above, except for INTERRUPT, where more parameters may have to be specified for execution.

(C) *Mode Control Instruction Format*

1	OP-CODE	ABS	SCS	EDG	SIZE	COLOR

OP-CODE: ENTER GRAPHIC MODE or ENTER CHARACTER MODE

Parameters: ABS = 1 means that all coordinates are absolute (not incremental).
SCS = 1 means that the analog scissoring circuit is enabled.
EDG = 1 means that a crossing of an edge of the viewing area causes an interrupt.
SIZE specifies the character size in the CHARACTER MODE.
COLOR specifies the color in the case of a color display.

(D) *Primitive-Generating Instruction Format*

0	SHORT POINT	SGN	ΔX	SGN	ΔY
0	SHORT VECTOR	SGN	ΔX	SGN	ΔY
0	SHORT BLANK	SGN	ΔX	SGN	ΔY
0	LONG POINT	TAG	SGN	X DATUM or Y DATUM	
0	LONG VECTOR	TAG	SGN	X DATUM or Y DATUM	
0	LONG BLANK	TAG	SGN	X DATUM or Y DATUM	
0	LONG DEFERRED	TAG	SGN	X DATUM or Y DATUM	
0	CHARACTER CODE (CH = 1)			CHARACTER CODE (CH = 2)	

Parameters: SGN means sign; TAG specifies a datum as the X or Y value.

of data words which contain the character code. Therefore, the display processor will interpret all words following a WRITE TEXT instruction as packed character code until a special "escape" character is recognized. The first case is less advantageous than the second, especially if a jump instruction requires two words.

The approach that we discussed in the previous section is of the second type. The ENTER CHARACTER MODE instruction plays exactly the role of the

WRITE TEXT command. Instead of an escape character at the end of the string, the system is switched back into the GRAPHIC MODE, serving the same purpose. Table 7.1 gives a synopsis of the formats of the instructions discussed above. We recognize that the first bit in the character code data word is a tag bit which is 0. Thus, the occurrence of a new ENTER MODE instruction (which always terminates a sequence of character code words) is easily detected, as this instruction belongs to the group that start with a 1 in the first bit position.

7.3.3 The Controller

A (simplified) block diagram of a typical display controller is depicted in Figure 7.8. Basically, a display processor has a number of registers and a bus system that provides data paths between these registers and the "outside world." All 16 bit words entering the display processor are at first sorted in the input data register (IDR). At this point, the system performs a first interpretation of these data in order to find out which of the following types the data are:

1. Jump addresses.
2. Mode parameters.
3. Control parameters.
4. Beam position data.
5. Vector specification data.
6. Character code.
7. Function keyboard lamp indication.

As a result of this interpretation, the data are routed to the respective part of the system. The only parts of the system where arithmetic operations can be performed are in the x and y increment adders and the address register adder. Mode instructions are transferred to the *mode register*. The content of this register defines whether the system is in the graphic or the character mode. Entity initialization instructions are routed to their respective destination (control register, index register, X position register via X increment adder, Y position register via Y increment adder). The control bits, which are stored in the *control register*, are interpreted and respective control commands are sent to the interrupt address generator, the vector generator, and the intensity control circuit.

The *X position register* and the *Y position register* are used to store the present position of the CRT beam. The use of the POSITION X or POSITION Y instruction will cause its data content to be placed in the respective position register via the input data register, main data transfer bus, increment adder, and ΔX (or ΔY) bus. When drawing vectors or characters, the content of the position registers is applied to the position generator. The content of these registers may also be read back into the display file memory via the main data transfer bus and the output bus.

The AX and DX data buses and the AY and DY buses, respectively, provide the various data paths between increment adder, position register, *destination register*, and back to the increment adder. The increment adder must be opened or closed to load the position register either in an absolute or in a relative fashion and increment its content in order to obtain the destination values of a vector. Note that the increment adders are actually adder/subtractors which serve two purposes:

1. Conversion of incremental data into absolute data in the incremental mode.

2. Conversion of absolute data into incremental data in the absolute mode. The necessity of converting incremental data into absolute data, and vice versa, stems from the fact that the positioning of the beam at a starting point of a vector requires absolute coordinates, whereas the writing of the line connecting this starting point with the end point, as performed by the vector generator, requires the ΔX and ΔY increments of that vector.

The *keyboard lamp register* stores the information which causes the illumination of certain function keys; the *keyboard register* stores the code which identifies a pressed function key for transfer to the host computer.

The constant stream of instructions from the refresh memory into the display processor that must be maintained for picture regeneration is controlled by the *address register*, i.e., the present content of that register defines which word has to be fetched next. After each step, the content of the address register can be incremented by 1 ("instruction counter") or transferred to the display file memory via the output bus. A new setting can be obtained from the memory via the input bus. The *interrupt address generator* responds to interrupts by issuing an appropriate memory request.

7.3.4 Analog Scissoring

In order to detect edge crossings, the wordlength of the increment adders must exceed that of the data word by at least 1 bit. Alternatively, an analog scissoring circuit may be employed. Provisions of that type are necessary to avoid display errors which stem from the fact that, in the incremental mode, the increments may assume values close to the maximal dimension of the screen raster. For example, if the start point of a vector comes close to the right-hand edge of the raster, and if the X-increment is positive and of maximal length, the end point of that vector has an X-value which is almost twice the raster dimension. If the adder wordlength were only matched to the raster dimension of, let's say, 1024 raster units, the adder would perform an addition modulo 1024, and whenever a calculated destination would exceed one of the four edges of the raster area, it would "wrap around," i.e., reappear at the opposite side, as illustrated by the following picture.

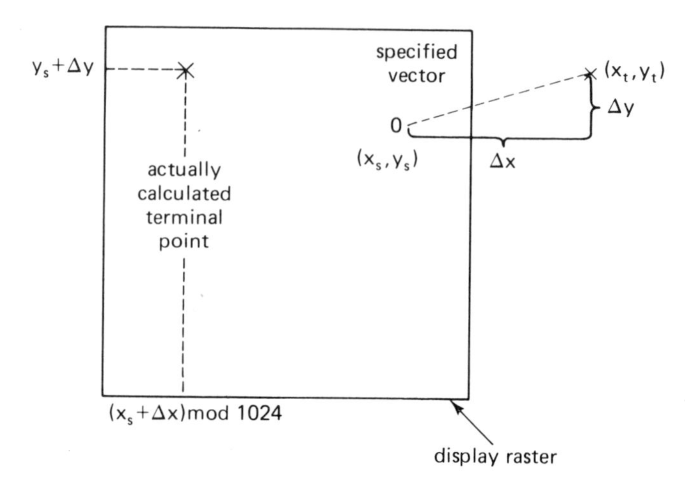

There is a simple remedy for this problem, which consists of providing a *drawing area* that has twice the dimension (or four times the area) of the raster area. That way the display raster will behave as if it were a window through which the larger drawing area can be seen (Fig. 7.9). To this end, the wordlength of the address registers in the controller and of the position generator is augmented by 1 bit (msb). Of course, this also means that it must provide twice the magnitude of the analog deflection voltage needed for deflecting the beam over the display raster proper. Whenever the beam is deflected beyond one of the boundaries of the raster area, it is switched off. Hence, only the parts of graphic objects lying within the raster area will be visible.

The following simple example shall illustrate this method, which is called *analog scissoring*. As depicted in Figure 7.9, two vectors, $\overline{P1P2}$ and $\overline{P2P3}$, are to be drawn. P2 lies outside the visible area (but within the drawing area). The deflection voltage is generated such that the beam would write the whole figure if it were not limited by the edges of the CRT screen. Actually, the beam is blanked out during the time it is traveling on the path $\overline{P2'P2}$ and $\overline{P2P2''}$. The reason for providing the deflection voltages for the whole drawing area is to let the beam leave the visible area at point P2′ and enter it again at point P2″-if we merely stalled the vector generation at point P2′, it would not reenter at the (different) point P2″. This method has the price of placing an additional burden of accuracy on the analog deflection system (or with a given linearity and repeatability it reduces the obtainable resolution by a factor of two). Furthermore, all the invisible parts of a drawing (the parts outside the window) must be actually generated, a process that consumes additional execution time.

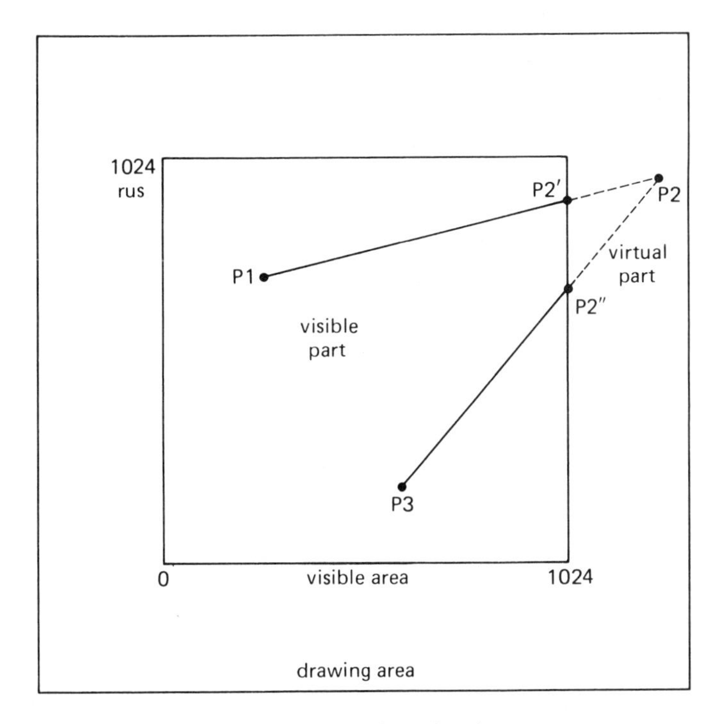

Figure 7.9 Analog scissoring.

7.4 HIGH-PERFORMANCE DISPLAYS

The display processor discussed so far is typical for a middle-of-the-road line-drawing display system. It was mentioned in Chapter 1 that display systems can be designed such that the constant stream of display processor instructions required for picture refresh is generated directly by executing the high-level language display program. It was pointed out that such an approach has two consequences:

1. The display processor must comprise hardware for picture transformations.

2. Programming is lowered to the "statement-for-each primitive" level; nevertheless, a multilevel picture structure can be introduced through the nesting of procedure calls.

The additional hardware for picture transformations consists typically of a "matrix multiplier" for performing the geometric transformations, combined with hardware for a three-dimensional clipping. A hardware stack may be provided for subpicturing or the composition of transformations [100]. The term "matrix multiplier" may be somewhat misleading. Actually, such a device typically comprises an array of registers and multipliers in connection with an accumulator. Hence, the multiplication of a coordinate vector with a row of the 3×3 matrix (for ordinary coordinates) or with a row of the 4×4 matrix (for homogeneous coordinates) can be executed in parallel. Some systems permit only the simple orthographic projection to avoid the division required for the more general perspective view (see Chapter 3). The coefficients of the transformation—which are primarily trigonometric functions of the angles of rotation—must be calculated by the host computer.

The first systems of that type used hybrid matrix multipliers given in the form of multiplying digital-to-analog converters. A multiplying digital-to-analog converter multiplies a digital value (the coefficient) with an analog value (the coordinate) and produces the result in analog form. Consequently, such a matrix multiplier is part of the display generator. Picture transformations are performed "on the fly"; i.e., their effect becomes visible on the screen, but the transformed data cannot be preserved for further use (e.g., for providing a hard-copy output or for additional processing of the obtained pictoral output).

With progress in semiconductor technology it became possible to replace the hybrid components by purely digital devices, thus permitting picture transformations at an earlier point in the processing chain from the display program to the display screen. Digital processing, with its higher accuracy and reliability, allows the transformed data to be stored back into memory, if their further use is intended.

Figure 7.10(a) shows the block diagram of a system with transformation hardware and buffered picture refresh (see Section 1.7), the PICTURE SYSTEM [26]. In this system, the "picture controller" is a general-purpose computer, typically a minicomputer. The picture controller contains the data base and the GPL program. The "picture processor" receives data sent by the picture controller in the form of two- or three-dimensional point coordinates. In either case, these points are expanded into 4-tuples of homogeneous coordinates, if necessary, by appending a dummy z-value. Simultaneously, relative coordinates are converted into absolute coordinates. Subsequently, the 4×4 matrix transformation is applied to the 4-tuples. Next, transformed points are windowed. In two dimensions, the window is a rectangular "viewbox"; in three dimensions it is a "viewing pyramid" (see Section 3.2). Subsequently, the transformed point is mapped into the two-dimensional screen coordinate system by the perspective transformation. The view port onto which the window is mapped may be considered as a three-dimensional space if a depth cue is introduced in the form of an intensity modulation, with intensity as the third coordinate. Transformed data can be read back into the picture controller memory, if wanted. A block diagram of the picture processor is depicted in Figure 7.10(b). A four-deep stack allows different transformation parameters to be stacked, thus providing for a fast switch between different transformations or the consecutive application of transformations on the same data, respectively. Special registers accommodate the parameters which define the window ("left," "right," "bottom," "top," "hither," "yonder").

As the "picture generator" can process data at a faster rate than the picture processor, a buffer memory called "picture system memory" is introduced to maximize the refresh rate or the number of elements in a picture, respectively. However, the buffer contains only data and control parameters; and no structural information is given that would allow the identification of graphic objects through a lightpen pick. On the other hand, the time link between program execution and object generation on the screen, which might also allow for an object identification, is severed because of the buffer. Therefore, a tablet is used in lieu of a lightpen.

At present, it is state-of-the-art to build with "off-the-shelf" standard components [47] multipliers which perform floating-point addition or multiplication in about 200 nanoseconds. A fast division can be performed in approximately 400 nanoseconds. In the near future, another increase in speed, maybe by as much as another order of magnitude, can be anticipated for such components. Therefore, very fast 4×4 matrix multipliers have become feasible in which a single high-speed multiplier/adder performs sequentially all required operations [131].

Unfortunately, there is not yet a minicomputer on the market which does contain such high-speed floating-point arithmetic. Fixed-point arithmetic, performed by a standard ALU (arithmetic and logical unit), is satisfactorily fast (60 nanoseconds for an addition). Picture transformations, however, require multiplications and, for perspective view, divisions. Therefore, it is at present not yet possible to use an off-the-shelf minicomputer as a display processor in a high-performance system. One possible solution is, as we have seen, to take an ordinary minicomputer and add special hardware (as is the "picture processor" in the previous example). Another possibility is to design an appropriate minicomputer that does have high-speed floating-point arithmetic.

Such a system is depicted in Figure 7.11. There is not the place to discuss all the details of this fairly sophisticated display processor. In a nutshell, the remarkable features of this system are:

- The display processor is microprogrammed (two-level quasi-horizontal or "diagonal" microprogramming [80]) and, with a writable control store option, microprogrammable by the user. Thus, the user may define and implement her/his own instruction set. Subpictures, which are very frequently used, may be generated as a microprogram. The manufacturer may supply various microprograms for different system configurations.

- The processor contains, in addition to a standard ALU, a high-speed floating-point multiplier/adder combination (MAD), destined primarily for use in picture transformations.

- A 256-word scratch pad is provided which may be used not only for storing the coefficients of transformation matrices, intermediate results, etc., but also as a stack (e.g., for subpicturing).

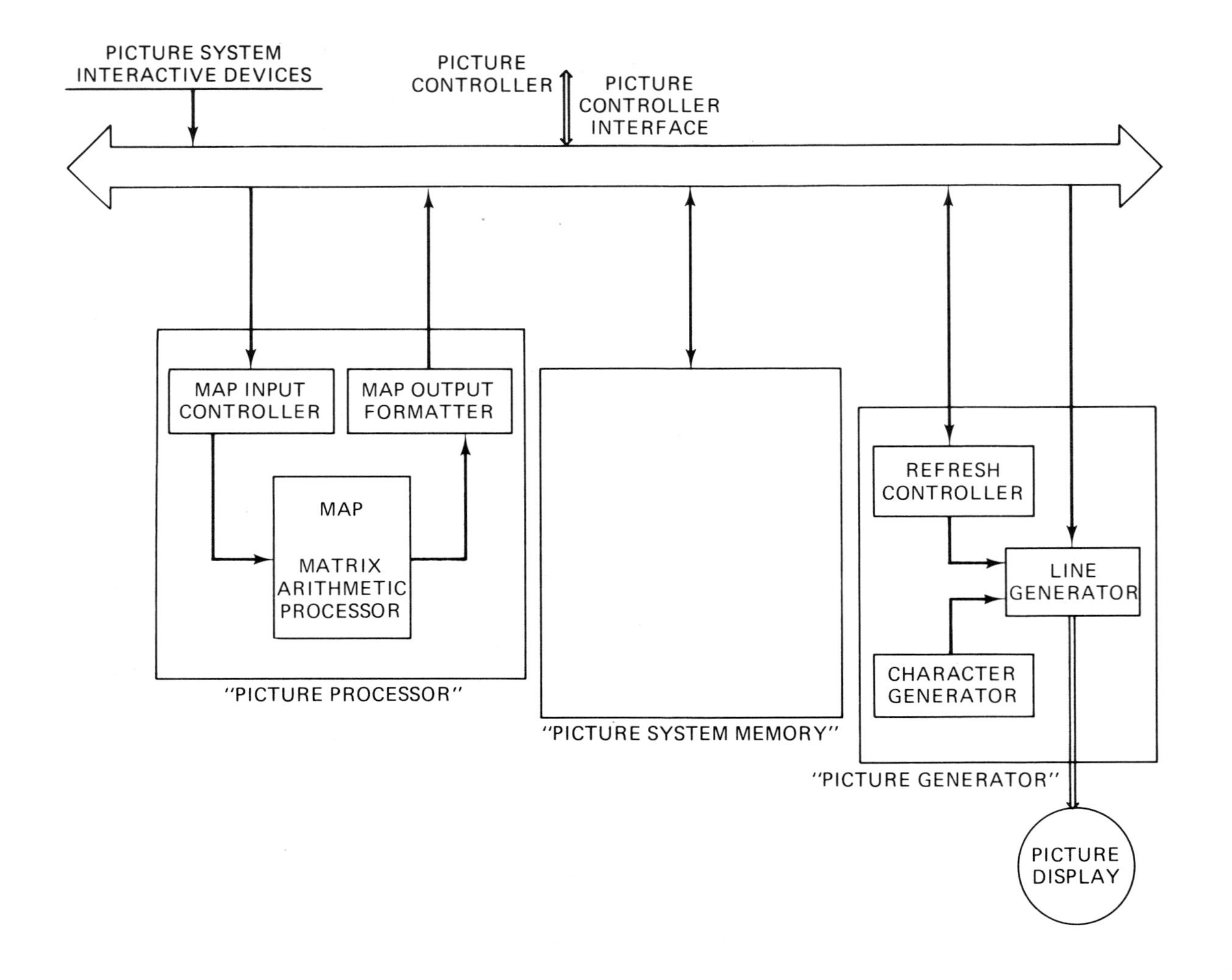

PICTURE SYSTEM INTERACTIVE DEVICES
PICTURE CONTROLLER
PICTURE CONTROLLER INTERFACE
MAP INPUT CONTROLLER
MAP OUTPUT FORMATTER
MAP
MATRIX ARITHMETIC PROCESSOR
"PICTURE PROCESSOR"
"PICTURE SYSTEM MEMORY"
REFRESH CONTROLLER
LINE GENERATOR
CHARACTER GENERATOR
"PICTURE GENERATOR"
PICTURE DISPLAY

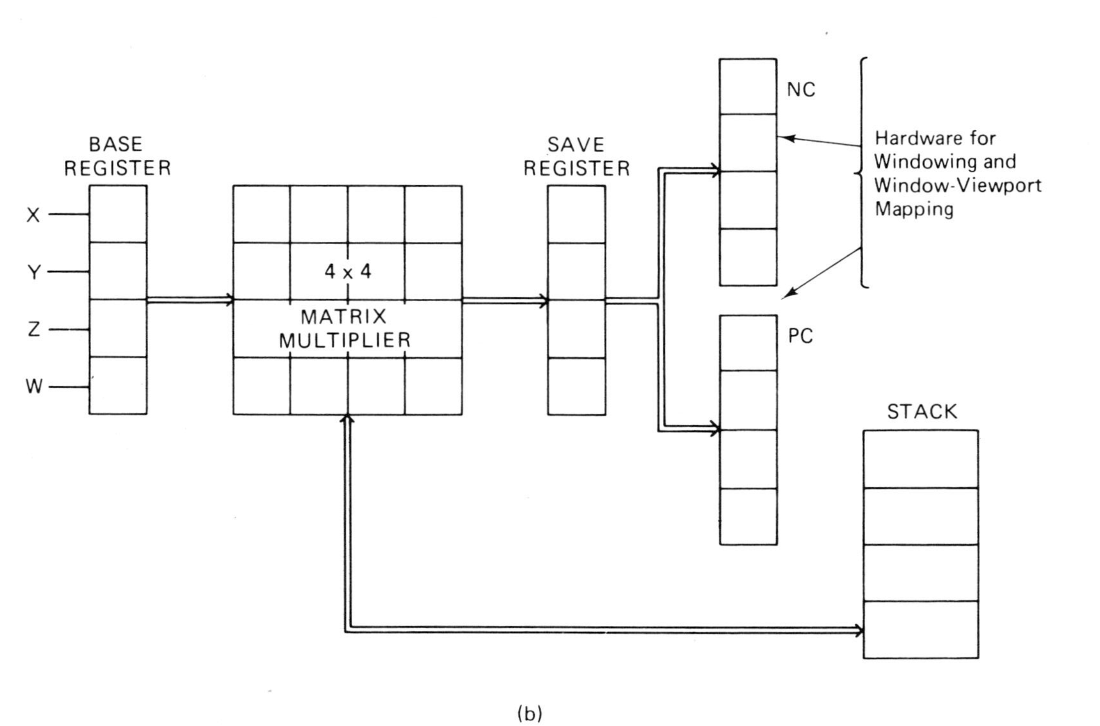

Figure 7.10 A and B (a) Block diagram of a high-performance system (The Picture System 2[37]); (b) block diagram of the picture processor (map).

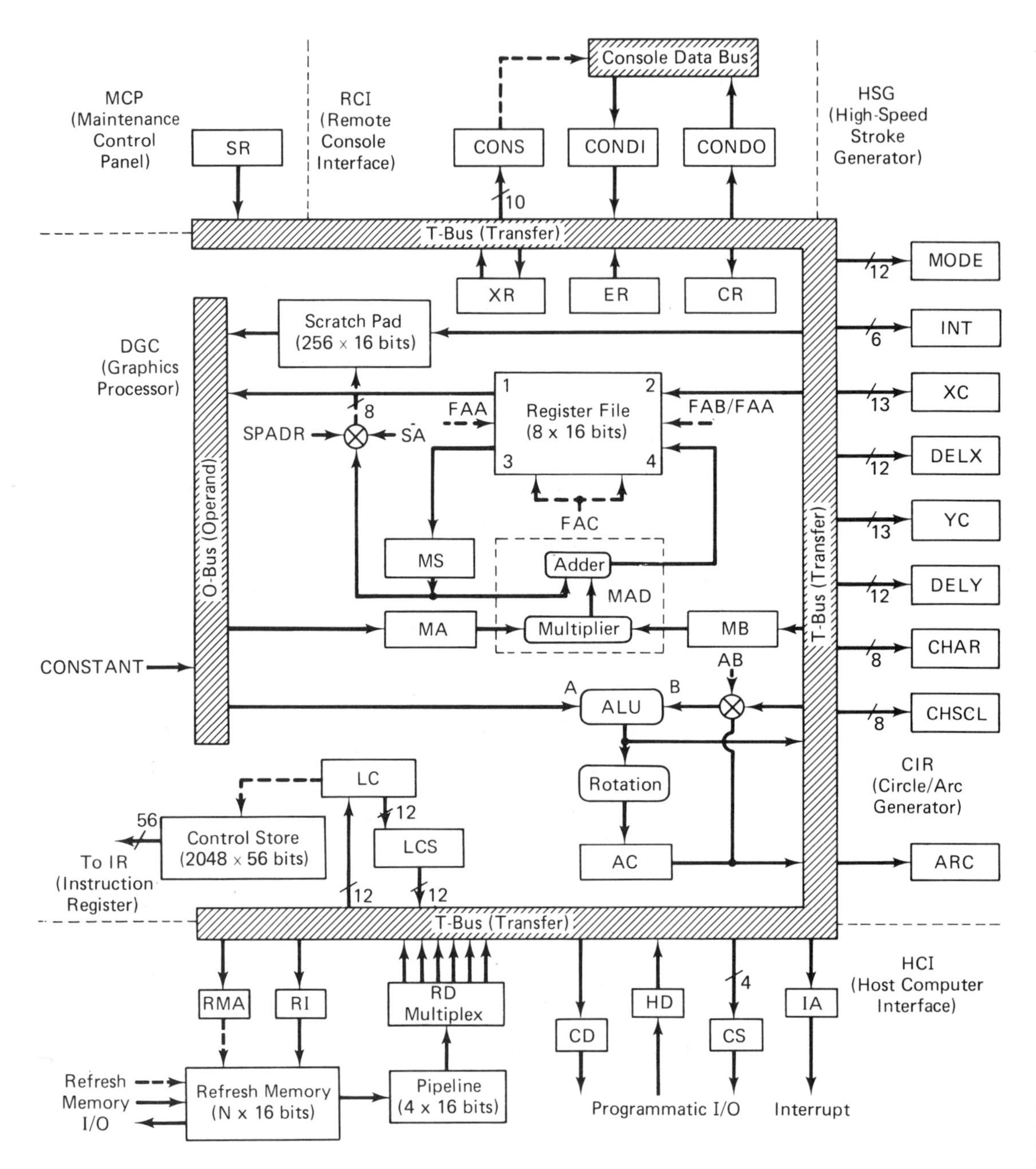

Figure 7.11 Block diagram of a modern microprogrammable display processor [80].

The block diagram is self-explanatory to everyone knowledgeable in computer architecture (MODE, INT, XC, DELX, YC, DELY, CHAR, CHSCL, and ARC are registers in the display generator).

Microprogrammable minicomputers with floating-point arithmetic, similar to the processor discussed above, will soon be on the market, allowing for the realization of high-performance display systems with off-the-shelf processors.

7.5 LOW-COST DISPLAY SYSTEMS

7.5.1 Storage Tube Displays and Plasma Displays

Storage tube displays are much less expensive than the vector display systems with picture regeneration that we have discussed so far. The cost reduction results from four main reasons:

1. As the storage tube screen itself has the capability of storing a picture for an arbitrary period of time, no picture refresh is necessary and thus no buffer memory is required.

2. Beam positioning and vector or character generation can be arbitrarily slow, since no time requirements must be met as in the case of regenerated picture displays (see Section 7.2). Hence, much less sophisticated—and thus much less costly—circuits for beam positioning and vector or character generation can be used. As already mentioned in Section 7.2, very inexpensive dot-matrix character generators can be applied when only one small character size is used.

3. Storage CRTs usually have a smaller screen size than the CRTs used in high-performance refreshed picture displays. Therefore, and as a result of the permitted low-speed operation, the deflection amplifiers are less expensive.

4. The display processor logic need not be fast and can thus be kept simple and inexpensive.

Storage tube displays are typically used as terminals of a remote computer. The communication link is usually a voice-grade telephone line, and the communication equipment may be either an acoustic coupler or a directly wired modem. Consequently, the data transmission rate may be 300 bauds (bits/sec) or less in the case of an acoustic coupler and 1200 to 9600 bauds in the case of a directly wired modem. A typical rating is 2400 bauds, as provided by the ubiquitous 201 type data set in connection with a voice-grade telephone line. If we assume in this case that the generation of a vector requires, on the average, 80 bits (five 16-bit words), the data transmission for the drawing of one single vector takes 33 ms. This explains why a vector generator that takes a drawing time of some milliseconds (this is a typical value in storage tube displays) is more than adequate. The lack of need for a picture regeneration as well as the fact that the recording of a picture may consume an arbitrary period of time leads to the consequence that no display file is needed. Any time a new picture is to be drawn, the picture definition as given in the data base (or in a particular picture file) is fetched, and a set of display procedures is employed to generate the appropriate display processor code and transmit it to the terminal.

A major disadvantage of storage tube displays lies in the fact that graphic entities displayed on the screen cannot be individually erased. Hence, a picture modification may require the very time-consuming reconstruction of the whole frame. This fact, the lack of a lightpen (see Chapter 6), and the reduced drawing capabilities (as compared to a high-performance display) is the trade-off for the much lower cost of such a system. Except for these different performance standards and the restricted interactive use, the display processor organization is principally the same in both cases. There are certainly many applications where the performance of the much more expensive refreshed picture display is not needed and the restrictions of the storage tube display are not significant. This holds true especially in time-sharing systems with a number of terminals used for the display of text and simpler graphics. On the other side, there exist other fields of application of computer graphics, mainly in the realm of computer-aided design, where the more powerful refreshed picture display systems are required with their capabilities of real-time person–machine interaction, of generating three-dimensional representations, of generating thousands of vectors in a high-resolution picture, or even of animating those pictures. In such cases the high cost of high-performance displays may amply be justified.

The plasma display consists of a very thin gas-filled chamber sealed between two parallel glass panes. A microscopically thin grid of orthogonal conductors is printed on the inner surfaces of the two panes, thus forming a matrix of small cells. A gas discharge can be fired or extinguished individually in each cell. Unlike a storage tube display, the plasma display thus allows the selective erasure of parts of a picture. Once a cell is fired, the gas discharge can be maintained for an arbitrary period of time. Hence, the plasma display has the ability to store a picture, and no refresh is needed. The typical resolution of a plasma display is presently 512×512, and a 1024×1024 raster seems to be the technological limit.

7.5.2 TV Raster Displays for Text and Simple Graphics

TV raster displays for alphanumeric text are currently at the very bottom of the cost range for computer display terminals. There are five main reasons why these systems can be marketed at a very low price:

1. As a standard TV raster is used for picture generation, any inexpensive standard TV set or—if one wants a higher quality—a studio-grade TV monitor can be used in the display console. These standard devices are mass produced and are, therefore, very inexpensive. No special deflection amplifiers are required.

2. The very inexpensive dot-matrix character generator available as an integrated circuit in MOS technology is ideally suited for the TV raster picture generation. Thus, the generation of alphanumeric characters and special symbols requires minimal expenses.

3. The currently available integrated MOS shift registers or MOS random-access memory devices provide an inexpensive way to construct memories at low cost.

4. All the symbols of a given representation on the screen are written in a fixed sequence, usually linewise. To give an example: the TV screen may be formatted into 32 lines with a maximum of 64 characters per line, totaling 2048 "character fields." Each character field is permanently associated with a memory cell in the picture-regeneration memory (i.e., the number of memory cells equals the number of character fields), and the character field address, as given by the line number (line index) and the character number in a line (column index), equals the memory cell address. In the course of the generation of a frame, the CRT beam moves consecutively through all fields (e.g., linewise) and writes a character whenever there is a nonblank character specified in the associated memory cell. Blanks on the screen are represented by the code for the blank character (that is, in the generation of a picture frame all 2048 characters are "generated," of which a certain percentage are blank characters). Because of the fixed sequence of character writing on the screen, no positioning data are stored in the regeneration memory. Thus, the capacity of this memory (and hence its cost) can be kept to a minimum.

5. The organization of the display processor, whose main task is to provide for the required editing functions, is simplified by the fixed-sequence character writing scheme.

Figure 7.12 shows a block diagram of a typical TV raster text display. The fixed sequence of character writing permits the use of a dynamic regeneration memory of the "delay line type," for which MOS shift registers proved to be quite adequate. However, random-access memory devices in MOS technology with a large storage capacity have recently become very inexpensive and, consequently, have begun to replace shift registers. Random-access memory devices have the advantage of requiring a less complicated organization of the display processor, and hence their slightly higher costs are offset by savings in the logic of the display processor.

Recent developments have successfully expanded the capabilities of such inexpensive TV raster display terminals for the display of graphical objects in addition to alphanumeric text. A first step in this direction is to equip the display system in addition to the alphanumeric character font with fonts of special symbols for various applications. These special symbols can be generated on the basis of the inexpensive dot-matrix principle [i.e., they require only the addition of some specially programmed MOS read-only memory (ROM) devices].

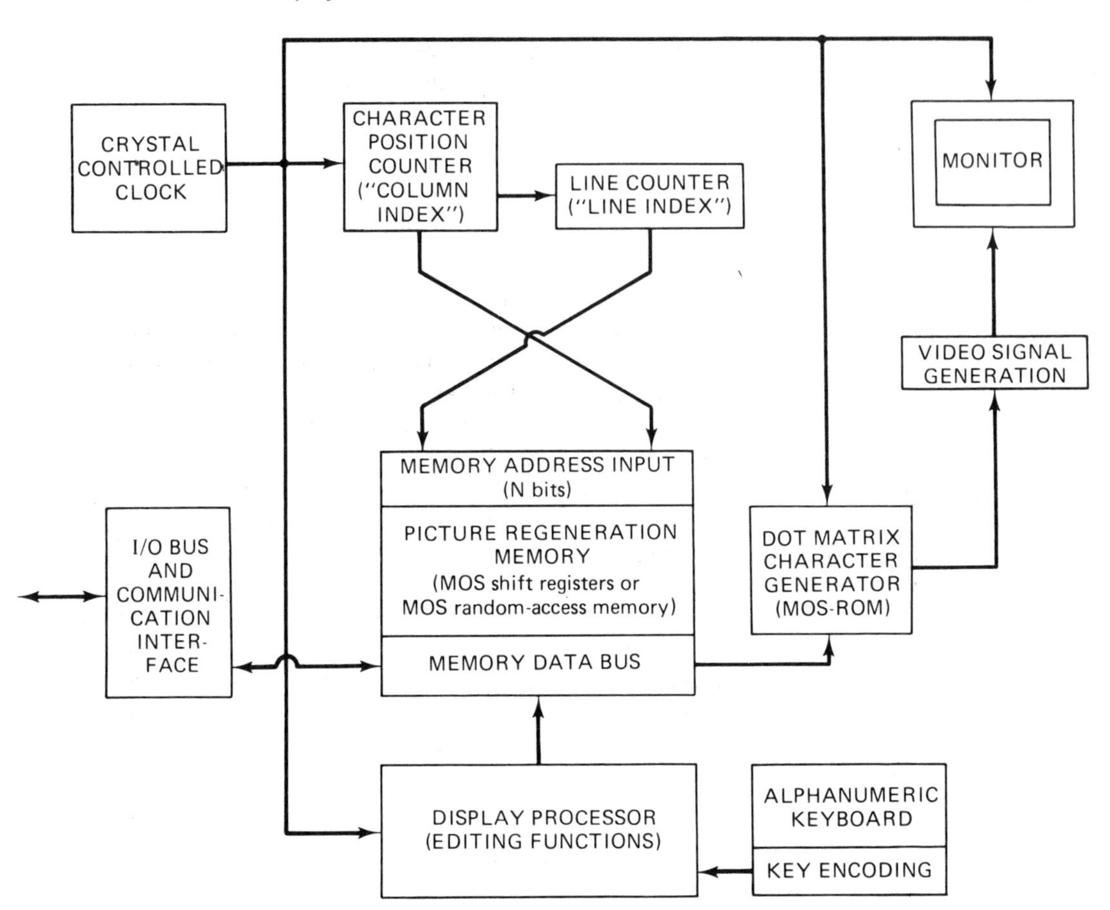

Figure 7.12 Block diagram of a TV raster edit display terminal.

The next step is to add line-drawing capabilities. The simplest way of doing that is to use a certain dot raster, say of 512×512 dots, and allocate a memory cell to each of these dots. In the simplest case, the memory cell capacity is 1 bit. This bit indicates whether the associated dot is bright or blank ("bit mapping" [53]). In our example, this requires a total memory capacity of $2^{18} = 256$K bits. If we want color, we need at least three bits per dot, which means to triple the figure above.

In the past, when the price of memory was the decisive factor, systems of that type were mostly based on a disk memory, with its considerable lower price per bit. The capacity of even a small disk allows the use of several bits for each raster dot in order to store gray levels or color components. The advantage of such a system is that solid surface patches can be as easily generated as lines and curves. The disadvantage is that the pictures must first be composed by the host computer, a procedure which may put a heavy burden on that computer. An even bigger disadvantage is that such a system cannot be connected to the computer via an inexpensive low-speed or medium-speed voice-grade communication line. Even for the simple bit mapping system with "only" 256K bits, a communication link using a standard 2400-baud (bit-per-second) modem would need approximately 110 seconds for the transmission of one picture. This unfavorable condition can be mitigated by transmitting addresses instead of the memory content, namely, the addresses of the cells that contain a logical 1. At least in line drawings, where the frequency of bright dots is relatively small compared to the total number of raster dots, a certain saving of transmission time may be accomplished by this measure. Nevertheless, transmission time remains a problem.

A much better approach than the bit mapping scheme is to use the "cell-organized" TV raster display. In this scheme, the whole raster area is divided into a number of cells (e.g., 64×64), as in the case of text displays. In addition to alphanumeric characters and special symbols, "elementary vectors" can be generated in each cell in order to construct line drawings. As in the case of pure text displays, a memory word is dedicated to each cell, in which the information for the object to be drawn in that cell is stored. These objects—characters or elementary vectors—are generated in a given dot raster. A raster of 8×8 dots per cell and of 64×64 cells for the usable screen area leads to a resolution of 512×512 raster units.

As in the case of the commonly used dot-matrix character generator, it is a favorable solution to use an inexpensive read-only memory for the generation of elementary vectors. Theoretically, this could be done by storing all segments that connect any of the $8 \times 8 = 64$ points of the cell raster with all other points. However, the large number of 1296 such segments (including the ones connecting themselves) prohibits this approach.

The solution to this problem [140] is to store only the 15 segments which connect the lower left corner with the 15 raster points on the right edge and the top edge, as depicted in Figure 7.13(A). Negative slopes are obtained by inverting the stored segments. These segments, plus a "no-segment" (blank), exist in a domain which we call the "vector field" and which can be translated relative to its cell. The decisive measure is now that only the part of the selected segment that lies in the intersection of the cell and the vector field is rendered visible [Fig. 7.13(B)]. This visible portion of the segment forms one of the elementary vectors out of which a picture is composed. Hence, by selecting the appropriate segment and by performing the appropriate translation, it is possible to connect any point on an edge of a cell with any arbitrary other point of the cell raster, regardless of whether the end points of the selected segment are in the cell.

The specification of an elementary vector requires three parameters: segment selection σ (4 bits plus sign) and translation parameters Δu and Δv (3 bits plus sign for each). The triples $(\sigma, \Delta u, \delta v)$ are internal parameters which can be calculated by a simple hardware circuit from the specified end points of an elementary vector. The translation mechansism can also be applied to characters, thus providing the

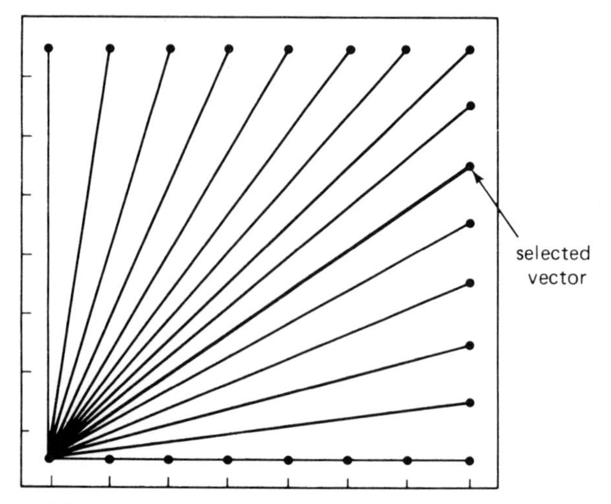

(a) Vector field with the set of 15 segments stored in a MOS-ROM. Negative slopes are obtained by inversion.

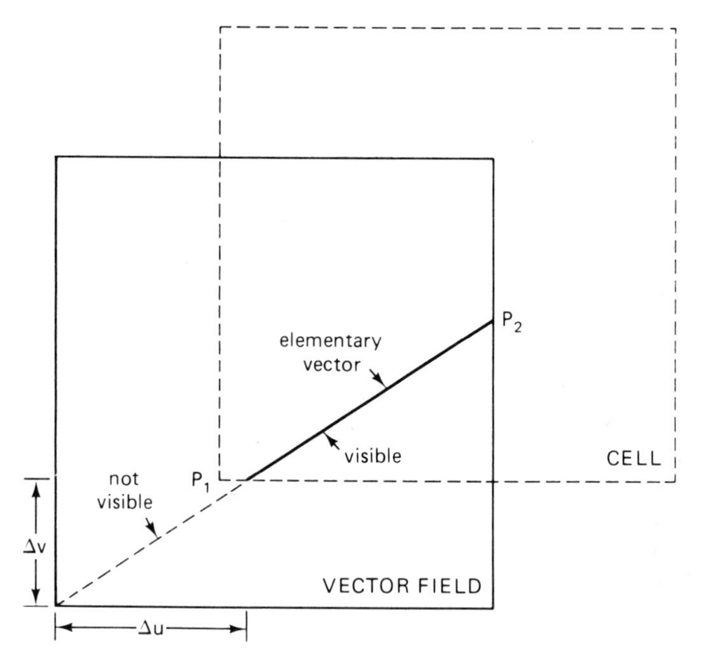

(b) A stroke has been selected and $(\Delta u, \Delta v)$ is chosen such that P_1 and P_2 are connected.

Figure 7.13 Principles of elementary vector generation.

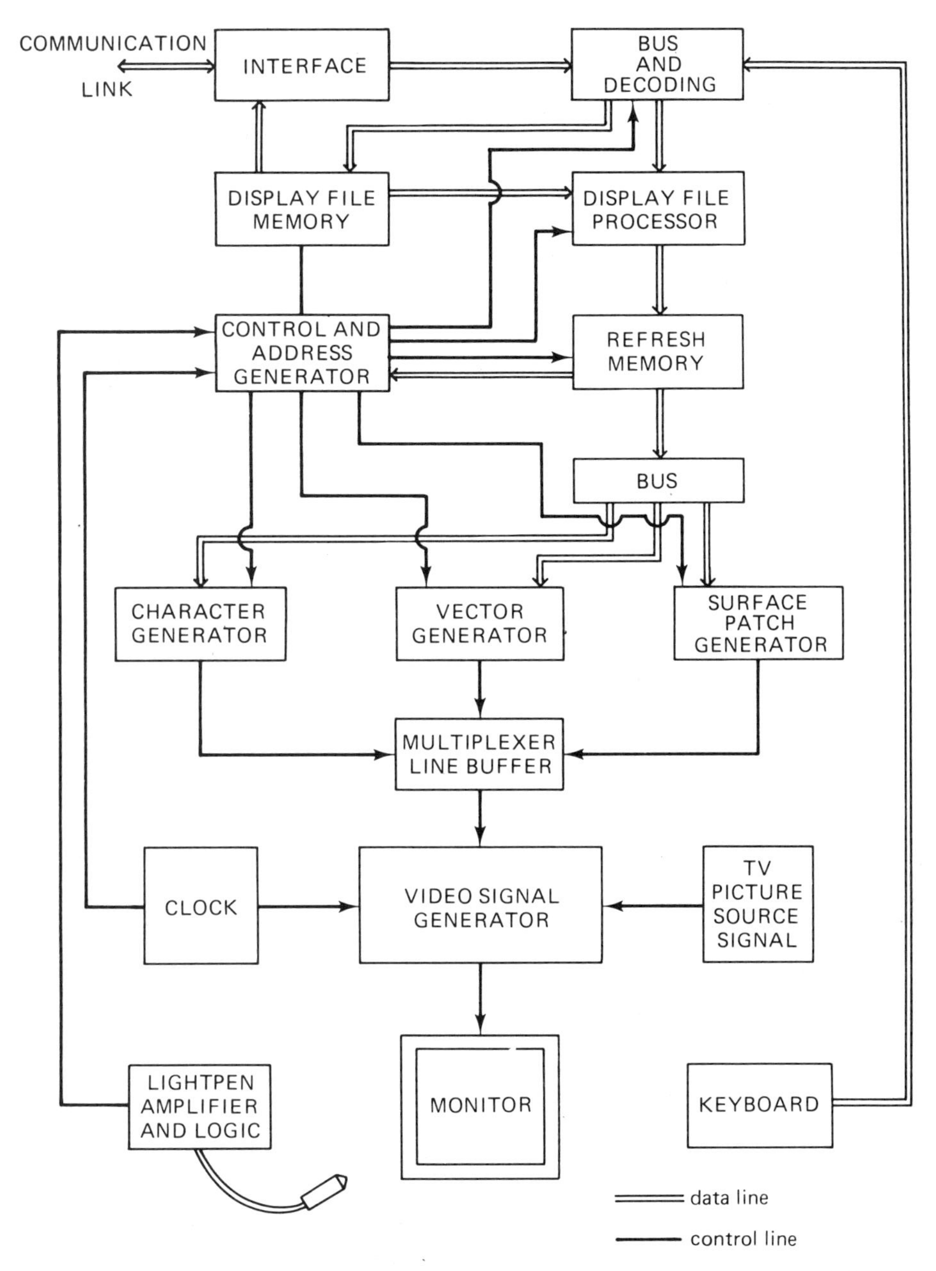

Figure 7.14 Block diagram of a TV raster display with (limited) graphics capabilities.

additional benefit of writing subscripts or superscripts. This is a very helpful feature in formula writing (e.g., in computer-aided instruction applications). Figure 7.14 depicts a block diagram of such a TV raster vector display.

The primary idiosyncrasy of this method is that, prior to their display, the given graphic objects must be decomposed into the appropriate sets of elementary vectors. This could be done by the host computer, but with the microprocessors now available it is more appropriate to have this carried out in the display terminal. In this case, we may have two memories in the display terminal, one for the customary display file and another for the picture refresh, the latter containing the end-point coordinates of an elementary vector or a character code word for each cell. To the host computer and to the programmer, such a system will not appear to be different from the directed beam vector display, except for a lower resolution. In contrast to storage tube displays, such a TV raster vector display can be interactively used without any limitation, including lightpen operation, etc. Furthermore, the display may use color, which adds a new dimension.

7.5.3 TV Raster Graphic Display Systems [54]

The advent of the microprocessor will lead in the foreseeable future to a new kind of low-cost graphics display. These displays will not be of the line-drawing variety, but picture generation will be based on the scan-line principle. So far, this principle has been used occasionally in line-drawing displays for generating gray shaded surfaces (see Chapter 5). However, it is more economic to use a TV monitor for this purpose rather than a line-drawing display, as a TV monitor (1) operates by nature on the basis of a scan-line raster, and (2) represents a very inexpensive and mature technology. However, one has to realize that these advantages may have to be payed for with a more limited resolution.

In television sets or monitors, the required bandwidth is halved by applying the *interlace* scheme; i.e., two subpictures with half the resolution are alternately generated, one consisting of the odd-numbered lines and the other consisting of the even-numbered lines. In the EIA standard used in the United States, the frame rate for each half-picture is 30 fps, resulting in 60 fps for the total picture. Furthermore, if the difference between the two subpictures is small enough, the half-pictures "fuse" in the viewer's perception into one picture that seemingly has twice the number of raster lines. This effect works well for photographic images, which, by nature, have a relatively small gray-level gradient, but it may not materialize in the case of high-contrast computer graphics pictures. Therefore, if a standard TV set or monitor is used as a graphics display, the interlace scheme tends to cause a rather intolerable flicker. There are two possible remedies for this problem:

1. The subpictures are made to be identical. This, in effect, eliminates the interlace, resulting in a genuine frame rate of 60 fps with half the resolution.

2. A picture tube with a phosphor of higher persistence is used, thus eliminating the flicker.

It is desirable to have a rectangular viewing area on the screen. In this case, only approximately 80% of the raster lines can actually be used in order not to have parts of a picture being cut off in the (rounded) corners of the picture tube. As a result, the obtainable resolution is further reduced. Table 7.2 lists some examples for the obtainable resolution of TV raster graphics displays. We recognize that the obtainable resolution is at least comparable with that of plasma displays and reaches in the best case, with special monitors, the resolution of high-performance displays.

Table 7.2 Obtainable Resolution in TV Raster Graphics Displays (approximate values)

Standard / Monitor	EIA (525 lines)	European (625 lines)	Industrial TV or high-resolution monitors
Standard, with interlace	200×500	256×512	400×1000
Special phosphor, no interlace	400×500	512×512	1024×1024 at 60 fps or 2048×2048 at 30 fps*

*This figure is based on a monitor that has 2500 lines at 30 fps and a bandwidth of 30 MHz. We realize that 30 MHz is not sufficient to distinguish 2500 points along a scan line (but only half that number). However, this is a similar problem as with the resolution of directed beam displays, where the obtainable minimal spot size may not really allow for 2048 distinguishable points either. Hence, it is safer to speak about "addressable points."

In the past, the dominating consideration in the design of display systems was to minimize the required capacity of the refresh memory. With the typical cost of memory having dropped to as little as 0.05 cent per bit, the postulate of memory minimization has become much less significant. It may now be more economic to spend more memory if this leads to a simpler logical structure of the display processor and, in particular, if the special-purpose, fast display processor can be replaced by general-purpose, "off-the-shelf" components. Such components have become available in the form of microprocessors.

One principle that helps to simplify a display processor organization is the separation of the refresh memory into a memory for characters and symbols and a memory for graphical elements. The symbol memory is a simple read-only memory (ROM) in the case of a fixed symbol set, and it is a random-access memory (RAM) if the symbol set is to be programmable. The data for graphic elements must be stored in a RAM. Graphic elements may be lines and gray shaded areas.

A block diagram of such a system is depicted in Figure 7.15. The core of the system is a microprocessor whose task is to interpret the display file and calculate the internal representation of graphic elements to be stored in the refresh memory.

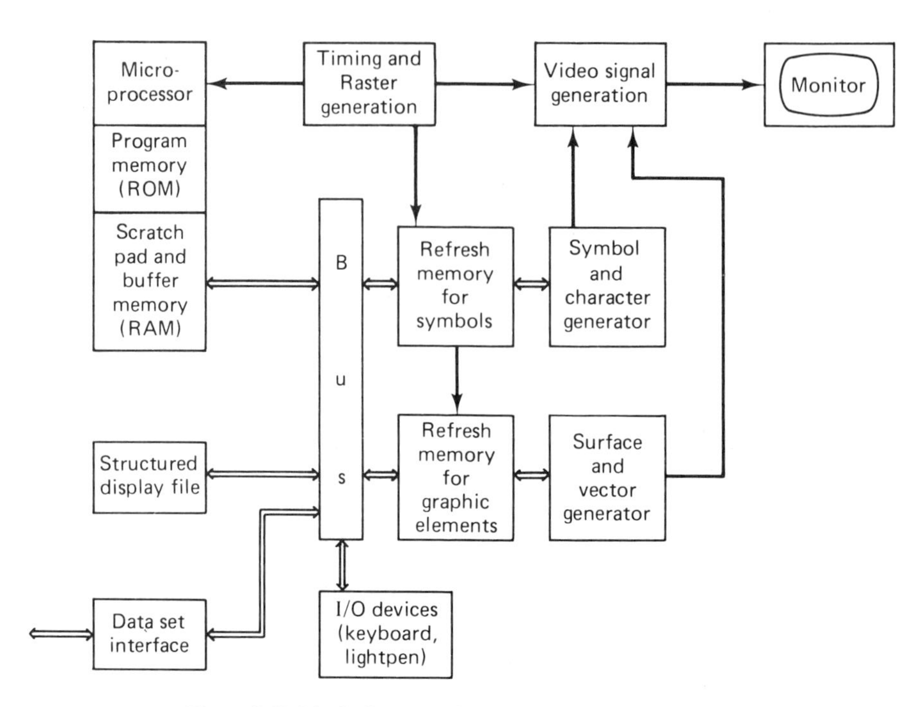

Figure 7.15 Block diagram of a TV raster graphics display.

No display file representation is generated for the character code of simple (nonidentifiable) output text strings, but this code is simply transferred to the symbol memory. Of course, all "pickable" objects must be adequately represented in the display file to allow for object identification. An outstanding feature of such a system is its modularity, providing for a maximum of flexibility. By programming the microprocessor that interprets the display file, the set of graphic primitives of the system can be extended or modified without any modification of the hardware. The display file interpretation may even be combined with picture transformations (e.g., scissoring). A fast execution of all these operations is guaranteed if a microprocessor of the *bit-slice* variety is employed.

The display file may be generated in the host computer and subsequently transferred to the display terminal, or the display terminal may be equipped such that it can process display instructions issued in a device-independent language (see Section 9.3). In the latter case, a second microprocessor may be added, to function as the (firmware-programmed) interpreter for the device-independent language. Since the execution time for program interpretation and display file generation is of secondary importance, the second microprocessor can be one of the less expensive, fully integrated types ("chip processor").

It may be of interest to have a closer look at the internal information structure of such a general-purpose TV raster display system, i.e., at the content of the display file and the refresh memories. Identifiable objects in the display file are: polygons, gray shaded (polygon-bounded) surfaces, light buttons, and special symbols (if the system comprises an appropriate symbol generator). Text strings not declared as light buttons are not identifiable. The character code for the text strings, light buttons, and special symbols is not stored in the display file but simply passed to the refresh memory for symbols. The general scheme for the definition of a graphic entity in the display file is to begin with an object definition, specifying the object type, its appearance (gray shade or color) and blink mode, as well as an object identifier (an integer number). Subsequently, the coordinates of the object origin are specified, either in terms of the character cell raster or the graphic raster, followed by a sequence of data specifications (graphic data or character code). The entity definition is delimited by an END statement. Test string definitions begin with a TEXT instruction which contains the appearance parameters but no identifier. Otherwise, the pattern is the same as for identifiable objects.

The refresh memory representation of graphic objects consists of a set of tuples, each tuple specifying the coordinates of the start point of a *sample span* (see Section 5.3.4), together with two parameters, a and b, which specify the appearance (gray shade or color) and the blink status of the respective line segment. A blank part of a scan line is also a sample span with gray level zero. For each scan line, the tuples indicating the start points of its sample spans are linearily ordered according to their x-coordinate value and grouped into a block (the end point of the last sample span is the end point of the scan line, any other end point is, as well, the start point of the following sample span). The blocks associated with the scan lines of the raster are, in turn, linearily ordered in the refresh memory according to the scan-line index. This has the disadvantage that, after insertion or deletion of a graphic object, the list of tuples must be reordered, requiring a sorting process. However, the time required for this is not so important, as the display may be halted during the sorting process. The decisive advantage, on the other hand, of the introduced ordering stems from the fact that in this case the refresh memory is simply sequentially scanned, and no search is required. Moreover, the y-coordinate values need not be stored in the tuples; i.e., it is sufficient to have a 4-tuple (x,z,a,b) for each start point of a sample span.

The sample span analysis and the generation of the ordered lists of 4-tuples, which necessitates among other operations the calculation of points of intersection between edges and scan lines (see the Watkins algorithm discussed in Section 5.3.4), is performed by the microprocessor (this is the main reason for the use of a fast microprocessor). The depth test which determines the visibility of overlapping scan-line segments, belonging to different objects is not performed by the microprocessor but "on the fly" by a hardware comparator which compares at the beginning of each sample span the z-values of the segments that coincide with the sample span (Section 5.3.4). This introduces a certain redundancy into the refresh memory but accelerates and simplifies the system organization (see the discussion

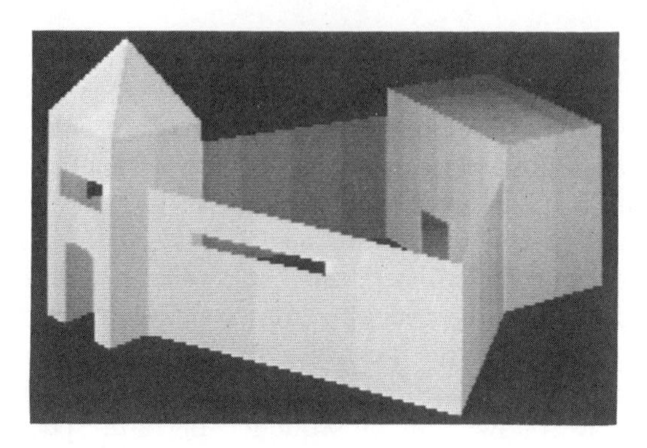

Figure 7.16 Example of a gray-shaded picture drawn by a low-cost TV raster display with a resolution of 256×256 points.

at the beginning of this section). Furthermore, in this case it is possible to erase one of the displayed surfaces by a lightpen pick and have immediately revealed what was hidden behind it. Thus, various layers of a picture can be displayed one after another. Figure 7.16 shows a picture generated on such a display [54].

EXERCISES

1. The purpose of a vector generator is to connect two given points, P_s and P_E, by a straight-line segment.

(a) Describe the vector connecting P_s and P_E in parametric form, i.e., as a function $v = v(u)$, $0 \leqslant u \leqslant 1$. Use the vector notation for the line equation.

(b) In a digital vector generator, the parameter u can assume only the discrete values $u_i \in [0:k]$. Consequently, the vector $v(u)$ is replaced by a sequence of discrete points $v_i = v(u_i) = P_S + i \cdot \Delta P/k$, $\Delta P = (P_E - P_S)$. How can the multiplication in the term $i \cdot \Delta P/k$ be avoided? Find a recursive equation that expresses a point as a function of its predecessor and the given parameters. How can the division in $\Delta P/k$ be avoided? Formulate an appropriate stipulation. What is the thus reduced set of possible values of k? What is, in general, the resulting number of computational steps in the recursion?

(c) In order to ensure equal beam intensity, all vectors must be drawn with the same velocity (constant-rate scheme). Since the vector generator is synchronized by an internal clock, independent of the vector length L [equation (7.6)], the stipulation of constant rate demands that the number of points of which the vector is composed be proportional to L. This sets a second condition for the proper choice of k. What is a satisfactory approximation for L, leading to the simplest possible calculation? A third condition for k stems from the stipulation that the generated "bead" of dots appears on the screen as a continuous line, i.e., that the spacing of the dots does not exceed the dot radius. Assume that the dot diameter is twice the raster unit. Formulate a condition for the value of k that satisfies all three stipulations with a minimum of computational steps.

(d) Flow-chart such a digital vector-generation algorithm.

(e) Design the block diagram of a hardware circuit that realizes this algorithm.

2. Some simpler display systems contain a vector generator that allows only the generation of short vectors (e.g., $\Delta X_{max} = \Delta Y_{max} = 31$ ru). Consequently, there are no LONG VECTOR instructions in the set of DPC instructions, but "long" vectors are composed out of an appropriate number of short vectors. Because of the lack of general-purpose computing capabilities of the display processor, the vector-generating GPL procedure must decompose a "long" vector into the appropriate sequence of short vectors and generate the respective instruction sequence. Flow-chart such a procedure.

3. Assume a TV raster graphics display system as discussed in Section 7.5.3. Assume that the screen raster comprises 64×64 cell addresses for characters and special symbols and 1024×1024 point addresses for graphic objects such as polygons (vectors) and shaded, polygon-bounded surfaces. Assume that the display processor instruction set encompasses eight instructions which are encoded by a 3-bit op code. Furthermore, assume that the instructions are 16 bits long (the display processor is a microprocessor). The instructions are

TEXT	ap, bm, cwm
ORIGIN	xc, yc
DATA	dt, dsp
BUTTON	ap, bm, bl, bid
GRAPHIC	ap, bm, et, gid
END	
JUMP	cond
DEST	addr.

The meaning of the parameters is

ap: *appearance*, alternatively eight gray levels or eight colors (including "blank")

bm: *blink mode*

cwm: *character write mode*, either normal (bright on dark) or inverted

xc, yc: *cell coordinates* of the 64×64 character-cell raster

dt: *data type*, either character data or graphical data
dsp: *data specification*: sk, tb, sh, cc for character data or
cd, coord for graphical data
sk: *skip* instruction for cursor
tb: *tabulator* mark for cursor
sh: vertical *shift* of characters or symbols, ranges from -7 to $+7$ ru
cc: *character code* for one character or special symbol
cd: *coordinate direction*, distinguishes x-value and y-value
coord: *coordinate value*
bl: *button length*, the number of characters in the button must be $\leqslant 7$
bid: *button identifier*, integer, $0 \leqslant \text{bid} \leqslant 31$
et: *element type*, either polygon (line drawing) or surface
gid: *graphic object identifier*, integer, $0 \leqslant \text{gid} \leqslant 127$
cond: *jump condition*
addr: *jump destination*

The parameterless instruction END delimits a sequence of DATA instructions which follow a TEXT, BUTTON, or GRAPHIC instruction. ORIGIN specifies the origin of a text string or a button in terms of cell coordinates. A special symbol (or string of special symbols) is either a TEXT (not pickable) or a BUTTON (pickable).

(a) Design the detailed formats of the instructions listed above.
(b) Indicate the instruction sequences that must be written for (1) a text string, (2) a light button, (3) a special symbol, (4) a polygon, (5) a polygon-bounded shaded surface.
(c) Develop flow charts for the procedures that are required for identification of buttons and graphic objects.

4. Usually, graphical data are directly accommodated in the DPC instructions. The alternative would be to reference in the instructions memory addresses and store the data in the referenced memory cells. Estimate the additional memory space which the latter scheme would in the average require (in percentage points). What would be the advantage of such a scheme?

5. Compare (1) a directed beam display, (2) a storage tube display, (3) a plasma display, and (4) a TV raster display with respect to the typical resolution and the typical refresh rate. What determines the maximally obtainable resolution in each case?

6. Is it possible to subject characters which are generated by a hardware character generator (a) to a rotation, (b) to a scaling, (c) to a translation?

7. Find an estimate for the maximal permissible average execution time of the arithmetic operations ADD/SUBTRACT/MULTIPLY/DIVIDE, if the 4×4 matrix operation is to be applied on a picture during continuing picture refresh.

The frame rate shall be 30 fps, and the picture definition may comprise the (homogeneous) coordinates of as may as 1000 points. Assume that the matrix components are precalculated and stored in memory.

8. Given a "scene" of plane-faced solids in 3-space. Each face is polygon-bounded. A perspective view of the scene shall be obtained by applying the simple orthographic projection. We assume that the object space is $\mathbf{N} \times \mathbf{N} \times \mathbf{N}$ and the projection space is $\mathbf{N} \times \mathbf{N}$, if $\mathbf{N}$ is the set of nonnegative integers.

(a) For the sake of simplicity, the projection of an object need not be identifiable. What is the data structure of an object, and what is the data structure of its projection? Design a procedure that performs the orthographic projection. Note that ordinary (not homogeneous) coordinates are used.

(b) Assume that the display system allows the generation of "halftone" pictures (i.e., the beam intensity is controllable over a wide range) proportionally to an intensity control variable v. This feature is exploited for the recording of gray-shaded surfaces in the following manner. A raster of 512 horizontal scan lines is generated. Over the time period during which a scan line intersects with a surface, the beam intensity is set to a constant gray level that is proportional to the average of the z-coordinates of all vertices of the bounding polygon. Therefore, two operations must be performed: (1) the average gray shade of a surface must be calculated, and (2) the intersections between the edges of the surfaces and the scan lines must be calculated ("segment analysis," see Section 5.3.4). Flow-chart the procedures for the two operations.

(c) Design an appropriate organization of the picture refresh file for each of the two alternatives: (1) the tuples of values associated with the beginning of each segment are ordered in the file exactly in the sequence in which they are to be processed; or (2) the tuples are stored in the file at random.

(d) Assume a picture refresh file organization according to (c1). Assume that during the processing of a tuple its successor tuple is already fetched and buffered by the display processor. The time for data transfer and processing of a tuple will be negligible in comparison to the memory cycle. Of how many scalar values must a tuple consist? If the memory cycle is 0.64 μ second and the drawing of a scan line takes 64 μ second, what is the maximal resolution of the rendering of shaded surfaces (measured in ru)? Would a "software solution" be possible for the real-time rendering of shaded surfaces, provided that the refresh memory were organized according to (c2)?

(e) The hidden-surface problem is to be solved as follows. If several surfaces overlap, the one with the highest gray shade determines in each instant the gray level along the scan lines. Does such a procedure also handle the problem of mutual covering of different surfaces? Flow-chart the algorithm by which the succession of gray-shade values along a scan line is calculated. Assume that the projection space representations of the objects are stored in a display file. How could this scheme be modified such that the gray shade of a surface is not constant but a linear interpolation of the gray values at its two boundary points? What undesirable effect could be avoided by such a measure? What price is to pay for it in terms of increased complexity?

8

Display File and Picture File Organization

8.1 DATA BASE AND DISPLAY FILE REVISITED

Let us return to the discussion of the roles of data base and display file in a computer graphics application system (Sections 1.5 and 1.6), as indicated by the general block diagram Fig. 1.7.

The considerations of the preceding chapters should have made clear that there exist no such thing as *the* data base organization for computer graphics. The fact alone that a data base, among other information, may contain graphic data is not specific enough to suggest a particular form of file management. Rather, the major determining factors stem from the particularities of the computer-aided design applications in which computer graphics is used as a tool. Certainly, hierarchical, relational, or set-theoretical data models could be applied as most general and powerful devices, the more so as such models allows the implementation of all given data structures. However, it would be wrong to advertise a data base management system that is based on one of those models as a standard system for computer graphics, as the price for the power and flexibility of such a universal system may be too high in many applications where a simpler data base organization is perfectly adequate. On the other hand, we are in a good position to propose a model solution [55] for a display file organization, as a display file structure is implied by the underlying picture structure. With the picture structure introduced in Chapter 3, we must represent in the display file a data structure consisting of a forest of three-level trees. From the functional point of view, two different parts of the display file can be distinguished. The first part consists of a list of all display processor instructions required for the generation of the actual *frame* (totality of all objects displayed on the screen). The second part provides for the structuring of the DPC list into code segments representing "segments," "items," and "primitives."

In order to justify our display file organization, let us first consider the order of magnitude of the number of segments, items, and primitives. With the presently given speed of the display generators (see Section 7.2), several thousand primitives can be contained in a frame at the minimal frame rate required for a flicker-free image. Consequently, the number of items may be in the hundreds, and the number of segments may be in the tens. The scanning and processing of the display processor program should be performed at maximum speed, as this determines the refresh rate. On the other hand, a search for an item name and/or a segment name is carried out while the display is interrupted. In this case, some milliseconds more or less will not count. Therefore, we can afford to perform a sequential search, the more so as the lists to be searched can be kept very small by a two-stage process: In the first step, the name list is searched for the segment name, and subsequently only the found segment block need be searched for the item name. Of course, this requires segment names and item names to be tagged accordingly. The implementation of this simple but very efficient scheme will be discussed in more detail.

8.2 DISPLAY FILE WITHOUT SUBPICTURE CALLS

As a result of the considerations above, we devise a display file in the form of a sequential list accessed via a directory (see Section 2.4). This list contains the DPC program, that is, the sequence of display processor instructions. Therefore, we call it the display processor code list (DPCL). DPCL is a one-dimensional array of dimension DPCL(L), if L is the total number of DPC instructions. The directory is a matrix with two columns, one containing pointers and the other containing data. We call it the *correlation table*, CT. Its dimension is CT(K, 2), with K = [number of segments] + [number of items] + [number of symbols (subpictures)]. Hence, CT is a two-dimensional array that has two entries for each such graphic entity. CT is linked to a name table NT of dimension NT(K) that contains the names of segments, symbols, and items. An entity name and its associated entry in the correlation table may be linked by having the same index in NL and CT. NT and CT could be combined into one array with dimension (K, n + 2), if n words are reserved for a name. The reason why we separate NL and CT is twofold. First, the data type in both lists may be different. CT contains solely integers, whereas NL may contain character strings (if mnemonic names can be used as identifiers rather than integers). Second, the name table need not be managed by the display processor but can be kept in the host computer. Hence, the internal identifier for an entity residing in the display processor may be the respective record number (row number) in CT.

When the name table NT is constructed (by executing a segment-generating program block), first the segment name and subsequently the names of all items belonging to that segment are entered. Primitives have no individual name but are identified by an index number. The order in which items are listed in a segment and primitives are listed in an item is the order in which they were created by the high-level program. This is also the order in which they are drawn by the CRT beam.

The two entries that we find for each entity in CT are in the case of segments and symbols two pointers to the display processor code list, delimiting the block of code for that segment or symbol. We call these pointers the start pointer SSP and the end pointer SEP. Symbol blocks are distinguished from segment blocks by the fact that the start pointer of a segment is entered as a positive integer, whereas the start pointer of a symbol is entered as a negative integer. The first of the two entries associated with each item is a start pointer to the code block in DPCL that represents the item. As item blocks are sequentially stored in the segment block they are part of, the start pointer of an item is also end pointer of its predecessor in the segment and, therefore, we can use the second cell in each item record for storing additional information about the item, given in the form of an item type code. The item type code consists of a negative integer: e.g., $-1=$lines; $-2=$ dots; $-3=$circles; $-4=$characters.

Such a scheme exhibits several advantages, such as:

1. Segments and symbols can be stored in an arbitrary order.

2. There need not necessarily be a fixed correspondence between the number of primitives in an an item and the number of display processor instructions for the generation of the item (such a fixed correspondence is not given if, in the DPC program, one-word instructions are used for short vectors and two-word instructions are used for long vectors.† (See Chapter 7.)

3. The search for a segment identifier as discussed in the following is facilitated, as a block in CT belonging to a segment is the only one containing two positive integers.

Figure 8.1 gives an example of a segment, named FILTER, which consists of 17 items: R1, R2, R3, C1, C2, L1, PS1, TS1, TS2, TS3, TS4, TS5, TS6, TS7, TS8, TS9, TS10. R1, R2, R3, C1, C2 are items of type line, L1 is an item that has only one primitive of type line, and PS1 is of type dot. TS1, TS2, TS3, TS4, TS5, TS6, TS7, TS8, TS9, and TS10 are text strings. The resulting display file layout is depicted in Figure 8.2.

We have to show that this scheme allows the identification of segments, items, and primitives on occurrence of a lightpen pick. Therefore, we consider briefly the fashion in which the display processor performs a picture refresh cycle. During picture refresh, the display processor code list is sequentially scanned from top to bottom, resulting in a stream of instructions flowing into the display processor and being there immediately executed. The scan is performed by a simple instruction

†However, it is advisable to restrict the use of "short vectors" strictly to symbols. This simplifies the code generation and facilitates the identification of primitives.

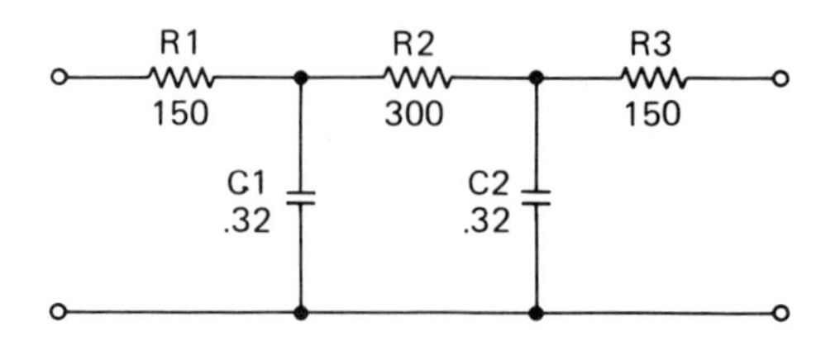

Figure 8.1 Picture of an *RC* filter.

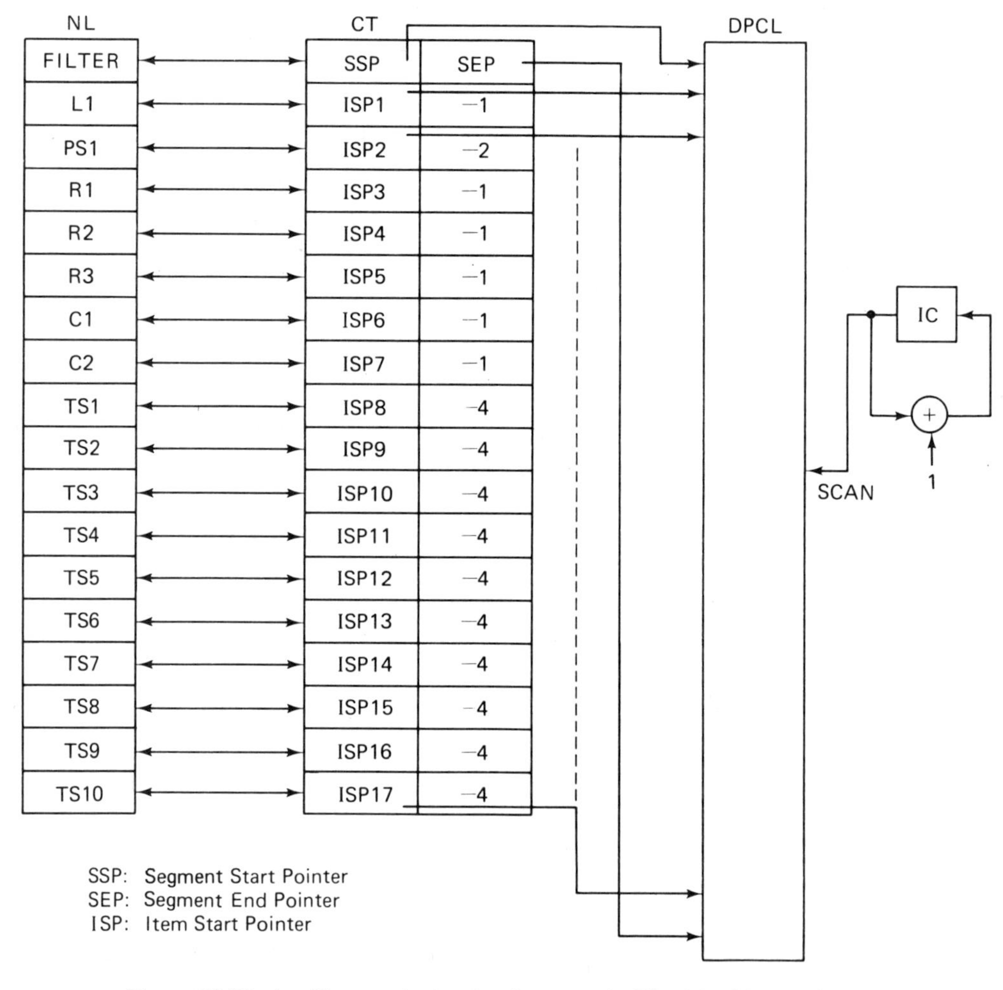

Figure 8.2 Display file organization for the example, Fig. 8.1 without subpicture calls.

counter, IC, which is incremented by 1 each time an instruction has been read from DPCL. At the end of the DPCL, a jump instruction causes the reset of IC to the beginning of the DPCL.

Let us assume that an interrupt occurs, caused by a lightpen pick. At this instant IC is "frozen," and its current content points to the successor of the cell from which the instruction has been read that caused the generation of the picked primitive. The first task is now to identify the segment to which this primitive belongs. Therefore, the (SSP, SEP) pairs of all segments are looked up in CT until the pair is found which bounds the given content of IC. We mentioned already that these (SSP, SEP) pairs can easily be recognized by testing the sign of SEP. Once we have found a pair (SSP, SEP) with $SSP \leqslant CONTENT(IC) \leqslant SEP$, the segment is known, identified by the index of the found record in CT. In case the user also wants the item to be identified, the system must now compare the content of IC with the ISPs of the items. If the condition $ISP_i \leqslant CONTENT(IC) < ISP_{i+1}$ is true, then the primitive belongs to the ith item. The ordinal number of a picked primitive cannot be found by a simple comparison, as there is normally no fixed correspondence between the number of primitives in an item and the number of instructions in the code segment for this item. One reason for this is that short vectors may take only one DPC instruction, whereas long vectors take two instructions; another reason is that the generation of a text string takes a variable number of instructions depending on the length of the string. Therefore, a primitive identification in such a case requires the analysis of the item-generating DPC segment.

8.3 DISPLAY FILE WITH SUBPICTURE CALLS

The display file organization as delineated above can easily be modified to allow symbol (subpicture) calls. Therefore, let us return to the example of Figure 8.1, this time with the proviso that the resistors and capacitors are stored in the display file as symbols, named R and C.

Symbols differ from segments in the respect that their corresponding display processor code segment does not contain any initialization instructions. We therefore call a symbol code segment the *symbol body*. The instructions for the initialization of a symbol form a separate code segment called the *symbol head*. A symbol body may be linked with several heads, leading to several *instances* of the symbol. While the symbol body is stored under the symbol name, the heads are listed under the name of the respective invoking segments. Figure 8.3 illustrates such an organization for the example of Figure 8.1.

In order to handle subpicture jumps, the display processor has been augmented by a special register RR that is used for storing the return address of a subpicture jump. If a JUMP TO SUBPICTURE instruction is encountered, the current content of IC is saved into RR and, subsequently, the jump destination address is loaded into IC, causing the DPCL scan to continue at this new location.

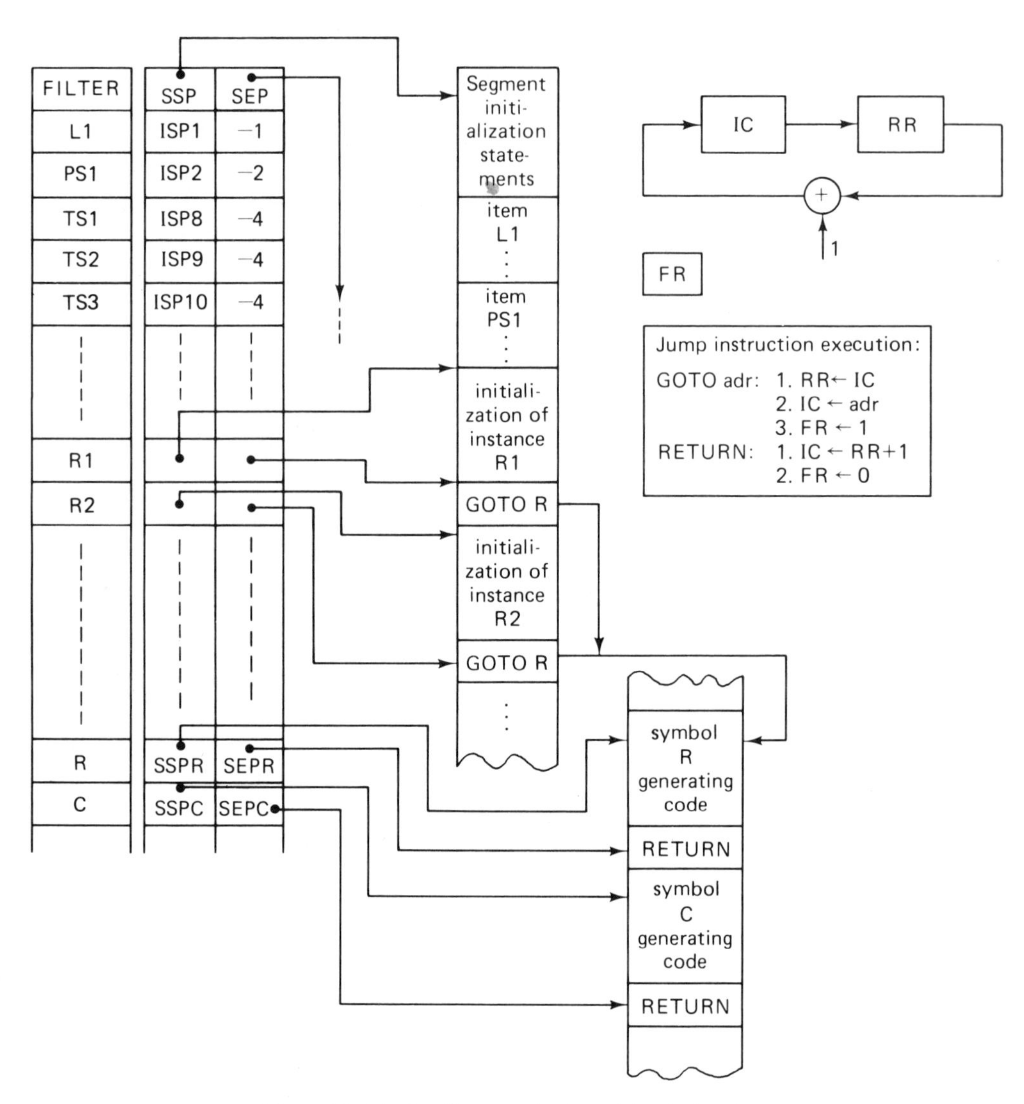

Figure 8.3 Display file organization with subpicturing.

Simultaneously, a flag register FR is set, indicating that the processor is now in the *subpicture mode*. The last instruction in the symbol body is a RETURN statement, causing the display processor to transfer the content of RR back to IC and reset FR subsequently. Thus the scan is resumed at the point following the point where the segment code sequence was left, and the system is back in the *picture mode*, as indicated by FR.

The flag in FR is needed for object identification after a lightpen pick. If, on occurrence of a lightpen interrupt, the system is in the picture mode, identification is carried out as described above. If the system is in the subpicture mode, the procedure is not changed, but it is now the content of RR instead of the content of IC that is compared to the pointers stored in the correlation table. Since RR contains the point of return in the invoking segment, it is the invoking segment rather than the invoked symbol that will thus be identified.

The possibility of nested symbol invocations (symbols, calling symbols) can be provided by expanding the "return register" RR into a deque (a double-ended queue). Return addresses are stacked as usual, i.e., pushed down on occurrence of a call and popped up on a return. However, the address to be compared to the pointers in CT for lightpen pick identification must always be the bottom element, as the bottom element is the point of return into the invoking segment.

8.4 DISPLAY FILE AND PICTURE FILE

We found it useful to separate the name list from the correlation table and to keep the name list only in the host computer. Since each name is firmly linked with a record in the correlation table, we may consider the combination of name and correlation table as a two-stage directory. The name list provides access to the correlation table, which provides access to the display processor code list. This address transformation sequence is traversed whenever a name is referenced by a display procedure. We have shown that the inverse transformation can also be easily performed whenever an entity is picked by a selector for identification.[†]

In cases where a fairly large name list (e.g., of some hundred names) must be handled, it may be worthwhile to organize the name list as a hash-addressed file in order to accelerate the search for a name. In this case, the fixed correspondence between name and list index cannot be maintained and, consequently, a pointer to the correlation table must be stored together with each name. Furthermore, provisions must be made for collision handling. As neither the number nor the size of the collision classes that may occur can be estimated, it may be the best approach to reserve a certain amount of "overflow" space outside the space accessible through hash addressing and to organize collision classes in the form of linked lists. To this end, each name record may encompass a special pointer field that contains either a particular NULL symbol as long as the name has not collided with another name or, in case of a collision, the link to the next record belonging to the collision class.

The implementation of a particular display programming system may necessitate additional entries in the name list. An example is the GRAP system (see the Appendix) which is an implementation of GRIP (see Chapter 10) in the form of a FORTRAN subroutine package. The simple FORTRAN subroutine mechanism does not provide variable name scope control facilities. Thus, the picture structure that is the basis for the GRIP philosophy can only be introduced

[†]Of course, a request for identification of a picked entity will now require a sequential search through the whole name list, with the pointers to the correlation table functioning as keys.

by a trick: the programmer may arbitrarily declare segment attributes as well as attributes of each individual item. Any item for which no individual attributes are declared assumes by default the declared segment attributes. Such a default mechanism, although simple in principle, is somehow complicated by the fact that GRIP permits the user to change the segment or item attributes dynamically. An efficient way of providing such a possibility is to associate the current attribute information with each entity, and the natural place for storing that information is the name list.

As was pointed out in Chapter 3, the process of primitive identification may be complicated by the fact that the image that is visible on the screen is an instance of an original picture, obtained after application of a windowing operation. The effect of windowing transformations on an item can be described as a relation between original and instance. If the identification of individual primitives is an important and frequently requested task, special provisions must be made to allow for backtracking from a primitive in an instance to the corresponding primitive in the original. A suggested solution consists in the addition of two more entries in the name table for each item. The first entry lists the number of primitives in the item, and the second entry is a pointer to a one-dimensional array called *item relation table* IRT.

Of course, the file we may thus obtain now represents more than just a name table. It contains all the pertinent information about the original picture, such as, for example, segment and item names, the structural relationship between segment and items, the number of primitives in each item, and the relation between the original items and their counterparts in the instance. Therefore, we call such an extended file in the host computer the *picture file*. Figure 8.4 depicts a picture file and the associated item relation table for the example given in Figure 3.8. For the sake of simplicity, we assume in Figure 8.4 that the picture file is not hash-addressed. For hash addressing, the one field in the picture file containing the name has to be expended into three fields, as indicated in Figure 8.4

Some explanation should be given as to the use of the item relation table. Each item that has been subjected to a windowing operation is represented in IRT by a vector of as many components as is the number of primitives in the windowed item. The entries in each such vector are the indices of the related primitives in the original. Hence, after the pick of a primitive of the transformed item, the IRT entry for that item is looked up, addressed by the index of the picked primitive, and the related primitive in the original is found (note that the user wants the primitives of the original to be identified and not the primitives of an instance).

We want to emphasize that the organization of a picture file as depicted above is not advertised as the only solution but merely as a suggestion. Depending on the tasks the system has to perform, only parts of it may need be implemented. Other modifications may be required for different problems.

At the end of Chapter 1 we mentioned the possibility of designing display systems whose computational power allows on-line execution of a high-level language display program at a speed sufficient for a continuous picture refresh. We

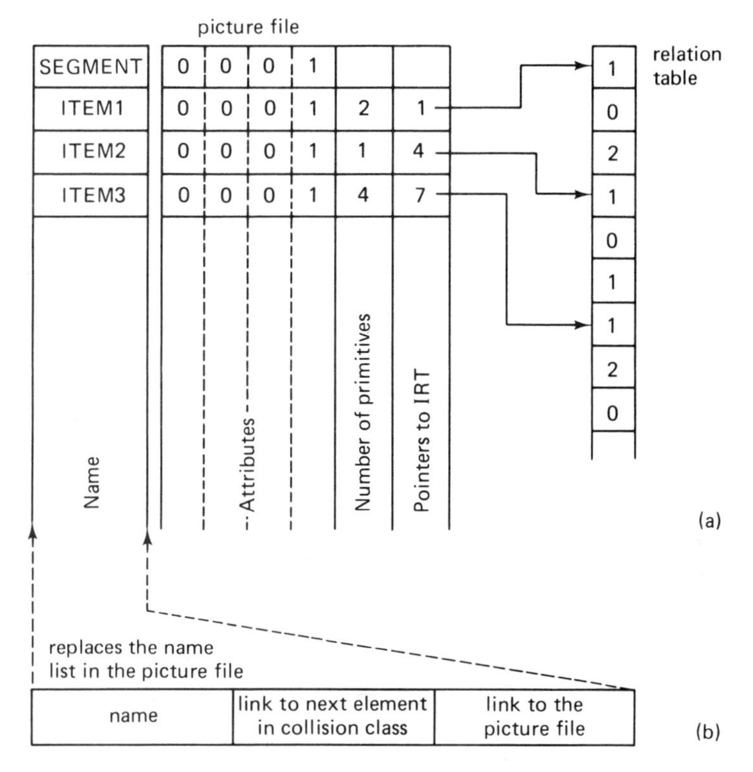

Figure 8.4 Picture file for the example of Fig. 3.8: (a) without hash addressing; (b) modification required for hash addressing.

mentioned also that the computing power of the display processor must include the capability of including on-line picture transformations in the refresh cycle.

According to our definition, this necessitates a picture file but no display file. However, this is only a question of terminology. Structure-wise, such a file corresponds very much with a display file as discussed in the previous section. The only major difference is that in lieu of the display processor code list, we have now to scan the high-level language program that is specifying the sequence of display procedure calls [102] whose interpretation (and execution) by the hardware leads to picture generation. As mentioned before, picture structuring can be accomplished by nesting entity-generating procedure calls. Figure 8.5 shows an example of the resulting file structure that can be found in such a display [152]. A built-in hardware stack will facilitate the nesting of procedure calls as well as the identification of an entity: If an interrupt occurs, the content of such a stack yields in connection with a correlation table all the information required for an identification.

In large picture files it may become rather tedious to retrieve data by specifying a record number (which may have to be looked up first in a directory). A better solution is in this case to enter and retrieve records associatively, that is,

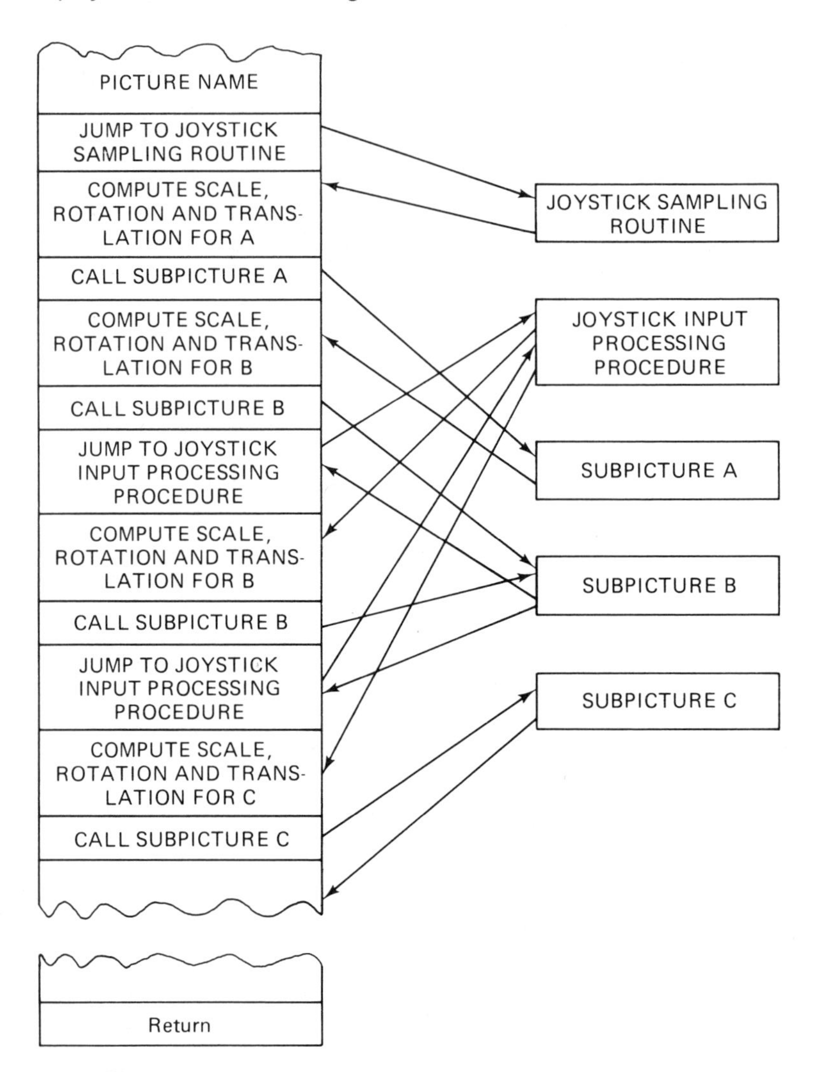

Figure 8.5 Picture file structure in an interpretative system [152].

through the reference of the entity name. This can be accomplished by hash coding.

Therefore, we need appropriate procedures for managing a hash-coded name table. These procedures, which may be of the function type, might be†

HASH '⟨name⟩'
FIND '⟨name⟩'
REMOVE '⟨name⟩'

†We use in this example the syntax of APL (in which these functions were implemented [56]).

All these functions operate on a "common" or *global* variable, a *h*ash-*a*ccessible *n*ame *t*able, called HANT. HASH enters the specified name into HANT and returns as its value the index of the cell where the name has been stored. If this name already exists in the table, it returns error code −1; if the name cannot be entered because of a table overflow, it returns error code −2. FIND retrieves a name in HANT and returns as its value the index of the cell in which the name is stored. If a name cannot be found, the error code 0 is returned. REMOVE deletes an entry from HANT. Two more functions, NEXT and LAST, are internally applied to the linked lists which represent the collision classes. NEXT I retrieves the pointer that is stored in the referenced cell I and points to its successor. LAST I starts from cell I and moves in the linked list from pointer to pointer until the terminal node has been found, the index of which is then returned. As this is a recursive procedure, we may write LAST as a recursive function, provided the language allows that.

HANT is structured as a matrix, accommodating the proper name list as well as an overflow space in which the collision lists will be stored. In FORTRAN we would declare this array as a common variable. In APL we represent it by a global variable that is initialized as an empty character matrix whose dimension originally is ρHANT↔N, L+M, if N is the maximal number of names to be stored, L is the maximal number of characters per name, and M is the number of digits in a pointer. Subsequently, the overflow space may be dynamically added by catenating additional rows to the matrix whenever needed. In order to satisfy the homogeneity constraint, pointers and tags must also be declared as characters (this assumes that the language allows the conversion of character strings, representing numerical constants, into numbers and vice versa). Each entry in HANT takes one row, in which the first L positions are occupied by the name and the remaining M positions are occupied by a pointer or a tag. A tag is one of two special symbols which either indicate that a cell (a row of HANT) is unoccupied or that it is the terminal node in a collision list.

Collisions may be handled as follows. If the cell indexed by the hashing operation is unoccupied, the name is entered into its name field, and the cell is tagged as a terminal node. If a cell, indexed by hash coding, is found occupied, the name to be entered and the name already residing in the cell is compared. In the case of a match, the error code −1 is assigned as output to HASH, and the function execution is terminated. Otherwise, HASH enters the name into the first free cell of the overflow space and tags this cell as a terminal node. The node thus created is linked to the appropriate collision list (i.e., the list whose first node is the hash-indexed cell that was found occupied).

The function FIND enters a collision list at the hash-indexed cell and compares the name found there with its argument. If the names do not match, it uses NEXT in order to proceed to its successor in the list, comparing the name stored there to its argument, and so on, until a match has occurred. The function REMOVE invokes first the function FIND to find the cell that contains the name to be removed. If this cell is in the proper name table, it is simply retagged as unoccupied. If it is a cell in the overflow space, it must be removed from the collision list it is part of and returned to the reservoir of available cells (in APL we simply let the overflow space dynamically grow or shrink).

EXERCISES

1. The objects to be displayed may be convex polyhedrons. [A convex polyhedron is a plane-faced three-dimensional object in which each face is bounded by a convex (plane, closed) polygon.] Thus, the subelements are faces, edges, and vertices. Two faces may have an edge in common, and several of the "contour polygons" of the faces may share some of the vertices.

(a) What is the resulting structure of the graphic data?

(b) Design a display file organization that implements the recognized data structure. The display file will be structured such that individual objects, faces, and edges can be identified, e.g. as the result of a lightpen pick. A detailed description of the display file implementation will be given. Illustrate with diagrams. Justify your approach.

(c) Outline the procedure that must be carried out whenever a graphical element is to be identified after a lightpen pick. Make sure that a pick of a part that is shared by several elements leads to the identification of all these elements.

(d) The identification problem always requires a search through the display file. Discuss the search techniques known to you and compare them in terms of efficiency and expenses (measure expenses in terms of required memory capacity).

(e) Identify subroutines that create and manipulate such a display file. Identify each subroutine by a name, list its input and output parameters, and briefly describe its function.

2. (a) Write a FORTRAN program for setting up a name table as part of a display file. The table will provide space for M records. There is one record for each segment and each item, accessed through the segment name or item name. A name must consist of four alphanumeric characters. Assume a font of 64 characters, which may be coded by the numbers 0 to 63 (there is no data type CHARACTER in FORTRAN). The table is to be declared as a matrix, each row of which corresponds with a record. The access to a record (i.e., the index of the respective row) will be obtained by applying the hashing function

$$\mathrm{I} = \mathrm{P1} + \mathrm{P2} \text{ modulo } \mathrm{P3}$$

to the associated name. Each record will consist of three information items: (1) the name, (2) a pointer to the corresponding row of the correlation table in the display file (see Fig. 8.4), and (3) a pointer that points to the next element of the same collision class. In case there is no next element, the value -1 is entered, denoting the record as a terminal element in the linked list that forms a collision class. Originally, on initiation of the table, all pointer fields are filled with 0s.

(b) What are the values of the parameters P1, P2, and P3? Write a subroutine HASH that performs the hash coding of a name into a record index.

(c) Somebody suggests solving the collision problem in the following way. A matrix with N rows is set up. If a collision occurs, a search is started for the first free cell, then the new record is entered there and linked to its predecessor in the collision list. Whenever this newly occupied cell should be indexed again, the respective entry will be made in another free cell and linked to the existing collision class. Hence, no dynamically growing overflow area is needed and the table is eventually filled to its full capacity (the problem of a "scatter memory" is circumvented). Provided that this scheme works, what would be the trade-off in terms of execution time for the obtained savings in memory space? Will this simple scheme work

(1) If the only objective is to access the correlation table properly by a hashing of a given name

(2) If the additional stipulation is made that it be possible to delete a name from the table and free the associated cell ("garbage collection")?

(d) Complete the package of FORTRAN subroutines required for the management of the name table. The complete package will consist of five subroutines:

CREATE:	Creates the name table as a matrix and initializes it.
PUT:	Enters a record into the table.
FIND:	Finds a record in the table and returns the associated correlation table index or an error code, respectively, if a name cannot be found.
DROP:	Removes a record from the table and frees the cell it occupied.
HASH:	Performs the hashing operation. HASH may be called by PUT, FIND, and DROP.

3. Consider the various forms of display file and picture file organization as discussed in this chapter. Assume that the following parameters are given L=total number of DPC instructions; K=[number of segments]+[number of items]; $M=K$+[number of symbols and symbol instances]; N=number of words reserved for a name. Assume that only long vector instructions (see Section 7.3.1) are used. List all tables required, together with their dimensions, for the following cases.

(a) Display file without symbols
  (1) What are the tables and their dimensions?
  (2) Flow-chart a procedure that identifies on a lightpen pick the picked primitive, the item, and the segment. The respective identifiers are to be returned as value of the three variables IDSEG, IDITM, IDPRM.

(b) Display file with symbols
  (1) What are the tables and their dimensions?
  (2) Does the object identification procedure as described in the text allow the identification of a picked primitive? If yes, how? If no, why not?
  (3) Assume that there is a hardware stack in the display processor. Can the stack be used for organizing a symbol return? Can it be used in the process of symbol instance identification?

(c) The index-addressed name table and the correlation table form a two-stage directory.
  (1) What are the tables and their dimensions?
  (2) Flow-chart a procedure that identifies a segment and returns its name as value of the variable IDSEG.

(d) The hash-addressed name table and the correlation table form a two-stage directory.
  (1) What are the tables and their dimensions?
  (2) Does the representation of collision classes by doubly linked lists have any advantage? If yes, in which cases? If not, why not?
  (3) Flow-chart a procedure that identifies a segment and returns its name as a value of the variable IDSEG.

4. A DPC program for the display of the RC filter diagram shown in Figure 8.1 will be written. The segment, called FILTR, consists of the five items R1, R2, R3, C1, C2; a set of connecting lines; and a set of text strings. An item is initialized by the instruction sequence.

CONTROL 047 ; XORG ⟨x-value⟩ ; YORG ⟨y-value⟩.

047 is the number whose binary representation is the bit string required for the desired appearance and status. A resistor is an entity consisting of seven short vectors; a capacitor consists of four short vectors and three short moves (in the right order). The DPC instructions for line generation, move, and character writing are:

LONG VECTOR ⟨x-value⟩
LONG VECTOR DEFERRED ⟨y-value⟩
SHORT VECTOR ⟨x-value⟩,⟨y-value⟩
LONG MOVE ⟨x-value⟩
LONG MOVE DEFERRED ⟨y-value⟩
SHORT MOVE ⟨x-value⟩,⟨y-value⟩.

The instruction for writing a character string is TEXT '⟨character string⟩'.

(a) List the DPC program segments for the two entities, resistor and capacitor.

(b) List the complete display file for FILTR, using an organization as proposed in this chapter, i.e., a file consisting of the display processor code table, the correlation table, and the name list. The type code is: -1 for vectors, -2 for dots, -4 for characters.

(c) Resistor R2 may be picked with the lightpen. How does the system identify it?

(d) Modify the display file such that the resistors and capacitors are generated as symbol instances. For this purpose, two more instructions can be used:

JUMP AND STORE ⟨adr⟩	A jump is performed to ⟨adr⟩ after the instruction counter (IC) content has been transferred to a special return address register (RA), i.e., after the register transfers RA←IC; IC←⟨adr⟩.
RETURN	Performs the transfers RA←RA + 1; IC←RA.

(e) How can this scheme be modified to allow nested symbol calls?

9

Language Concepts for Interactive Computer Graphics

9.1 HIGH-LEVEL GRAPHIC PROGRAMMING LANGUAGES

9.1.1 An Example

In 1968, Kulsrud published a paper [88] in which the requirements for a general-purpose graphic language were outlined and a model language was defined. This paper—although it was not the first attempt to define a graphic language—has a certain pioneering character because of the rigorousness and cleanness of the presented concept of high-level graphic programming. This is reflected by the fact that Kulsrud's paper rpobably is the most frequently cited paper in the realm of graphic languages. As the semantics of Kulsrud's model language represents a concept that is still valid, we present it here as an example for a high-level, general-purpose graphic programming language.

Kulsrud's postulates for a general-purpose graphic language are:

1. The language must provide its user with the ability to describe, generate, and manipulate pictures. To describe and generate a picture, the language must have commands by which certain graphic primitives or collection of primitives are specified as well as certain attributes determining their appearance on the display screen. Manipulations require geometric and windowing transformations and the ability to rearrange, merge, or delete subelements of a picture.

2. A graphic language should exhibit dynamic flexibility. Procedures are essential.

3. A general-purpose graphic language should allow for the analysis of pictures. Analysis requires commands for the location and examination of special features in the picture. Also, analysis may be concerned with the relations between subelements (see the discussion in Sections 1.1 and 3.1).

From our hindsight, one correction of these postulates must be made. Kulsrud felt that a general-purpose graphic language (the word "universal graphic language" might even be more appropriate) should encompass the generative as well as the cognitive aspects of computer graphics (see Section 1.1). The trend of the past decade, however, has not led to the development of such universal graphic languages. Generative (interactive) computer graphics and cognitive graphics (image analysis) turned out to be two rather separate fields, and how these two areas could be brought closer together is still being discussed.

In our presentation of the Kulsrud language as a model for a high-level graphic programming language, we are not concerned with syntactic details (to a certain degree, the syntactic forms of a language are arbitrary and reflect the taste of its designer). Therefore, we explain in Table 9.1 only the semantics of the commands.

The reader will notice that the Kulsrud language has no commands such as DRAW (for the generation of a single line segment) or MOVE (for beam positioning), commands that can be found in many of the graphical languages or programming systems designed subsequently. In fact, commands such as DRAW or MOVE correspond with the respective display processor instructions and hence are low-level. Furthermore, note that the language processes, in addition to the common data types of high-level programming languages, the following graphic data types: points, line segments, pictures, subelements, and regions.

9.1.2 Language Extensions vs. Subroutine Packages

The number of graphic languages or extensions of existing programming languages proposed over the past decade is legion, the reason for this being twofold. First, graphic data types and operators usually are not encompassed in general-purpose, high-level programming languages. Therefore, if graphic data types† are desired, the general-purpose programming language must be appropriately augmented. Second, the conceptual design of a programming language is an intriguing intellectual pastime and thus self-stimulating (this author does not exclude himself from that observation). However, it is one thing to design a language conceptually, and another to implement such a design. Language extension means to add new syntactic constructs to an existing "host" language. Some existing programming languages lend themselves toward an extension and thus are called "extensible." In general, however, a language extension necessitates major

†Note that we can only then speak of a particular data type of a programming language if there exist as well operators specifically defined on that data type.

Table 9.1: KULSRUD LANGUAGE

A. Commands for Description (Generation)

SCALE	::	Determines the range of the viewing area and generates a grid in a specified spacing. Also deletes the current content of the display file.
POINT	::	Accepts as parameters a name and a list of points. As a result, the points are plotted as dots and identified by the given name.
LINE	::	Accepts as parameters a name and a list of points. As a result, a polygon is generated and identified by the given name whose vertices are the points, taken in the order in which they occur in the list.
ARC	::	Accepts as parameters a name and three points. As a result, an arc is generated connecting two of the specified points, the center of which being the third point.
FUNCT	::	Accepts as parameters a name and two expressions representing the functions $y = f_1(x)$ and $x = f_2(y)$. As a result, the expressions are evaluated.
CURVE	::	Accepts as parameters a name and numerical data specifying the end points as well as a desired breakpoint spacing of a segment of the functions identified by the name. The segments are plotted as a curve.

B. Commands for Manipulation

PRINT	::	Produces an output version of the picture currently being generated. A list of parameters may be added specifying entities to be superimposed on the output picture but not kept in memory (e.g., coordinate axes, a grid, etc.).
STORE	::	Causes the picture currently being worked on to be stored in a stack of pictures under a specified name.
GET	::	Replaces the picture in the display file by the picture previously stored under the specified name.
NAME	::	Accepts as parameters a list of names and provides for the labeling of subelements in the picture currently in the display file. The entities identified by the listed names are linked together and fetched under a common name specified in the list.
COPY	::	Accepts as parameters a list of names referencing entities in stored pictures and merges these entities into the picture currently in the display file.
ROTATE	::	Accepts as parameters a list of subelement names and rotation parameters and rotates the named subelements as specified by the data.
ERASE	::	Accepts as parameters a list of names referencing stored pictures or subelements in the currently displayed picture and deletes the named entities.

C. Commands for Regional Analysis

REGION	::	Identifies and names a region containing a specified point. Additionally, a property can be specified which the region must satisfy.
CONECT	::	Checks whether two specified points are in the same region and whether a curve connecting the points is in the region.
ADJAC	::	Checks the adjacency of two specified entities and produces a respective boolean output.
INTERX	::	Determines the point of intersection of two lines.
DIST	::	Determines the number of steps between two points. A step is made by a move to a nearest neighbor.

D. Commands for Topological Analysis

WITHIN	::	Checks whether a subelement is contained in another subelement.
SEPAR	::	Checks whether two subelements are separate (disjoint).
SIMPLY	::	Checks whether a picture or a subelement of a picture is a simply connected region.

modifications or a rewriting of the existing compiler and hence represents a rather tedious and time-consuming effort. Furthermore, it creates a major obstacle to the portability of the language dialect thus obtained.

A common method to avoid modifications of the host-language compiler is to write a precompiler. In this case the source program comprises two kinds of statements: (1) statements of the host language, and (2) statements of the language extension; the latter to be translated by the precompiler into statements of the host language.

However, new data types and operators on these data types can also be introduced without any syntactic extension of the programming language. As a matter of fact, programmers are doing this all the time. Let this point be illustrated by the following example. In ALGOL 60, the operators of the language are defined solely on scalars (of the data type integer, real, or boolean). Arrays exist only as a data structure into which a number of scalar data items of the same type can be grouped, but there are no array operations (as, for example, in APL or, to a much smaller extent, in PL/I). In other words, no access is provided to an array as a data entity, only to its scalar components (through indexing). However, the user may write procedures that transform the elements of an array (e.g., apply a component-wise scaling) or rearrange them (e.g., by transposition). Or he/she may write procedures that connect the elements of two arrays arithmetically, forming as a result a new array (e.g., matrix multiplication). Hence, she/he has introduced operators on arrays, and arrays have genuinely become a data type.† This approach raises no portability problems. However, it has the disadvantage that the newly introduced data types are only implicitly given and cannot be explicitly declared. Hence, domain and range of the operators represented by procedures are not explicitly declared either. Let us illustrate this by an example.

Over the years, a number of picture description languages have been developed for the purpose of describing the composition of a picture out of a small set of primitive elements. One of the first languages of that type is Shaw's "Picture Description Language" (PDL) [126]. In PDL, graphic primitives are line elements (e.g., vectors or arcs) whose characteristic property is to have a "head" and a "tail." Additionally, blank (invisible) and don't-care objects are introduced to connect disjoint parts of a drawing or to specify the geometrical relationship between parts of a picture. On these objects the set of operators $\{+, \times, -, *\}$ is defined as follows. Let A and B be two objects and let head(A) and tail(A) denote the head and the tail of A, respectively. Then the meaning of the operators is:

- Concatenation operators:
  - $a+\mathrm{b}$ head(a) linked to tail(b)
  - $a-b$ head(a) linked to head(b)
  - $a\times b$ tail(a linked to tail(b)
  - $a*b$ tail(a) linked to tail(b) and head(a) linked to head(b).

†The generality and flexibility of the procedure concept provides the potential for a host of "hidden operations" [124]. Some authors feel that this potential should be drastically restricted (see [124] and the references cited there).

- Monadic operators: $\sim a$ head($\sim a$) = tail(a) and tail($\sim a$) = head(a) (head–tail reverser)
  $\neg a$ the operator $\neg$ blanks out the structure to which it is applied.

In a language extension, for example, we could introduce the additional data type GRAFEL for "graphic element"; and the variables A and B could be declared in the program as of such type, e.g., by the statement DCL (A, B, C) GRAFEL.[†] Subsequently, we could write program statements such as the assignment C=A+B, where the operator + could be given the meaning defined above. That is, C would now identify a graphical element formed by the concatenation of tail(B) to head(A). Instead of extending the language, however, we might as well simply write a number of external procedures (subroutines) such as, e.g., a procedure for the "addition" of GRAFELs. In this case, the creation of element C might read, for example, CALL GRAFADD (A, B, C). Note that A, B, and C are now of the data-type array (e.g., in 2-space we may use a 2×2 array to represent the head and tail coordinates of a graphic element). The interpretation of such a 2×2 array as data associated with a graphic element is restricted to the special procedures written for the processing of graphic elements. Outside these procedures, A, B, and C are just arrays.

The language that lends itself more than others toward an augmentation of its set of operators through special procedures is APL, for two reasons. First, APL is an "attribute examining language"; i.e., there exist no name-attribute declarations, but the system interprets a variable on the basis of its current value and the context given by the expression in which it occurs. Declarations are only required for constants. Hence, the lack of explicit type declarations in the "subroutine package" approach does not hurt in a language that has no variable declarations anyway. Second, APL procedures are strictly "function-type procedures" [150] with a maximum of two formal input parameters (function arguments). Hence, a function is invoked by its name like a keyword operator. In our example, the appropriate APL statement may read C←A PLUS B, where PLUS is the function performing the head-to-tail concatenation.

Nevertheless, if special graphic operations are to be performed on variables of a language, the cleaner and more secure approach is to extend the language appropriately. This holds true for all languages of the "picture-processing" type. However, if the purpose of graphic programming solely is to interactively generate and manipulate pictures, then special graphic data types are not mandatory. For example, the commands for description (generation) and manipulation of graphic entities have names, points (coordinate tuples), and other numerical data as parameters. Only the commands for "regional analysis" and "topological analysis" necessitate graphic data types. Coordinates are of data type integer or real (if the

[†]In these examples, we assume a PL/I-like syntax.

host language makes such a distinction) but do not constitute a special graphic data type. Operators for the generation of graphic objects receive numerical data and produce, accordingly, a certain output: a plot or a picture displayed on the CRT screen. In principle, that does not distinguish them from other output procedures.

9.1.3 The "Prefabricated-Structure" vs. the "Building-Block" Concept

The designer of a high-level programming language is confronted with the following basic question: Should he/she be guided primarily by the principle of providing "prefabricated" data structures, operators, language constructs, etc., or should he just furnish the "building blocks" that enable the user to build his own data structures, operators, and language constructs. The early algorithmic languages strictly followed the first pattern.

Prefabricated structures offer the advantage that their use can be very convenient, elegant, and of high level. On the other hand, it is, of course, impossible to provide such structures for every conceivable application of the language. Therefore, we deem it a good policy to provide, in addition to these structures which may be suitable in the bulk of cases, building blocks that allow the user to tailor certain tools to his needs.

An obvious case in point for such a philosophy is the realm of data structures. Here, readily applicable structures are typically given in the form of homogeneous, rectangular arrays—early languages, such as, FORTRAN or ALGOL60 provide nothing else—as well as inhomogeneous structures such as, for example, STRUCTURES in PL/I and ALGOL68 or LISTS in PASCAL. These structures may be sufficient in standard application programming, albeit the system programming aspects must also be considered. Denying a system programmer the building blocks for the realization of more complex data structures may lead to a less efficient coding.

Naturally, building blocks belong to a lower level than prefabricated structures. Actually, all that needs to be added to a language is the data type POINTER. In Chapter 2, we discuss PL/I as the example of a language that offers both readily applicable structures (in the form of ARRAYS and STRUCTURES) and building blocks (in the form of POINTERS in connection with the BASED memory allocation mode). If a general-purpose programming language that will be used for graphic programming does not provide for the required special building blocks, then either a language extension becomes inevitable, or one has no other choice but to resort to the prefabricated structure solution. Most general-purpose programming languages do not enable the user to define his/her own operators, control constructs, and data structures. In the following section we consider two cases in point, one for the prefabricated structure and one for the building-block concept.

9.2 HIGH-LEVEL GRAPHIC LANGUAGES: TWO CASES IN POINT

9.2.1 LEAP—An ALGOL60 Extension Based on the Prefabricated-Structure Concept

The LEAP language found its realization in connection with the LEAP data base management system [37] discussed in Chapter 2. It is primarily a language for establishing associations (relations) between information items and entering them into or retrieving them from the data base on the basis of these associations. Therefore, the data base manipulated by LEAP comprises a universe of *items* and a universe of *associations*.

The universe of items is divided into three subclasses, differing in the way an item enters the universe of items:

1. A DECLARED item results from each declaration of an ⟨identifier⟩ to be of type *item*. The declaration is processed once at translation time; items are not declared dynamically.

2. A CREATED item results from the execution of a *new* expression.

3. An ASSOCIATION item results from the execution of a ⟨bracketed triple⟩, [A∘O=V] (read: "Attribute of Object=Value").

An item of type 1 or 2 may have an associated DATUM of algebraic type which can be used or altered like an ALGOL variable. An item of type 2 or 3 can be deleted from the universe of items by a delete statement; no item is automatically deleted at block exit.

An ITEMVAR is a variable whose value is an item. A LOCAL is an itemvar which obeys special building rules and is used only in a restricted set of contexts. A SET is a finite (unordered) collection of items containing at most one occurrence of an item.

An ASSOCIATION is an (ordered) triple (A, O, V) of items which are called the COMPONENTS of the association. The totality of associations created during execution of a program (and not subsequently removed by *erase* statements) is called the universe of associations. Whenever an association is created from its component items (by execution of a ⟨bracketed triple⟩, or of a *make* statement), the universe of associations is searched for an association with the same components, and a new association is constructed only if the search fails. Thus, the universe never contains more than one association with the same components. Once an association has been created, its component items cannot be changed.

In the following we give the syntactical description (in BNF) and the semantics of DATA TYPES, EXPRESSIONS, and STATEMENTS which are added to the regular ALGOL 60 data types, expressions, and statements.

DATA TYPES OF LEAP

The augmentation of ALGOL for associative processing requires the addition of four new type declarators, corresponding to the notions *item, itemvar, local,* and *set*. A declaration is formed from ⟨type⟩ declarators as in ALGOL 60.

Syntax:

⟨simple type⟩: : = *real* | *integer* | *boolean*
⟨algebraic type⟩: : = ⟨simple type⟩|⟨simple type⟩ *array*
⟨leap type⟩: : = *item*|*itemvar*|*local*
⟨type⟩: : = ⟨algebraic type⟩|⟨leap type⟩|*set*|⟨algebraic type⟩⟨leap type⟩

Examples of declarations:

set	sons
real item	pi
local	*x*
boolean	married

Semantics: An ⟨identifier⟩ declared to be of ⟨algebraic type⟩ behaves exactly like an ALGOL 60 variable. An ⟨identifier⟩ declared with *item* (*itemvar, local*) denotes an item (itemvar, local). An item whose declaration is qualified by an ⟨algebraic type⟩ has an associated datum of that type. All items are treated as if they were declared in the outermost block. If an itemvar (local) has its declaration qualified by an ⟨algebraic type⟩, the item which is the value of the itemvar (local) is assumed to have a datum of that type for algebraic operations. An ⟨identifier⟩ declared with *set* denotes a variable that takes on sets as values.

EXPRESSIONS OF LEAP

The additional expressions required for LEAP are divided into those which yield algebraic values and those whose values are LEAP types. The use of τ in a production is a shorthand notation for two productions, one with τ replaced by "retrieval" and one with τ replaced by "construction."

Syntax:

⟨additional arithmetic primary⟩: : = *datum* (⟨item primary⟩)|*count* (⟨retrieval set expression⟩)
⟨additional boolean primary⟩: : = ⟨retrieval item expression⟩ *in* ⟨retrieval set expression⟩|⟨retrieval set expression⟩ ⊂ ⟨retrieval set expression⟩|⟨retrieval set expression⟩ = ⟨retrieval set expression⟩|*istriple* (⟨item primary⟩)|*datum* (⟨item primary⟩)

⟨algebraic expression⟩::=⟨arithmetic expression⟩|
  ⟨boolean expression⟩
⟨item primary⟩::=⟨item identifier⟩|⟨itemvar identifier⟩|⟨local identifier⟩
⟨retrieval item expression⟩::=⟨item primary⟩|⟨selector⟩ ⟨item primary⟩
⟨construction item expression⟩::=⟨item primary⟩|*new*|*new*(⟨algebraic
  expression⟩)|⟨bracketed triple⟩
⟨bracketed triple⟩::=[⟨item primary⟩·⟨item primary⟩≡⟨item primary⟩]
⟨τ set primary⟩::=∅|⟨set identifier⟩|{⟨τ item expression list⟩}|
(⟨τ set expression⟩)|(⟨τ derived set⟩)
⟨τ item expression list⟩::=⟨τ item expression⟩|⟨τ item expression list⟩,
  ⟨τ item expression⟩
⟨τ set factor⟩::=⟨τ set primary⟩|⟨τ set factor⟩−⟨τ set primary⟩
⟨τ set term⟩::=⟨τ set factor⟩|⟨τ set term⟩∩⟨τ set factor⟩
⟨τ set expression⟩::=⟨τ set term⟩|⟨τ set expression⟩∪⟨τ set term⟩
⟨retrieval associative expression⟩::=⟨retrieval item expression⟩|⟨retrieval
  set primary⟩|*any*
⟨construction associative expression⟩
  ::=⟨construction item expression⟩|⟨construction set primary⟩
⟨τ derived set⟩::=⟨τ associative expression⟩ ⟨associative operator⟩ ⟨τ associative
  expression⟩
⟨τ triple⟩::=⟨τ derived set⟩≡⟨τ associative expression⟩
⟨associative operator⟩::=.|'|*
⟨selector⟩::=*first*|*second*|*third*

Examples:

count (sons) − *datum* (bill) + 3
{bill, tom} ⊂ sons ∪ (father'don)
number.[part.hand≡finger]≡ *new* (5)

Semantics: The distinction between ⟨construction item expression⟩ and ⟨retrieval item expression⟩ is primarily semantic and gives rise to the further (construction, retrieval) distinctions. In LEAP, an expression is sometimes used to create new items or associations (construction type) or to retrieve information about existing items or associations (retrieval type).

Any ⟨item primary⟩ yields an item as its value. If the item has an associated algebraic datum then *datum* will yield its value; otherwise, *datum* is undefined. The unary operator *count* returns an integer equal to the number of elements in the set specified by its argument. The notions of membership, set containment, and set equality have their usual meaning.

The predicate *istriple* returns *true* if the item specified by its argument is an association item. A ⟨selector⟩ applied to such an item will yield the corresponding component; otherwise, it is undefined. A ⟨construction item expression⟩ may be the operator *new*, which causes a new item to be created and assigned an internal name. If *new* is given a parameter of ⟨algebraic type⟩, the created item has an associated datum of that type.

The execution of a ⟨bracketed triple⟩ causes a new association to be created and placed in the universe of associations (unless it is already present) and then produces an association item which denotes this association.

Any set construct can be of either retrieval or construction type, depending on its context. The symbol $\varnothing$ denotes the empty set, a list of item expressions denotes the set containing the values of the item expressions, and the set operations $-$, $\cap$, $\cup$ have their conventional meanings. A ⟨τ derived set⟩ is defined in terms of the universe of associations. If the two operands are both ⟨τ item expressions⟩ with values P and Q, then P.Q is the set of items X such that $P.Q \equiv X$ is in the universe of associations, P′Q is the set of items X such that $P.X \equiv Q$ is in the universe of associations, and P*Q is $(P.Q) \cup (P'Q)$. If the ⟨τ associative expression⟩ S has a set $\bar{S}$ as its value, the value of P.S is the union of P.Q for all Q in $\bar{S}$; P′S, P*S, S.Q, S′Q, and S*Q are defined analoguously.

The ⟨τ triple⟩ is a generalization of the 3-tuple $A \circ O = V$, which is the LEAP representation of an association. The meanings of a ⟨construction triple⟩ and of a ⟨retrieval triple⟩ are quite different.

PROCEDURES OF LEAP

Statements: There are relatively few statement types in LEAP because the control structure of ALGOL is already largely adequate.

Syntax

⟨additional statement⟩::=⟨set statement⟩|
⟨associative statement⟩ ⟨loop statement⟩
⟨set statement⟩::=⟨set identifier⟩←⟨construction set expression⟩|
put ⟨construction item expression⟩ in ⟨set identifier⟩|
remove ⟨retrieval item expression⟩ *from* ⟨set identifier⟩
⟨associative statement⟩::=*delete* ⟨retrieval item expression⟩|
⟨itemvar identifier⟩←⟨construction item expression⟩|
datum (⟨item primary⟩)←⟨algebraic expression⟩|
make ⟨construction triple⟩|*erase* ⟨retrieval triple⟩
⟨loop statement⟩::=*foreach* ⟨associative context⟩ *do* ⟨statement⟩
⟨associative context⟩::=⟨element⟩|⟨associative context⟩
and ⟨element⟩
⟨element⟩:=⟨retrieval triple⟩|⟨boolean expression⟩

Examples:

```
put tom in sons
sons←sons {tom}
make father.tom≡bill
foreach x in sons do datum(x)←datum(x)+2
foreach father.x≡bill do put x in sons
```

Semantics: The assignment statements to an ⟨itemvar identifier⟩ and to the datum of an ⟨item primary⟩ are discussed earlier. The set-assignment statement assigns the value of the ⟨construction set expression⟩ to the ⟨set identifier⟩. The *put* (*remove*) operation is simply a more efficient way of doing union (subtraction) of a single element. An item that was created (e.g., by a *new* expression) can be destroyed by *delete*. The internal name associated with this item will be reassigned, and it is the user's responsibility to assume that there are no uses of a deleted item.

The *make* statement causes the association represented by a ⟨construction triple⟩ to be placed in the universe of associations if it is not already there. This may entail the creation of new items and associations, as the following example will show.

TWO SAMPLE PROGRAMS

```
1. begin comment This program will find the polygon in some figure which has the largest
   perimeter and will write its name, number of sides, and its perimeter. We assume that there
   are items representing figures, objects (components of figures), and lines, and that there are
   associations in the universe of the form part.figure≡object, type.object≡polygon, and
   side.object≡line. We also assume that the number of sides of a polygon is stored as its
   datum and the length of a line is stored as its datum, the data base is assumed to be built in
   an outer block:
   item part, side, type, polygon;
   real sum, max;
   local x;
   real local y;
   integer itemvar largest;
   itemvar figure;
   read (figure);
   max←0;
   foreach part.figure≡x and type.x≡polygon do
       begin sum←0;
         foreach side.x≡y do
           sum←sum+datum(y);
         if sum ≥ max then
           begin largest←x; max←sum end
       end;
     write(largest, datum(largest),max)
 end
```

2. *begin comment* This program will determine if Mary is related to Joe by virtue of the fact that Mary's paternal aunt is married to Joe's paternal uncle and, if so, will record that fact;

```
item father, mary, joe, sex, married, male, related, reason;
local x, y;
set uncles;
boolean switch
uncles←∅; switch←false
foreach father.x≡(father.(father.joe)) and
     sex.x≡male do put x in uncles;
  foreach father.x≡(father.(father.mary)) and
  married*x≡y and y in uncles do
    begin
      make related.mary≡joe
      make reason.[related.mary≡joe]
      ≡[married.x≡y];
      switch←true
    end;
    write (switch)
end
```

9.2.2 EX.GRAF—A FORTRAN Extension Based on the Building-Block Concept

An excellent example for a realization of the building-block philosophy is EX.GRAF, an extensible language for graphical purposes [153]. Williams, who designed EX.GRAF, sees the prime requirement of a general-purpose graphic language in the provision of a sufficient set of data types and operations on these data types. He devised EX.GRAF as an experimental tool. Consequently, the language provides an initial set of data types and operators and allows the user to extend this set at his discretion.

Data types in the initial set are: BIT (for boolean variables), CHAR (for characters), INT (for integers), REAL (for floating-point numbers), and PTR (for pointers). The data type POINTER actually is the building block for the construction of data structures. Variables can be declared to have one of these data types. It is an unusual, but for an experimental language quite appropriate, feature of EX.GRAF that operators can also be declared. Another interesting feature is the existence of a "prefabricated" data structure called *block*. A block is an inhomogeneous list of records. By declaring the structural mode (called SMODE) of a block, the user creates a certain block type. Subsequently, he may declare any number of actual blocks of that type: e.g., a block-type definition may read

```
SMODE HEAD, CHAR(20); BODY1, INT; BODY2, INT; TAIL, INT; TYP1.
```

This creates a list consisting of four records, HEAD, BODY1, BODY2, and TAIL. Connected with each record name is a data-type declaration for this record: HEAD is of type CHAR, and the other records are of type INT. The declaration at the end assigns the name TYP1 to this structure. Subsequently, variables may be declared as having the structure TYP1 by writing, for example,

DECLB BLOK1, BLOK2, TYP1.

This statement declares the variables BLOK1 and BLOK2 as being blocks of the structure TYP1. The possibility of declaring individual data types for the records of a block allows the construction of inhomogeneous data structures.

As an example for the use of the data type pointer as a building block of data structures, we define a structure RING consisting of a backward pointer, an integer entry, and a forward pointer, as illustrated by the following symbol.

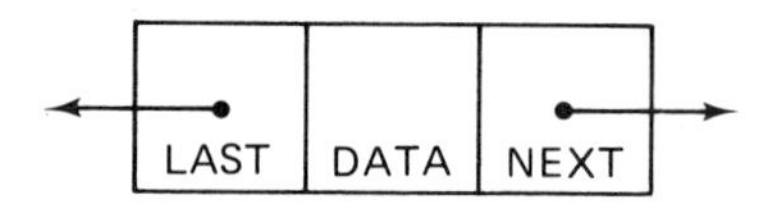

The declaration of the structure type is, in this case,

SMODE LAST, PRT; DATA, INT; NEXT, PTR; RING.

From now on, the structure type RING can be used as a building block for constructing ring lists. A record in a block is accessed by the name of the block and the name of the record or by the name of the block and the list index of the record.

Operators are declared by a statement of the form

NEWOP OPND1, OPND2, OPER, NAME,

where OPND1 amd OPND2 are the types (or SMODE names) of the operands, OPER is the operator that may be named by any arbitrary character string, and NAME is the name of a routine to be invoked whenever the operator and these two operand types occur together in a subsequent statement.

9.3 L^4—AN INTERMEDIATE LANGUAGE FOR DEVICE-INDEPENDENCE AND INTRASYSTEM COMMUNICATION

In Section 1.3 we modeled an interactive graphics system as a stratified hierarchy of *processing systems*, a processing system consisting of a language and an interpreter for the language. We introduced as one of the layers in the stratification the processing system

(Intermediate Language, Abstract System)

as the link between the two systems: (GPL, GPL translator) and (DPC, display system) (see Figure 1.3).† The insertion of this system yields a high-level programming system (GPL) that is device-independent, as the object language for the GPL compiler now is the intermediate language and not the "machine language" of a particular display processor.

A language that will serve as an intermediate language for any type of graphic device—from a simple plotter to the highly sophisticated directed-beam display with transformation hardware and a variety of interactive devices—must satisfy the following basic conditions for its syntax:

1. It must be simple, versatile, and easily extensible.
2. It must be free of rigid format conventions.

The first condition stems from the fact that commands in this language may range from a simple control command (e.g., "pen down") or a primitive-generating command (e.g., "draw line") to sophisticated transformation commands (e.g., "perform the combined operation of rotation, translation, scaling, and perspective projection"). Furthermore, it is impossible, at the time of a first design of this language, to anticipate all possible future uses of it, and therefore its extensibility becomes mandatory. The second condition reflects the fact that the number of parameters associated with a graphic command may range between zero and a large number.

As the intermediate language will also serve as a communication language in a host computer–satellite system, we stipulate as a third condition that the language should be as concise as possible, in order to minimize the data flow in such a network (see the discussion of Section 1.7). All these conditions may be ideally met by the intermediate language we call L^4 [49].‡ L^4 instructions have the format

⟨command⟩ ⟨parameter⟩ ⟨pointer⟩.

⟨Command⟩ stands for an operation code that may correspond with a primitive operation in the display system or with an elaborate procedure, however complex.

Any number of parameters may be furnished for the execution of a command. These parameters are forming a block in a list that is associated with the list of L^4 instructions and is thus called the *associated data list* (ADL). The beginning of each block is indicated by the pointer in an associated instruction, and the end is indicated by the pointer in the subsequent instruction. It is appropriate to have, in addition to the pointer, one parameter directly contained in the instruction. This parameter may be used for indexing the display file, for specifying the number of instructions that will be skipped in the case of a branch command, and other purposes. As we have no labels, a branch destination is specified relative to the position of the branch instruction, i.e., by a (positive or negative) integer specifying

†GPL, graphic programming language; DPC, display processor code.

‡L^4 stands for "linear low-level language."

the number of instructions that will be skipped (in a forward or backward direction).

Hence, L^4 commands present arbitrary multiadic operations, in combination with any required number of parameters. A program in this communication language consists of a *linear list of instructions*, terminated by a special delimiter command. A DATA statement provides the carrier for data transfers. Consequently, this language may serve the purpose of carrying information in both ways, from the host to the satellite, and vice versa. Hence, it may function as the carrier of protocols and, simultaneously, as a device-independent intermediate language. Only the interpreter in the satellite that translates L^4 statements into appropriate DPC routines is device-dependent. This interpreter must be tailored for the individual display system. It consists of a package of interpreting routines that may grow with a growing configuration. Seen from the outside, such a display system can be considered as an abstract machine whose language is L^4.

The advantages of this solution are the following:

1. We obtain a most concise, yet totally universal language (one L^4 instruction may represent a large segment of the display processor program). Thus, the amount of information to be exchanged between host computer and satellite is minimized.

2. All statements have the same format, regardless of whether they represent very simple or very complex operations or procedures. Moreover, the information types of this language can be better matched to the one used in the high-level programming language than the information types of a typical display processor code. A 4-byte format for an L^4 instruction is adequate: One byte is sufficient for the op code and one for the parameter. The other two bytes represent the pointer. Thus, we have a perfect match to the word format of minicomputers as well as of byte-oriented machines.

3. The introduction of this intermediate language results in high-level display programming that is device-independent and thus not only portable but also uniform for any given variety of computer systems and graphical peripherals.

4. The separation of instructions and data in L^4 facilitates greatly the execution of picture transformations after compilation. Hence, picture transformations can be continuously applied, without the need for executing a HLL program again and again. This feature supports strongly the generation of animated pictures.

Table 9.2 gives an example for a possible set of L^4 instructions. Note that the mnemonic notation for the instructions is introduced only for the sake of a better understanding by the reader, as L^4 is not a programming language but an intrasystem communication language. Furthermore, we want to emphasize again that L^4 is extensible, so that a user can tailor certain instructions to his needs. Consequently, the instructions listed in Table 9.2 are only examples, and we do not claim that this list is anything like complete.

Table 9.2: Examples of L^4 Statements*

Mnemonics	Code	Semantics
		Group I: Display processor code generating commands
SET CONTROL	00	Control of blink status, beam intensity, line style lightpen pick status, or color (whatever is applicable)
SET ORIGIN	01	Moves beam to a certain position
DRAW POINTS	02	Generates $N \geqslant 1$ dots
DRAW LINES	03	Generates $N \geqslant 1$ connected lines
SET MODE	04	Control of character write mode
WRITE TEXT	05	Generates a text string
DRAW CIRCLE	06	Generates $N \geqslant 1$ circles
DRAW GRAPH	07	Generates a graph
JUMP TO SUBPICTURE	08	Subpicture jump
SET LAMP.REGISTER	0A	Controls illumination of the keys of a function keyboard
PENDOWN	0F	Plotter pen control
		Group II: Display file manipulating commands
DELETE	10	Deletes an object from the display file
INSERT SEGMENT	11	Inserts a segment into the display file
INSERT ITEM	12	Inserts an item into a picture already existing
INSERT SYMBOL	13	Inserts a symbol into the display file
NEWSTATUS	14	Changes the attributes of a segment
NEWORIGIN	15	Changes the origin of a segment
NEWDATA	16	Replaces data of an item by new data
INITIALIZE	17	Initializes the display file
DATA	18	Transmission of a data block of specified length
END DATA	19	Delimits a data block
SUPPRESS	1A	Suppresses a primitive of an item
		Group III: Interaction handling commands
WAIT FOR PICK	20	Causes the terminal to wait for an interrupt and subsequently to report the user's action
WAIT FOR HIT	21	Wait for key activation
WAIT FOR POINT-IN	22	Wait for input of coordinates by lightpen, cursor, etc.
SKIP	23	Skips the next k commands
SKIP IF HIT	24	Skips the next k commands conditioned by key identifier
CREATE QUEUE	26	Creates a queue with a specified length and name for a specified device
ENABLE	29	Enables a specified attention source
DISABLE	2A	Disables a source
DO TRACKING	2B	Starts tracking. (X, Y) values are placed into identified queue
DO INKING	2D	Same as TRACKING but points are connected by straight-line segments
REPORT	2F	Transmits content of MT and ADL and subsequently clears the two files

*Note that L^4 provides a format in which the user may implement any command of his choice. Thus, the table may be arbitrarily extended.

Group IV: Report Commands		
PICK RETURN	30	Enters identification of picked object into MT+ADL
HIT RETURN	31	Enters identification of hit function key into MT+ADL
POINTIN RETURN	32	Enters pair of coordinate values into MT+ADL
REPORT ID	33	Returns terminal ID plus a 64-bit boolean vector indicating the status of the L^4 implementation in the terminal
QUEUE RETURN	33	Terminates CREATE QUEUE and enters content of queue into MT+ADL

The simplest way of organizing a dialog between computer and terminal would be to let only the application program in the host manage the display file. In this case, the user's attention would have to be served by appropriate attention-handling programs in the host, resulting in modifications of the display file, similar to modifications requested by the application program. Each time a change has been made, a copy of the updated display file could be sent to the terminal in order to replace the current copy. Such a simple system, however, would violate the two rules for a satisfactory performance of interactive graphics in a time-sharing environment: not to transmit any redundant information and not to handle single attentions in the time sharing loop with its prohibitively slow response (see the discussion in Section 1.7). Therefore, we generate the entire display file just once, at the beginning of a dialog, and update it from then on successively. Only the irredundant information concerning a picture editing need be exchanged. As picture editing may be done from either side, we need a vehicle that will carry the required information back and forth. This carrier consists of a list of L^4 instructions, called the *message table* (MT), combined with an *associated data list* (ADL).

In this network, each satellite will contain a preprocessor and display processor. The preprocessor must perform the following tasks:

1. Interpretation of L^4 instructions and translation into appropriate DPC program segments.

2. Creation and updating of the display file copy residing in the terminal.

3. Identification of picked objects, recognition of attention sources, and formulation of appropriate messages to be sent to the application program in the host computer.

4. Acceptance of input from the alphanumeric keyboard or the function keyboard.

5. Execution of subprograms that handle certain interactions locally.

10

High-Level Language Implementation of Display Programming Systems

As pointed out in Chapter 9, the realization of a graphical programming system in the form of a subroutine package is an adequate approach, as long as the tasks of such programs are restricted to generative graphics. Consequently, many such subroutine packages for interactive, generative computer graphics were implemented over the years, most of them in FORTRAN. The primary justification for the dominant use of FORTRAN is its ubiquitousness. Besides COBOL, FORTRAN was and still is the most commonly used programming language. In the case of stand-alone graphic systems, the small-scale to medium-scale computers used as the dedicated sole processor frequently had only a FORTRAN compiler available.

Rather than FORTRAN, we consider PL/I, APL, and PASCAL to be more eligible host languages for a graphic programming system. In this chapter, a model programming system for interactive and generative computer graphics is presented and the suitability of some of the most commonly used high-level, general-purpose programming languages for the implementation of such a system is discussed. In the course of this discussion, the fact must be considered that in actual computer graphics applications, the suitability of a language for constructing and managing data bases is at least equally important. The reason we shall, nevertheless, dwell more on the picture generation than on the data base aspects is that in the latter case the language capabilities are fairly well understood. Hence, we consider primarily the interrelationships among picture structure, data structure, and language structure as well as the mechanism of picture specification in order to identify the required capabilities.

10.1 TASKS OF A GRAPHICAL PROGRAMMING PACKAGE

The process of picture generation may be classified into two categories; program-created pictures (plotting) and interactively created pictures. Whereas the first case is rather straightforward, the second case includes the "attention-handling" techniques (see Chapter 6), and the associated program structure is affected by the choice as to whether all steps of an interactive procedure are directly controlled by the main program or whether we have an "intelligent" terminal with local computing power. In the latter case the program in the host computer may give only some global directives, whereas the individual steps of an interactive procedure are controlled and executed locally.

As was pointed out in Chapter 3, a programming package for generative graphics must provide tools for the structuring of pictures into entities and subentities. A graphic entity may have its own name, attributes, and coordinate origin. The program segment for the definition of an entity can conceptually be divided into two portions, one for an entity initialization and one for an entity specification. The entity initialization consists in the assignment of name, attributes, and origin to an entity; the entity definition consists in a sequence of object-generating statements. The entity attributes determine the appearance and the status of an entity (see Chapter 3). In a practical system, it is convenient to concatenate the attribute values into one array, which we may call the "BLIP vector," as it consists of the four components B, L, I, P (e.g., as defined in Table 10.1).

Table 10.1: EXAMPLE FOR POSSIBLE ENTITY ATTRIBUTES

Component	Usage	Values	Meaning of values
B:	Blink status	$B=0/1$	
L:	Line style *or* character size	$L=-1/0/1/2$	Dot-dashed/solid/dashed/dotted; small/normal/large/extra large
I:	Line boldness (intensity) *or* gray levels *or* color code	$I=-2/-1/0/1$ *or* $I\in[0:N-1]$ *or* $I\in[0:M-1]$	Invisible/dim/normal/bright (N = number of gray levels; M = number of colors)
P:	Lightpen *or* character overwrite control	$P=0/1$	Lightpen disabled/enabled *or* overwrite disabled/enabled

In the case that certain picture transformations are to be performed after the execution of entity-generating procedures, the transformations may also be considered as part of the picture-generation mechanism. The structure of the program segments for picture generation depends on how the data structures and language constructs available in the high-level programming language are related to the picture structure. Besides operators for picture generation, a graphical programming system must also contain operators for creating and editing the display file as well as operators for reporting the results of the user's attentions to the application program. The block diagram, Figure 10.1, represents the tasks of a graphical programming package and their interdependence.

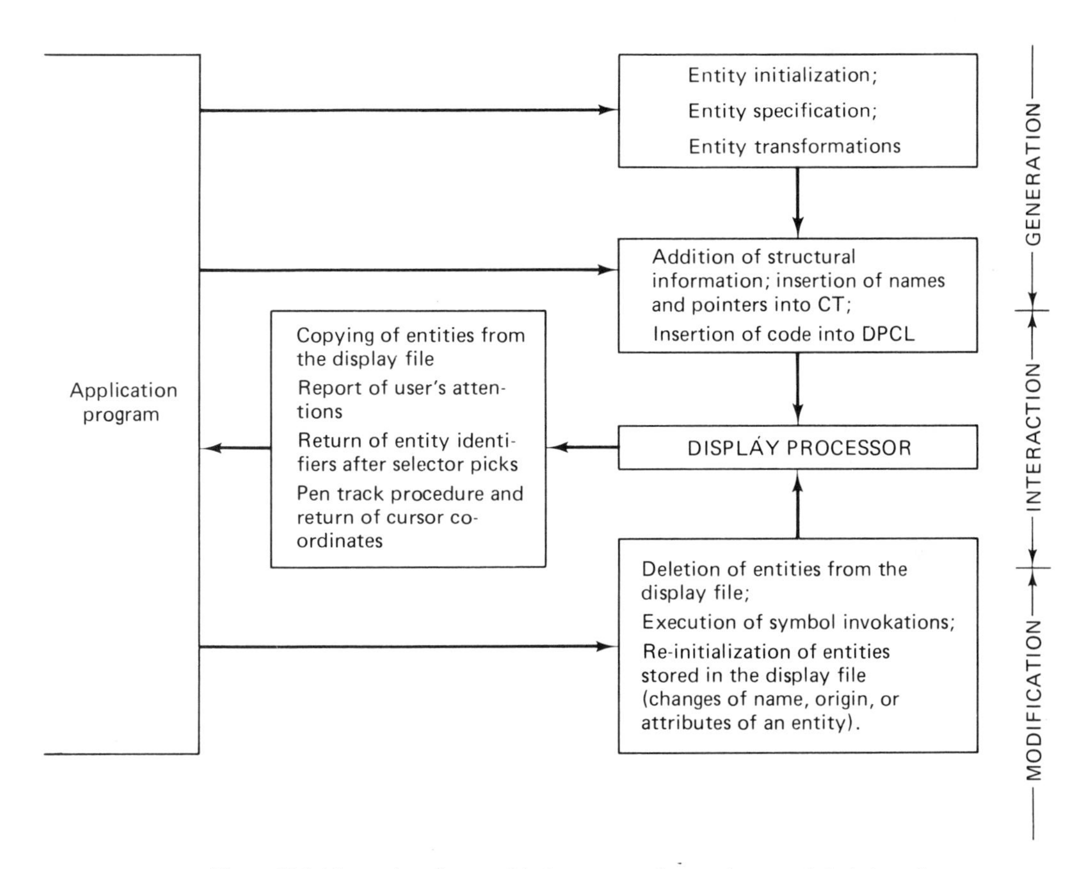

Figure 10.1 The tasks of a graphical programming package and their interdependence. The tasks centered around the display processor actually interface between application program and display.

Display file editing is primarily a problem of searching for the names of the entities to be modified or deleted. The same holds true for the copying of an interactively created entity. The problem of entity identification through a "pick" is also primarily a search problem; however, this time a search must be performed on the list of pointers by which the display processor code segments of the entities are correlated with the names.

10.2 LANGUAGE STRUCTURES

Naturally, we have at the display processor level a one-to-one correspondence between a primitive and a program statement, since the display processor code is a low-level language that offers no possibility for further structuring. Many suggested language extensions and existing subroutine packages maintain the same one-to-one correspondence between statement and primitive in the high-level graphical programming system. Such an approach somehow contradicts the idea of a high-level language. For the picture structure introduced in Chapter 3, it is certainly more logical and appropriate to reserve the statement level for *items* and the procedure or program block level for *segments*. In that case primitives occur only implicitly as the elements of items. Procedure calls may be nested, as segments may invoke *symbol instances*, and so on.

One situation where single primitives might be generated in a statement-by-statement fashion occurs in the case of interactive picture creation when the display system is a stand-alone unit in which all the computing power is concentrated in one dedicated computer. A statement-wise generation of primitives may here be indicated to provide immediate feedback to the user about the results of his actions. For example, the user may locate a point on the screen with the lightpen, and the immediate response of the system may be the generation of a vector that connects this new point with the previously located point. The adequate control construct for such a process is the iterative loop, and a special device must be provided to group into items the primitives thus generated. Furthermore, event-driven routines must exist for the handling of each of the given input devices (see Chapter 6).

A language structure, well suited for the structuring of display programs, exists in the PL/I-type language, i.e., a language that has internal procedures (or blocks) as well as external procedures of the function type and the statement type [150]. In contrast to such a highly structured language, APL and FORTRAN provide only structurally independent external procedures (subroutines). A common measure that compensates for the lack of a block structure may be the introduction of "pseudo-blocks" through appropriate subroutine calls. Of course, such a measure does not create new name scope rules, and the passing of parameters from one such pseudo-block to another can only be carried out within the given subroutine parameter passing mechanism.

On this basis, we may have the following mechanism for entity definitions:

1. *Items*: Items are created by appropriate item-generating library procedures with reserved names.

2. *Segments*: The proper construct for a segment definition is the program segment, given in the form of internal procedures or of BEGIN...END blocks. The program segments comprise invocations of item-generating library procedures and/or symbol-generating program segments. Segment and/or symbol definitions may be nested. The common name scope rules, where in a nesting of blocks or procedure calls an invoked procedure inherits the variable access rights of the invoking procedure unless the variables of the invoked procedure are declared as local, may be exploited to let a symbol instance or an item either have its own individual attributes or assume the attributes of the invoking segment. In languages without block structure and/or internal procedures, a segment or a symbol may be delimited by a pair of special library procedures. The function of these procedures is to enter the entity name into the name list, the appropriate pointers

into the correlation table, and the generated display processor code into the DPC list.

3. *Symbols*: Like segments, symbols are defined by a program segment. However, there are two decisive differences between segments and symbols: (a) a symbol is a nondivisible entity (i.e., all its items have the same attributes and are not individually identifiable), and (b) the execution of the DPC segment for a symbol is deferred until the symbol is invoked, then causing the creation of a symbol instance. Attributes are assigned to symbol instances, not to symbols.

It may be useful to again emphasize the difference between a symbol invocation and the repeated call of a picture-generating procedure. A picture-generating procedure may be connected with various transformations which modify the picture from call to call. Hence, such a sequence of calls generates a sequence of different instances of a picture. Each invocation results in a new segment of display processor code and, therefore, from the viewpoint of the DPC program, the effect is that of a *macro*. Contrastingly, a symbol resides in the display file and may be called during the execution of the display processor program. From the viewpoint of the DPC program, its effect is that of a *subroutine*. This implies that a symbol instance cannot be subjected to picture transformations other than translation, unless the display processor comprises special hardware for picture transformations. The result of our discussion is summarized in Table 10.2.

The first draft of the CORE recommendations [1] by the ACM/SIGGRAPH Standards Planning Committee proposes a two-level picture structure consisting of primitives and segments. However, the appropriateness of having a third structure type, more complex than a single primitive but less complex than a segment, was also acknowledged. The inclusion of such "advanced primitives," on the other hand, contradicts the objective to have a minimal set of primitives. Nevertheless, the committee decided to introduce the polygon as a primitive, and it was recognized that the addition of other "advanced primitives" may be indicated as well.

Table 10.2 PICTURE STRUCTURE, DATA STRUCTURE, AND LANGUAGE STRUCTURE

Data structure	Picture structure	Language structure	Comment
n-tuple	Primitive	Library subroutine (function in APL)	Generated by display processor hardware
Homogeneous array	Item	Library subroutine (function in APL)	Defined as an ordered set of primitives of the same type
Set of arrays	Symbol	External or internal procedure* or block	Nondivisible entity
Tree	Segment	External or internal procedure* or block	Defined as a set of disjoint items

*In FORTRAN and APL only external; in ALGOL 60 only internal.

In our opinion, defining sets of primitives as new primitives is not a clean solution. This dilemma is solved and the principle of minimality is restored by the three-level picture structure as defined above, where primitives are only those elementary graphical objects which are generated by the hardware display generator. The introduction of homogeneous sets of primitives as a new elementary structure type constitutes a very natural extension that provides the desired advanced structures in a clean way. If the programmer wants to stay with a two-level structure, he/she can exercise the option to declare such an "item" to be a single element set. Furthermore, such a structure enhances the efficiency of the attribute declaration and visibility control mechanisms.

10.3 INTERACTION HANDLING ROUTINES

In the case that pictures are constructed interactively (i.e., primitive-wise), we need a device for grouping primitives into items. This can be accomplished analogously to the grouping of items to segments. The (virtual) devices for command and data input and object identification (selector, valuator, locator, button; see Section 6.4) must be serviced by appropriate routines. One of the most popular media for man–machine interaction is the display of light buttons by the program and the "pick" of such light buttons by the user (see Section 6.1). Consequently, we need a number of library procedures for the output and input of information through these devices. A good control structure of the language is very helpful in supporting such routines and aids in distinguishing the interactive operation from the straightforward picture-generation operations.

As an example, let us consider the following problem. At some point in an interactive program, four light buttons, LINE, DOT, ARC, EXIT, are to be displayed. By picking one of them, the user will decide which item type she/he wants to construct interactively on the screen: POLYGON, DOTSET, or ARCS. We assume that we have two library procedures available, a subroutine MENU(...), whose call results in the display of a number of specified light buttons at specified locations with specified identifiers, and a procedure BUTTON(ID), whose call causes the system to enter a WAIT state until a light button has been picked, upon which event it returns the identifier of the picked button (see the following section). Let us assume that we have the following identifiers ID: LINE=1, DOT=2, ARC=3, EXIT=4. The next step in the program is a multibranching to one of four different procedures, conditioned by the BUTTON return. ALGOL and PL/I provide for branching by the "IF B THEN S1 ELSE S2" clause. Such a binary branching is not very appropriate, as the general case in interactive graphics programs requires a multibranching, in which case we must resort to the less-than-elegant approach of using the IF...THEN clause several times. A more efficient device would be the CASE clause suggested by Hoare, Wirth, et al. If we had this available, we could write, for instance,

```
SEGMENT(SKETCH, MXA, MYA, IAV)
          . . . . . . . . . . . . . . . . . .

          MENU(....);      BUTTON (ID);
          CASE ID          OF
          1:     BEGIN POINTIN(IX,Y);
                           MXR:=IX -MXA; MYR:=IY-MYA;
                           ITEM(ORIGIN,0,0,0,IDAT); DOT(IX, IY);
                           ITEM(POLYGON,MXR,MYR,IAV,IDAT); I:=0;
                           WHILE    ID=1 DO POINTIN(IXP, IYP); I:=I+1;
                                    IDAT [1,I]:=IXP-MXA; IDAT[2,I]:=IYP-MXA;
                                    LINE(IXP, IYP); IX:=IXP; IY:=IYP; BUTTON(ID);
                           END;  DROP(SKETCH,ORIGIN,1);
                 END;
          END;
          2:. . . . . . . . . . . . .
          .
          .
          .
          END;
END;
```

Comments: The outermost block is a segment definition block. After some other statements, the menu with the four light buttons, LINE, DOT, ARC, EXIT, is displayed and the identifier of the picked button is returned by the subroutine BUTTON(ID). The value of ID is used in the case construct, and we assume that LINE was picked. The next line begins the program block for ID=1, beginning with the call of the subroutine POINTIN. POINTIN returns as the values of IX and IY a pair of absolute coordinates which are transformed in the next line to relative coordinates. In the following line, a dot is generated to indicate the picked point. As we want to later delete the dot, we define it as an item, named ORIGIN. The definition of this item may not need to be closed by an END, as the following statement is again an item definition, this time for the interactively constructed polygon. The construction takes place in an iterative DO loop which goes on as long as the user continues to pick the button LINE. The first action is to call POINTIN, i.e., to let the user specify a new point. The absolute coordinates of this point are made relative and stored in the array IDAT. Subsequently, the created line is displayed. A call of BUTTON returns the information describing the desired action of the user. On completion of the polygon, that is, when the DO loop exits, the initial dot is erased. The reader may have noticed that the language which is used is a PASCAL-type language [154], augmented by some of the GRIP statements of Section 10.4.

We stated before that certain routines (e.g., POINTIN) must wait for the input of a pair of coordinates through the process of positioning a cursor interactively by the lightpen or a similar device.† We assume that the completion of such

†For the sake of simplicity, in the following we use the word lightpen consistently for a selector device. The reader may replace in her/his mind "lightpen" by "selector".

a process is indicated by the user by pressing (or releasing) a certain control button (e.g., the lightpen switch), thus causing an interrupt. We assume furthermore that, as a result of that interrupt, the system sets a boolean variable B to 1 (or TRUE). In PL/I we could now write the POINTIN procedure as follows

```
POINTIN: procedure (IX,IY)
         .............................................
         WAIT(B)
         .............................................
         END
```

Here, B is the name of the event which causes the procedure, which was halted at the occurrence of the WAIT statement, to continue. If the host language does not have such a WAIT construct, it must be added (e.g., in the form of an interrupt-driven routine). Our example program lacks elegance, inasmuch as after each input of a coordinate point, the light button LINE (or DOT or ARC, respectively) must be picked again for the sole purpose of indicating to the system that the user does not want yet to exit, but wants to continue drawing lines (dots, arcs). In our example program, we are checking this condition by a repetitive "polling" of the ID value. In principle, however, our program could be organized such that, after initiation, the process of item construction would go on until the light button EXIT is hit. The organization of the program in such a way requires the capability of detecting a light button pick (or any other attention caused by the user) whenever it occurs; i.e., it requires the capability of interrupt handling (see Chapter 6).

PL/I offers certain limited facilities for interrupt handling in the form of the ON statement [76]:

```
ON ⟨condition-name⟩ ⟨action-specification⟩.
```

Thus, the user can specify what action shall be taken if the named condition arises. Such conditions may occur in the process of

- Computation.
- Input/output.
- Program checkout.
- List processing.
- System actions.

The user is thus given a list of conditions to choose from. However, what is needed are conditions derived from external events (e.g., caused by attentions). PL/I does not provide for such a possibility. Although the programmer is allowed to name his/her own condition—for example, one could write ON CONDITION (LIGHTPEN) BEGIN...END—the occurrence of the named condition can only be simulated by inserting a SIGNAL statement at a certain point into the program. [In the example above, this SIGNAL statement may read SIGNAL CONDITION(LIGHTPEN).] Such a construct is helpful for testing ON statements but is not a measure for handling external interrupts. However, PL/I may be rather easily extended such that it would accept external attentions as conditions [77], as the ON statement already provides the syntactic form for such an extension. This distinguishes PL/I favorably from other languages. Moreover, PL/I provides for multitasking and task synchronization through the WAIT(⟨event⟩) statement. This may be very useful in interactive graphics, as it permits the simultaneous execution of an interrupt-driven procedure and one or more regular procedures: e.g., while the system must wait for a user's response, other parts of the program could already be executed. By synchronizing these tasks it can be ensured that all steps are kept in the right order. Task synchronization is accomplished by halting at a given point a procedure execution by a WAIT(event) statement and creating the event by a second procedure the first one must wait for, and so on.

Zahn suggested for future programming languages event-driven constructs, for example, in the form [157, 86]:

```
BEGIN UNTIL ⟨event⟩_1 OR...OR ⟨event⟩_n:
        ⟨statement list⟩_0;
END;
THEN ⟨event⟩_1 = >⟨statement list⟩_1;
                  :
     ⟨event⟩_n = >⟨statement list⟩_n;
FI;
```

Such a construct would be well suited for programming interrupt-driven routines.

As long as we do not have constructs of that type available in the high-level language which we use for display programming, we must resort to programming the interrupt handling at the assembly-language level, preferably in the fashion of writing code procedures that can be called as library subroutines.

Hence, in addition to the building blocks for picture generation, given in the form of item and primitive generating library procedures, we need to have building blocks for the interactive execution of a display program. The latter come in two groups: a group of library procedures for interactive display file management (change of attributes and origin of graphic entities; erasure of segments, items, or primitives; augmentation of existing entities) and a group of library functions for attention handling. Finally, we may devise some standard procedure for inking and sketching to be executed in an "intelligent" terminal.

All the conclusions drawn from our discussion are reflected in a model language outlined in the following section. This model language can be thought of as a blueprint for either the extension of a host language or its augmentation by a procedure package. In the latter case, modifications will become mandatory, owing to the idiosyncrasies and limitations of the respective host language.

10.4 GRIP (GRAPHIC PROCEDURES FOR INSTRUCTIONAL PURPOSES)

GRIP [51] was designed as an instructional vehicle to clarify the mechanism of structured, high-level graphical programming rather than as an extension to a particular programming language. Consequently, idiosyncrasies of secondary importance within existing languages were not taken into consideration. For the sake of convenience, multiparameter operators were introduced, a rather unusual construct in most programming languages (except for PL/I, where the more complex file-handling functions are of a similar type). Nevertheless, GRIP has already found two implementations, one in the form of a FORTRAN subroutine package (see the Appendix) and the other in the form of a package of APL "defined functions." GRIP has been used and is being used every year by scores of students with great success. We are particularly pleased to notice that most of the conceptual framework of GRIP can also be found in the CORE concept of the ACM/SIGGRAPH Graphics Standards Planning Committee, which GRIP predates. On the other hand, GRIP comprises the features set forth by Kulsrud for the generative and manipulative parts of her language.

The reader will notice that GRIP is very procedure-oriented. In fact, GRIP is the concept of a procedure package without the special syntactic conventions of a particular language. This fact makes its implementation very straightforward. In the following, we define the set of GRIP commands in the form of a table. The reader may notice that the naming of parameters is (1) mnemonic and (2) declares them as integers in the case of FORTRAN or PL/I (by default).

GRIP

Group 1: Object-generating statements

segment level

SEGMENT '⟨Segmentname⟩; MXA; MYA; IAV'

Parameters: ⟨Segmentname⟩ identifies the segment

MXA,MYA: Segment origin in absolute coordinates ("move to X(Y) absolute")

IAV: Segment or item attribute vector

IAV=B,L,I,P

B: blink status

L: line writing style or character size

I: boldness (intensity) or gray level or color

P: lightpen pick status

GRIP

END Closes a segment definition block and inserts the code generated since the last END or SHOW command (see group 2) into the display file.

Note: The scope of IAV extends over all items or symbol instances of a segment unless their attributes are individually specified.

Item level

Common Parameters:
- ⟨Itemname⟩: Identifies the item
- MXR, MYR: Item origin, relative to the segment origin (absolute if MXA = MYA = 0)
- IAV: Item attribute vector (see above) If IAV is not specified, the system assigns a default value (e.g., the value 0, 0, 0, 0).

POLYGON '⟨Itemname⟩; MXR; MYR; IDAT; IAV'

Parameters: IDAT: Array of coordinates, relative to (MXR,MYR), and control parameters. Dimension(IDAT) = 3,N (N = number of primitives. The control parameters are either 1 = visible or 0 = invisible.

DOTSET '⟨Itemname⟩; MXR; MYR; IDAT; IAV'

Parameters: IDAT: Array of coordinates relative to (MXR,MYR). No control parameter; dimension(IDAT) = 2,N.

TEXT '⟨Itemname⟩; MXR; MYR; MODE; MESSAGE; IAV'

Parameters:
- MESSAGE: Character vector of the text
- MODE: Character write mode

ARC '⟨Itemname⟩; MXR; MYR; IDAT; IAV'

Parameters: MXR; MYR are the center coordinates of the circle that contains the arc. IDAT specifies the end points of the arc. The points are connected in the mathematically positive sense.

Symbol level

INSTANCE '⟨Symbolname⟩; ⟨Instancename⟩; MXR; MYR; IAV'

Parameters ⟨Symbolname⟩ identifies a symbol to be activated in the segment definition block in which the INSTANCE command occurs. The created symbol instance is identified by ⟨Instancename⟩; it has the (relative) origin and the attributes as specified by MXR, MYR, and IAV, respectively.

SYMBOL '⟨Symbolname⟩'

Parameters: ⟨Symbolname⟩ identifies the symbol (subpicture) created by the statements following the SYMBOL statement. Items and primitives of a symbol are not individually identifiable. Any origin and attribute declarations for these items will be ignored.

GRIP

END Closes a symbol definition block. Furthermore, END inserts the code generated since the last END or SHOW command (see group 2) into the display file.

Group 2: Display file manipulation and object identification

SHOW : Inserts the code generated since the last SHOW command into the display file

---

CLEAR : Erases the entire display file

DELETE '⟨Segmentname⟩; ⟨Itemname⟩' : Deletes an item or a segment

Parameters: If an item is to be deleted, it must be identified by its name and the name of the segment it belongs to. If ⟨Itemname⟩ is omitted, the entire segment identified by ⟨Segmentname⟩ is deleted.

DROP ⟨Segmentname⟩;⟨Itemname⟩;IN' : Removes the identified primitive from view

Parameters: IN: Index number of the primitive in the identified segment and item to be removed from view.

ADDITEM '⟨Segmentname⟩;⟨Itemname⟩' : Adds an item to an existing segment

---

CATENATE '⟨Segmentname⟩;⟨Itemname⟩;IDAT' : Concatenates additional primitives to an existing item

Parameters: IDAT: Data array for the additional primitives

PICK 'PSNAME;PINAME;IN' : Identifies a displayed entity through a pick by a *selector*

Parameters: IN : Index number of the picked primitive

PSNAME, PINAME : Output parameters identifying the item and segment to which the picked primitive belongs

The search through the segment tree can be accordingly truncated if PINAME and/or IN is not listed.

NEWSTATUS '⟨Segmentname⟩ ; ⟨Itemname⟩ ; IAV' : Assigns new attributes to an item

Parameters: IAV: New attribute vector that replaces the one currently assigned to the specified item

---

NEWORIGIN '⟨Segmentname⟩ ; ⟨Itemname⟩ ; IPX ; IPY ; IIX ; IIY' : Assigns a new origin to an item

Parameters: ⟨Segmentname⟩ identifies a segment existing in the display file and ⟨Itemname⟩ identifies an item in that segment

IPX, IPY: Increments to be added to the current segment coordinates (MXA, MYA) in order to change the segment origin

GRIP

IIX, IIY: Increments to be added to the current item coordinates (MXR, MYR) in order to change the item origin;
IIX=0, IIY=0: No change of the item origin.

NEWNAME '⟨Oldname⟩ ; ⟨Newname⟩' : Changes the name of a segment

Parameters: ⟨Oldname⟩ is the current segment identifier that is to be changed into ⟨Newname⟩.

(*Note*: PICK is an event-driven construct and includes the invocation of SHOW.)

Group 3: Input commands

POINTIN 'IX ; IY' : Input of a coordinate point through a *locator*

Parameters: IX, IY: Output parameters specifying the absolute coordinates of a point on the screen marked by the locator

---

TEXTIN 'MESSAGE MAX' : Input of a text

Parameters: MESSAGE: Output vector containing the characters typed in by the user
MAX: Maximal number of characters that can be typed in

KEYIN 'ILLUM ; KEYCODE' : Activation of a control key (*button*)

Parameters: ILLUM: Boolean vector whose length equals the number of control keys. 0=key not illuminated, 1=key is illuminated
KEYCODE: Output parameter identifying the control key pressed by the user.

VALUEIN 'N ; DAT' : Input of an *n*-tuple of numerical values by a *valuator* (e.g., the keyboard or a "digital potentiometer")

Parameters: N: Number or values to be put in
DAT : Output parameter representing the array of typed-in values. The values must be separated by one or more blanks.

(*Note*: POINTIN, TEXTIN, KEYIN, VALUEIN are event-driven constructs. They include the invocation of SHOW.)

Group 4: Interactive picture generating statements

NEWITEM '⟨Itemname⟩; MXR; MYR; KIND; IDAT; IAV'

Parameters: ⟨Itemname⟩: Identifies the item created by the subsequent primitive-generating statements
MXR, MYR, IAV: see above
KIND: Input parameter declaring the type of the interactively created item
IDAT: Output parameter returning the data of the interactively created item

GRIP

Primitive level

DOT 'IX; IY'
LINE 'IX; IY'
NOLINE 'IX; IY'
Parameters: IX; IY: Absolute screen coordinates specifying a dot or the end point of a line (or no line, respectively) whose start point is the current beam position

---

CIRCLE 'IDAD'
Parameters: IDAD: Array containing the coordinates of the points specifying a circle

---

END : Closes an item-generating block. END includes an invocation of SHOW.

Menu technique

MENU '⟨Menuname⟩ ; NC; ID; MESSAGE; LOC; IAV' : Creates a menu
Parameters:
NC: Number of characters in each light button (must be the same for all light buttons of a menu)
ID: Array of integers identifying the light buttons
MESSAGE: Character vector of the light buttons (padded up by blanks to equal length)
LOC: Array defining the location of the equally spaced light buttons by its four components, which specify the absolute coordinates of the first character in the first button (from left to right and from top down) as well as the *x*- and *y*-increment from button to button

---

BUTTON 'ID' : Waits for the pick of a button by a *selector* and returns the identifier of the picked button. BUTTON is event-driven and includes the invocation of SHOW.

Pen-track and inking routines

PENTRAK '⟨Segmentname⟩; ⟨Itemname⟩; MX; MY'
Parameters: ⟨Segmentname⟩ and ⟨Itemname⟩ identify the object that can be attached to a *selector* (e.g., a lightpen) and moved to a new position. If ⟨Itemname⟩ is not specified, the whole segment is moved.

MX, MY: Output parameters that yield after termination of the procedure the origin of the moved entity (in absolute coordinates in the case of a segment, in relative coordinates in the case of an item).

GRIP

SKETCH '<Itemname>; MXR; MYR; MAX; NP; IDAT; IAV'

Parameters:

- <Itemname>: The figure created by freehand sketching (inking) is declared as an item of type POLYGON and identified by <Itemname>.
- MXR, MYR: Origin coordinates
- MAX: Maximum number of line segments by which the drawn figure will be approximated
- NP: Actual number of sampling points of the figure. The procedure yields a value of NP for which is $MAX/2 \leq NP \leq MAX$.
- IDAT: Output array containing the coordinates of the sample points. Dimension(IDAT) = 2NP.
- IAV: Attribute vector of the generated item

Comments

1. A segment definition block must be delimited by SEGMENT and END. A symbol definition block is formed by the delimiters SYMBOL and END. Symbols may consist of an arbitrary number of items or other symbols which, however, cannot be declared as entities in their own right. Symbol definition blocks may be nested in other symbol definition blocks or in segment definition blocks, but segment definition blocks must not be nested in other segment definition blocks.

Example:

```
SYMBOL      'FILTER'
POLYGON     'LINE;0;0;IDAT1'
DOTSET      'NODES;0;0;IDAT2'
INSTANCE    'C;C1;0.2;0.2'
INSTANCE    'C;C2;0.4;0.2'
INSTANCE    'R;R1;0;0.2'
INSTANCE    'R;R2;0.2;0.2'
INSTANCE    'R;R3;0.4;0.2'
TEXT        'RNAME;0.1;0.22;IDAT3'
TEXT        'RVAL;0.1;0.18;IDAT4'
TEXT        'CNAME;0.17;0.1;IDAT5'
TEXT        'CVAL;0.17;0.08;IDAT6'
END
SYMBOL      'R'
POLYGON     ...
END

SYMBOL      'C'
POLYGON     ...
END
```

```
SEGMENT    'CIRCUIT;0;0;IAV'
INSTANCE   'FILTER;F1;0.5;0.9'
TEXT       . . . . . . . . . . . .
SHOW
. . . . . . . . . . . . . . . . . . . . . . .
END
```

(*Note*: An item consists only of primitives of the same type, but these primitives are individually identifiable. A symbol may consist of items and/or other symbols, but these items or symbols and their primitives, respectively, are not individually identifiable. Symbols are also indivisible entities with respect to the attributes.)

2. NEWSTATUS reassigns attributes to a segment or an item; NEWORIGIN can translate either one. NEWNAME allows the interactive creation of a modified version of a segment while the old version is being displayed. The new version is at first created under an intermediate name. After completion, the old version is deleted and, subsequently, its name (which is now free) is assigned to the new version.

3. The pair MENU...BUTTON provides a very convenient mechanism for creating a light button menu and handling light button picks. MENU has the status of a segment. The restriction of this technique lies in the fact that the light buttons can only be arranged in one column and must all have the same attributes. If light buttons with different attributes are wanted, they must be declared as individual items. MENU will usually contain an escape facility (e.g., a light button EXIT) for termination of a process by the user. MENU includes the invocation of SHOW.

4. NEWITEM inserts an interactively created new item as part of an existing segment into the display file. PENTRAK initiates a procedure for lightpen tracking, and SKETCH initiates a procedure for freehand drawing with the lightpen.

5. The system may further be augmented by procedures for the generation of special items such as arcs, squares, or perhaps even certain conic sections. Other routines may perform rubberbanding, spline approximation of curves and surfaces, and picture transformations. We do not claim any completeness of the GRIP package as presented here. What we want to demonstrate is the "mapping" of the picture structure and picture-generation mechanisms introduced above into appropriate language constructs.

10.5 THE INSTANT OF OBJECT CREATION

One of the questions to which the designer of a graphical programming system must find an answer is the question of at what instant a generated object will become visible. One possible solution is to apply the principle of *instantaneous visibility* [1]. Instantaneous visibility demands that the smallest graphical entity becomes visible immediately after its generation. In the GRIP picture structure, that rule would apply to items and symbol instances. The principle of instantaneous visibility is quite appropriate for the interactive process of object creation. However, such a solution aggravates the data-transmission problem in graphical systems which are not stand-alone systems, i.e., where data are exchanged between the display terminal and its host computer at a relatively low rate. In this case it is more efficient to transmit larger data blocks by generating the code for larger graphic entities prior to the initiation of a data transmission. The scheme employed in GRIP is a reasonable compromise between the contradicting demands for instantaneous visibility and efficiency. It works as follows

1. Segments, symbols, and interactively created items are entered into the display file immediately after their generation. Hence, a newly created segment or an interactively created item of an existing segment becomes instantaneously visible (a symbol is never visible, only its instances are). This is accomplished by letting the block delimiter END invoke an operator, SHOW, that causes the entry into the display file of the code generated thus far.

2. If the user wants certain parts of a segment (e.g., certain items) to become visible before the entire segment has been generated, he/she may explicitly issue the SHOW command.

3. The displayed image should be updated prior to a "prompting" of an attention. Therefore, all event-driven constructs, such as PICK, POINTIN, TEXTIN, KEYIN, VALUIN, and BUTTON, include an invocation of SHOW. SHOW is also invoked by MENU, as a menu has the status of a segment.

This mechanism, and the scope control mechanism for attributes, rule out the possibility of nested segment definitions. However, this is no deficiency, as a nested segment definition would spoil the clean structuring of a graphical program anyway, without offering any particular advantage. Contrastingly, the nesting of symbol definitions is a very useful feature. As symbols are never visible, their nesting can be allowed.

10.6 COMMAND LANGUAGES

Whether a host language is truly extended or only augmented by a subroutine package, it must provide for the real-time environment required for the interactive use of computer graphics; i.e., it needs to respond to attentions issued by the user during run time. Moreover, the user must be provided with a language that enables her/him to exercise the person–machine dialog that is the very core of interactive computing. Such a language, in which the user can formulate requests to the system, is called a *command language*. GRIP encompasses the necessary tools for the application programmer to set up any appropriate command language, provided that the language also comprises the appropriate program flow control constructs (see Section 6.3). Hence, it is left to the user's discretion whether commands are associated with light button picks, function keys, or character strings forming keyword instructions.

If the host language is a "batch language," augmented by a graphic subroutine package, all that needs to be done is to identify a given command and generate a respective subroutine call with the specified parameters as arguments. The invoked subroutine is immediately executed ("compile and run" mode). If the host language is interpretively executed, the mechanism is basically the same, the only difference being that no compilation is involved. It should also be mentioned that one normally does not write one monolithic program for command processing, but that one rather prefers an implementation in the form of a modular system. This has a number of practical advantages. First, the overloading of the screen with light buttons can be avoided, for only the light buttons that are relevant for the current module need to be displayed. Second, and this is very important for small-scale systems, the common overlay technique may be applied so that only the modules needed at any given moment need to be in memory. The advantages that modularity generally offers with respect to writing, testing, debugging, and maintaining a program hardly need be mentioned.

10.7 IMPLEMENTING THE GRIP PHILOSOPHY IN EXISTING HIGH-LEVEL LANGUAGES

In the following, we shall discuss the suitability of existing high-level languages for being the host of a GRIP-like programming package.

10.7.1 FORTRAN

FORTRAN lacks any structuring facilities and therefore provides no support for a structured programming as introduced above. Standard FORTRAN offers no tools for bit manipulation and string handling, nor does it have any facilities for building data structures and handling files. A particular disadvantage is that only static arrays can be declared. The control structure of FORTRAN is awkward, to say the least.

In the case of a graphics subroutine package, the required structuring of the program can be introduced as a pseudo-block structure by respective subroutine calls. As we have no scope control facilities, the passing of attribute values from a segment to its items cannot be automatically performed but must be explicitly called for by the programmer (e.g., by inserting a respective code in lieu of IAV in the item-generating subroutine call). With these restrictions, all GRIP operators listed in Section 10.4 can be directly transposed into the form of FORTRAN subroutine calls. Such a GRIP implementation shall be discussed in more detail in the Appendix.

The FORTRAN subroutine concept is general enough to offer ultimately a solution for any problem—and this has been the key to the success of the language despite its many shortcomings. This holds also for the file management systems which have been programmed in FORTRAN for all conceivable data models, although the language itself provides only scalars and (homogeneous) arrays as readily available structures. However, it must be emphasized that the price to pay for the use of FORTRAN is a poor program structuring. Hence, for a genuine extension, FORTRAN is certainly the least appropriate candidate.

10.7.2 ALGOL 60

A positive feature of ALGOL is the block structure and the possibility of dynamic array declarations. On the negative side, we find the same deficiency as in FORTRAN, namely an overly restrictive set of data types and operators (no operations on bit strings and character strings). The often semantically unnecessary need of elaborate declarations is sometimes bothersome. This declarative overhead may be especially annoying in the case of display programming, if pictures are defined in the screen domain (i.e., all numbers are integer by nature). Real numbers and boolean variables which are logically connected cannot be packed into one array. The control structure of ALGOL 60 is inadequate. The language was designed strictly as an algorithmic language and is therefore rather inadequate for the programming of file management systems.

10.7.3 PASCAL

PASCAL is a language that combines a concept of outstanding clarity and simplicity with a modern control structure and is, therefore, of high pedagogical value. PASCAL has the block structure of ALGOL 60, but exhibits in comparison with the latter a number of improvements. Examples are: a better control structure; the existence of data structures other than only homogeneous arrays; input/output operators; the existence of character strings as a data type and of certain string-handling capabilities. We consider PASCAL as a very desirable environment for an implementation of the GRIP philosophy, and we are quite sure that with the increasing use of PASCAL, such an implementation will materialize.

10.7.4 PL/I

PL/I has almost all desirable features: a block structure, reasonable control constructs, and all required data types and operators for bit manipulation, string handling, and file handling (especially the type STRUCTURE is here very useful). A negative feature of the language is its arbitrary and voluminous features, albeit this is mitigated by built-in defaults which allow the use of the language even if not all of its facets are completely understood. From a practical point of view, PL/I is a strong competitor for the title of the "best-suited" language for graphical programming.

PL/I excels all other languages considered here by its facilities for building more complex data structures. This is primarily attributable to the existence of a data type called STRUCTURE. Contrasting to arrays, STRUCTURES need not be homogeneous. Furthermore, whereas array elements are referenced by a location index relative to a base, structure components are referenced by name. As an example, let us consider a simple PL/I structure declaration

```
DECLARE 1 PART
          2 NAME
            3 PART CHARACTER(10)
            3 SUBASSEMBLY CHARACTER(10)
            3 ASSEMBLY CHARACTER(10)
          2 PARTNUMBER FIXED BINARY
          2 QUANTITY FIXED BINARY.
```

PART is a structure with 3 components (the level-2 entries): NAME, PARTNUMBER, and QUANTITY. NAME again is a substructure with 3 components (the level-3 entries): Part name, subassembly name, assembly name (e.g., PISTON-RING, ENGINE, MB450SE). Such a structure can be depicted as a tree, as shown in the following figure. Another important mechanism for building data structures is the existence of a data type *pointer* which can be used to allocate various values of a variable in storage (see Section 2.4.2).

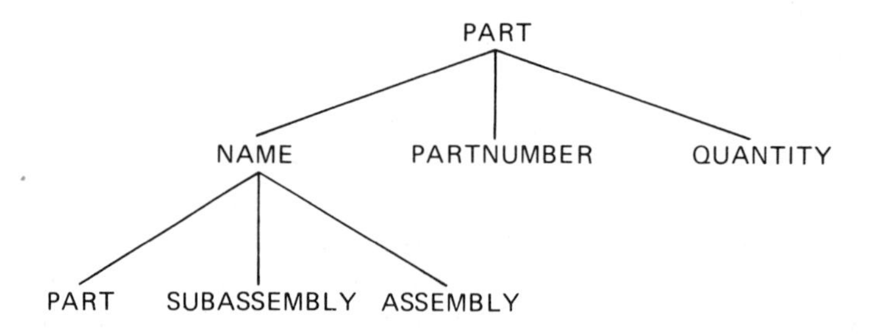

10.7.5 APL

Among the languages considered here, APL is the only genuinely conversational language and thus it is predestined for interactive computer graphics carried out in time-sharing systems [55]. Additionally, it is an asset of APL to have operators for all matrix operations applicable to picture transformations. APL is declaration-free (except for the required distinction between numerical (including boolean) and character values). One most useful property of APL for graphics purposes is the total freedom in using arrays dynamically. Since the name-value binding of variables is deferred until the first occurrence of a name in an assignment statement, at which point the system derives the type specification from the type of the assigned value, arrays can be arbitrarily redefined, restructured, or concatenated. This makes it very simple to perform domain transformations, to augment items by adding more primitives, or to connect items.

Objections voiced against APL are basically: (1) APL has only scalars and homogeneous arrays as data structures but no lists; (2) its control structure lacks the iterative control clauses of modern languages which help to reduce the use of GOTOs; and (3) APL programs are hardly intelligible for anybody but the one who wrote the program.

APL shares the first deficiency with most other languages. However, there is always the possibility of setting up a file management system in an APL workspace that can be used in connection with APL programs and that provides the required data base management capabilities [56]. The second argument is certainly a valid point. However, it has to be considered that in the case of "intelligent" terminals most of the iterative loops of a display program are handled locally. In this case, the "sequentialization" of the main program (which is the main purpose for abolishing GOTOs in favor of clauses such as WHILE and REPEAT) is automatically obtained. Furthermore, it is possible to introduce all desired control constructs (including synchronization constructs for concurrent processes) by an appropriate combination of special functions and labels [58].

The tremendous power of APL stems from the quasi-mathematical construction rules according to which a relatively small number of operators can be combined in a very regular way, resulting in a large variety of different operations. Once one has really become familiar with the language, APL expressions are as easy to read as mathematical expressions, if the programmer abstains from forcing too many and too heterogeneous terms into elaborate compound expressions. The extreme liberalism of APL, the possibility of multiple assignments, of catenating a branch destination with other expressions, etc., made it possible that a phenomenon could arise known as the "one-liner," and, alas, some of the APL literature encourages more than discourages this syndrome. In short, the extreme liberalism of APL requires from its user more self-discipline than other languages, but whenever such discipline is exercised, APL programs can be as readable as other programs. A particular asset of APL is that it enables its user to define his/her own operators (in the form of keyword instructions) and even her/his own control constructs.

The main restriction which APL imposes on an implementation of the GRIP concept is that the language has no internal block structure and that the APL "defined functions" have at most two explicit input parameters and one explicit output parameter. An item definition, however, requires the specification of three parameters, NAME, ATTRIBUTES, and DATA. This problem may be resolved by using a pair of dyadic functions for the creation of an item, one with name and attributes and one with name and data as arguments [57].

Another method is to use only monadic functions, with a character string as argument representing the list of parameters. Inside the functions, the character string is analyzed and the various substrings as delimited by the separator symbol (;) are processed: i.e., names are preserved as such, whereas substrings representing expressions are converted into expressions and evaluated. Such a possibility is provided for by a special APL operator called EXECUTE. Hence, the user is in this case entirely unrestricted as to the syntactic form of a parameter list and the number of parameters contained in it. Actually, as the reader may have noticed, the GRIP syntax was chosen such that it allowed us to literally realize GRIP in the form of a library of APL functions.

The usual APL systems are designed to be used in a conversational mode of operation, carried out via teletype or typewriter terminals. Consequently, the only I/O operator available is the "quad" function (or the "quote-quad," respectively) for the input and output of data or program code on the terminal. In interactive computer graphics, however, we have to deal with an additional I/O medium, the display console. This problem can also be handled by special functions at the APL end of the communication link and by an appropriate organization of the interpreter in the terminal.

APL lends itself ideally toward an implementation of the GRIP philosophy because of its absolutely dynamic handling of arrays. In the case of a matrix, for example, additional rows or columns may be arbitrarily added or "catenated." Hence, the fact that a windowing operation may increase the cardinality of the affected items causes no problems whatsoever. In the process of an interactive construction of items, the corresponding data array may initially be created as an empty matrix. Subsequently, whenever a new primitive is created, a corresponding row is simply added to that matrix. Hence, a high-level graphic programming system can be obtained in which windowing does not cause a program modification.

EXERCISES

1. Write a GRIP program that accomplishes the following actions.

(a) Set up a menu that contains the light buttons:

FACE: Allows the drawing of a circle. When this button is picked, the comment DRAW FACE should appear on the screen. The circle is specified by two points picked with the lightpen.

EYES: Allows the drawing of two circles. Comment on the screen: DRAW EYES.

NOSE: Allows the drawing of a line. Comment on the screen: DRAW NOSE.

EARS: Allows the drawing of polygons. Comment on the screen: DRAW EARS.

MOUTH: Allows the drawing of a polygon. Comment on the screen: DRAW MOUTH.

HAIR: Allows the drawing of a polygon. Comment on the screen: DRAW HAIR.

BEARD: Allows the drawing of a dotset. Comment on the screen: DRAW BEARD.

The coordinate points of all drawings should be specified with the lightpen. When the drawing of an object is finished, the corresponding comment should be deleted.

Why do you need the additional light button EXIT?

(b) The menu should contain another light button ORDER, which allows the user to type in a character string. When this light button is picked, the system should display the comment TYPE: the typed-in text should be displayed following this comment.

(c) The program should allow the drawing of the following face:

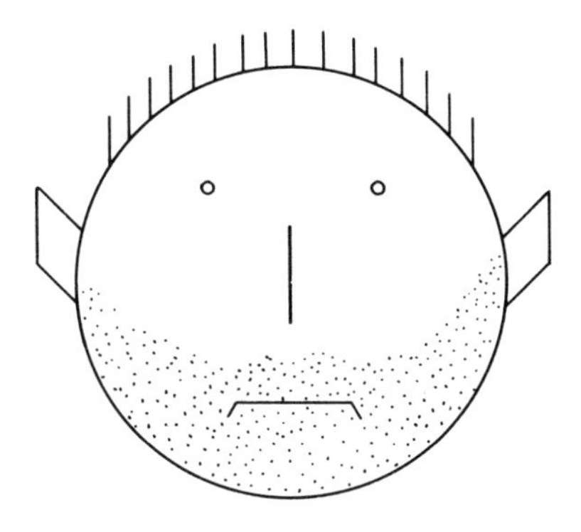

TYPE: I NEED A SHAVE!

(d) The computerized barber is activated by an additional light button SHAVE in the menu. When this button is picked, the face on the screen is shaved and appears as shown below. The invoked modification deletes the beard and the text I NEED A SHAVE! and replaces the original mouth by a grinning one.

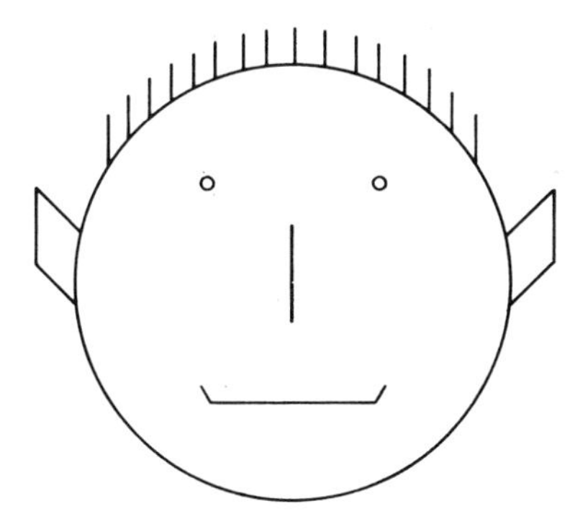

2. (a) Write a GRIP program that creates an aerial view of a carrier as shown below. Assume that the picture is defined in a [0: 1000]×[0: 1000] screen domain. The points P1 to P4 should not be displayed. The center coordinates of the carrier are (500, 500). All other measures can be taken from the picture.

(b) Write a symbol for the airplane shown. Display 4 instances of that subpicture at the points P1, P2, P3, and P4.

(c) Expand the program such that another airplane is initially displayed at the location (500, 800), heading east. Subsequently, this plane should circle the carrier clockwise. Can the fifth airplane be generated as an instance of the symbol? If not, what other language construct may be appropriate for its creation?

(d) Modify the program so that you can land the fifth airplane on the carrier or let it take off from the carrier. What is the most adequate input device for controlling the airplane's movements?

Assume that GRIP is embedded into FORTRAN in the form of appropriate subroutine calls.

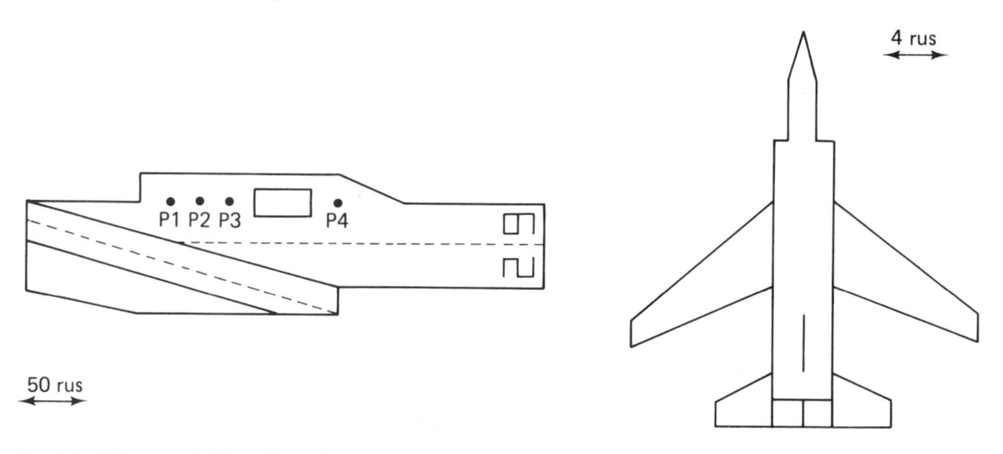

3. (a) Write a FORTRAN subroutine

MATRIX4(ALPHA,BETA,GAMMA,SX,SY,SZ,XT,YT,ZT,ZO,PIN,POUT)

which performs the 4×4 matrix transformations on homogeneous coordinates. The coordinates and parameters are given in a virtual, real-valued domain. The original coordinates are represented by the array PIN, and the obtained result is found in the output parameter POUT. See Fig. 3.10 for the meaning of the other parameters.

(b) Write a FORTRAN subroutine for the windowing of lines.

LWINDOW(WCC,ICC,RDAT,IDAT).

The subroutine allows the definition of a rectangular window in the virtual domain as specified by the coordinates of its lower left and its upper right corner point. These coordinates are represented by the array WCC. The array ICC represents the coordinates of a similar viewport, onto which the window is mapped. RDAT represents the end points of the original lines, specified in terms of virtual domain coordinates. IDAT represents the end points of the "windowed" (clipped and scaled) line, specified in terms of screen coordinates. LWINDOW may be invoked following the MATRIX4 procedure.

(c) Write a GRIP program, assuming FORTRAN as host language, which displays a cube. The edges of the displayed cube have a length of 100 raster units. In the world space the length of the edges may be 10. The point (0,0,0) in the three-dimensional world space should correspond with the point (500,500) on the screen. This point is the center point of the cube. The normalized screen domain is assumed to be $[0:1000]\times[0:1000]$. Apply a central projection with the parameters $z_O=100$ and $z_I=0$. Call the subroutines MATRIX4 and LWINDOW for the projective transformation of the cube in the world domain into the screen domain. Is there a particular order of invocation of MATRIX4 and LWINDOW?

(d) Set up a menu that contains the following light buttons. Integrate the corresponding actions into your program developed under point (c).

ROTATE: When this button is picked, the cube is rotated in increments of $\pi/18$ until the cube has reached its original position again. The rotation can be performed with respect to all 3 axes as specified by the user. Therefore, the system displays the message SPECIFY ROTATION, to which the user responds by the input of a vector (r_z, r_y, r_x). Each component of this vector may take one of the following values:

- -1: rotation in mathematically negative (clockwise) direction with respect to the axis indicated by the subscript;
- 0: no rotation with respect to the indicated axis;
- 1: rotation in mathematically positive direction.

SCALE: When this button is picked, the system prompts the user to input a vector of scale factors, (S_x, S_y, S_z), by displaying the message SPECIFY SCALE FACTOR. Subsequently, the cube is displayed in the modified form [in perspective view as defined under (c)].

TRANSL: When this button is picked, the cube is to be translated *in the screen domain*. The system prompts the user to specify the new center point of the cube by displaying the message SPECIFY NEW CENTER. Subsequently to the input of the new center coordinates, the cube is moved into the new position.

EXIT: When this light button is picked, the screen is cleared and the program terminates.

(*Note*: Displayed messages should disappear as soon as the prompted input has been completed.)

4. A special application program may allow its user to construct logical diagrams interactively. The diagram may contain an arbitrary number of gates. The gate types are AND, OR, and NOT, as denoted by the following symbols.

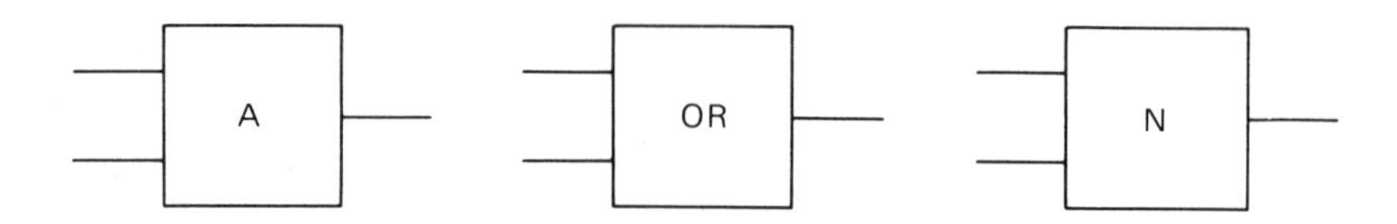

The generated block diagram should be defined as one segment. Gate symbols can be connected by polygons whose edges are strictly horizontal or vertical. In order to make these connections exactly match the input and output leads of the symbols, these leads should be lightpen-pickable. The pick of an input lead automatically defines the terminal point of the connecting polygon. The pick of an output lead automatically defines the start point of the connecting polygon. No connecting polygon must have more than three segments. This leaves the vertex between start point and terminal point to be specified (e.g., through a pen-track procedure). For the sake of simplicity, we assume (unrealistically), that each gate has only a "fan out" of 1, i.e., each output is connected with not more than one input.

Can the symbols be generated as symbol instances (make sure that the leads can be picked individually)? If not, what other language construct may be appropriate?

(a) Write a GRIP program for the interactive construction of logical diagrams. The gates should be created by picking the appropriate light button in the menu listed below, followed by a pen-track procedure in order to specify the center coordinate of the gate to be created. The gate symbol will then appear at that location. A output/input connection should be drawn by a pick of the light button LEAD, followed by a pick of the output lead of the first gate (in the sense of signal flow), followed by a pick of the input lead of the second gate, followed by the indication of the intermediate vertex through a pen-track procedure.

OR	Initiates the insertion of an OR gate into the diagram.
AND	Initiates the insertion of an AND gate into the diagram.
NOT	Initiates the insertion of an inverter into the diagram.
LEAD	Initiates the drawing of a connecting polygon.
TRACK	Invokes the pen-track procedure for the location of a point.
NAME	Initiates the input of a name via the alphanumeric keyboard. Subsequently, a gate in the diagram is identified by a lightpen pick, and the name, which is limited to not more than 3 characters, appears inside the symbol.

ERASE	Initiates the deletion of a symbol of the diagram. The symbol to be deleted must subsequently be identified by a lightpen pick.
EXIT	The pick of this light button terminates the design process.

(b) Add a light button INSERT to your menu, which inserts an inverter into a horizontal segment of a picked connecting polygon. The replaced part of the horizontal segment must disappear.

Bibliography

[1] ACM/SIGGRAPH Graphics Standards Planning Committee, First Report of the CORE Definition Subgroup, Preliminary Draft, Feb. 1977.

[2] Adage Inc., *FORTRAN IV Language and Programming System*, Programmer's Reference Manual.

[3] AFIPS NCC 1974, Panel session on "Manufacturing Control Systems."

[4] Ahuja, D. V., An Algorithm for Generating Spline-like Curves, *IBM System J. 7*, 3/4, 1968, 206–218.

[5] Appel, A., The Notion of Quantitative-Invisibility and the Machine Rendering of Solids, *Proc. 1967 ACM Nat. Conf.*, 387–393.

[6] Ash, W. L., and E. H. Sibley, TRAMP: An Interactive Associative Processor with Deductive Capabilities, *Proc. 1968 ACM Nat. Conf.*, 144–156.

[7] Bézier, P., *Numerical Control*. Wiley, New York, 1972.

[8] Birkhoff, G., and S. Maclane, *A Survey of Modern Algebra*. Macmillan, New York, 1941.

[9] Blaser, A., and H. Schmutz, Data Base Research: A Survey, IBM Germany, Heidelberg Scientific Center, *Tech. Rept. 75.10.009*, Nov. 1975.

[10] Bouknight, W. J., A Procedure for Generation of Three-Dimensional Halftone Computer Graphics Representation: *CACM 13*,9, Sept. 1970, 527–536.

[11] Brayer, J. M., and K. S. Fu, Some Properties of Web Grammars, Purdue University, *Tech. Rept. TR-EE 74-19*, Apr. 1974.

[12] CALLAN, J. F., The Architecture of the Picture System, *Proc. 2nd Ann. Symp. Computer Architecture*, IEEE Computer Society Publ. 75CH0916-7C, 13–16.

[13] CHAMBERLIN, D. D., Relational Data-Base Management Systems, *Computing Surveys 8,* 1, Mar. 1976, 43–66.

[14] CHILDS, D. L., Feasibility of a Set-Theoretic Data Structure: A General Structure Based on a Reconstituted Definition of Relation, *Proc. IFIP Congr. 1968*, North-Holland, Amsterdam, 1968, 162–172.

[15] CHILDS, D. L., Description of a Set-Theoretic Data Structure, *Proc. AFIPS FJCC 1968*, 557–564.

[16] CODASYL DEVELOPMENT COMMITTEE, An Information Algebra, Phase I Report—Language Structure Group, *CACM 5,*4, Apr. 1962, 190–204.

[17] CODASYL DATA BASE TASK GROUP, Apr. 1971 Rept., ACM, New York, 1971.

[18] CODD, E. F., A Relational Model of Data for Large Shared Data Banks, *CACM 13,*6, July 1970, 377–387.

[19] CONTROL DATA CORP., *CDC DIGIGRAPHICS—User's Manual*, CDC Product M 007*2-E 006*2.

[20] COONS, S. A., Surfaces for Computer-Aided Design of Space Forms, *MIT Project MAC-TR-41*, 1967.

[21] COTTON, I. W., AND F. S. GREATOREX, JR., Data Structures and Techniques for Remote Computer Graphics, *Proc. AFIPS FJCC 1968*, vol. 33, pt. 1, 533–544.

[22] CRICK, M. F., AND A. J. SYMONDS, A Software Associative Memory for Complex Data Structures; IBM Cambridge Scientific Center, *Rept. G 320-2060*, Aug. 1970.

[23] CUTMULL, E., A Subdivision Algorithm for Computer Display of Curved Surfaces, University of Utah, Department of Computer Science, *UTEC-CSc-74-133*, Dec. 1974.

[24] DAVIS, J. R., Device Independent Graphics Software—Is It Possible?, *Computer Graphics 9,*1 Spring 1975, ACM-SIGGRAPH Publ., 232–245.

[25] DAVIS, M. R., AND T. O. ELLIS, The RAND Tablet: A Man–Machine Graphical Communication Device, *Proc. AFIPS FJCC 1964*, 325.

[26] DAVIS, P. J., *Interpolation and approximation*, Blaisdell, New York, 1963.

[27] DODD, C. G., APL—A Language for Associative Data Handling in PL/I, *Proc. AFIPS FJCC 1966*, vol. 29, 677–684.

[28] DUDA, R. O., AND P. E. HART, *Pattern Classification and Scene Analysis*, Wiley, New York, 1973.

[29] ECKERT, R., Implementierung eines weitgehend allgemeinen Visibilitätsverfahrens für Funktionen zweier Variablen, Diploma thesis, Technical University of Berlin, Institute for Information Processing, 1970.

[30] ELSON, M., *Concepts of Programming Languages*, Science Research Associates, Chicago, 1973.

[31] ENCARNACAO, J., Computer Graphics, *Eine Einführung in die Programmierung und Anwendung von graphischen Systemen*, Oldenbourg, Munich, 1975.

[32] ENCARNACAO, J., W. GILOI, J. SANITER, W. STRASSER, AND K. WALDSCHMIDT, Programmierungs- und geraetetechnische Realisierung einer 4×4 Matrix fuer Koordinatentransformationen auf Computer-Bildschirmgeraeten; *Elektron. Rechenanlagen 14*, 5, 1972.

[33] ENCARNACAO, J., Survey of and New Solutions for the Hidden-Line Problem, *Proc. GC Symp. Delft*, Oct. 1970.

[34] ENCARNACAO, J., AND W. GILOI, PRADIS—An Advanced Programming System for 3-D Display, *Proc. AFIPS SJCC 1972*, 985–998.

[35] ENGLEBART, D. C., AND W. K. ENGLISH, A Research for Augmenting Human Intellect, *Proc. AFIPS FJCC* 1968, vol. 33, pt. 2, 395–410.

[36] EVAND & SUTHERLAND COMPUTER CORP., Picture System 2 Literature, 1977.

[37] FELDMAN, J. A., AND P. D. ROVNER, The LEAP Language and Data Structure, *Proc. IFIP Congr. 1968*, vol. 1, 579–585.

[38] FOLEY, J. D., AND V. L. WALLACE, The Art of Natural Graphic Man–Machine Conversation, *Proc. IEEE 62,* 4, Apr. 1974, 462–471.

[39] FOLEY, J. D., An Approach to the Optimum Design of Computer Graphics Systems, *CACM 14,*6, June 1971, 380–390.

[40] FORREST, A. R., Computational Geometry, *Proc. Roy. Soc. Lond., A321*, 1971, 187–195.

[41] FORREST, A. R., Mathematical Principles for Curve and Surface Representation, *Proc. Curved Surfaces Eng.*, IPC Science and Technology Press, London, 1972, 5–13.

[42] FORREST, P. M., Computational Geometry—Achievements and Problems, in R. E. Barnhill and R. F. Riesenfeld (eds.), *Computer-Aided Geometric Design*, Academic Press, New York, 1974, 17–43.

[43] FORREST, A. R., Interactive Interpolation and Approximation by Bézier Polynomials, *Computer J. 15,*1, 1972, 71–79.

[44] FORREST, A. R., On Coons and Other Methods for the Representation of Curved Surfaces, *Computer Graphics and Image Proc. 1*, 1972, 341–359.

[45] FRY, J. P., AND E. H. SIBLEY, Evolution of Data Base Management Systems, Computing Surveys 8,1, Mar. 1976, 7–42.

[46] GALIMBERTI, R., AND U. MONTANARI, An Algorithm for Hidden-Line Elimination, *CACM 12,*4, Apr. 1969, 206–211.

[47] GHEST, C., Multiplying Made Easy for Digital Assemblies, *Electronics*, Nov. 1971, 56–61.

[48] GILOI, W. K., Moving the Hardware–Software Boundary Up in Hierarchical Multistage Image Analysis, *Proc. Milwaukee Symp. Automatic Computation and Control*, 1976, 439–443.

[49] GILOI, W. K., J. ENCARNACAO, AND S. L. SAVITT, Interactive Graphics on Intelligent Terminals in a Time-Sharing Environment, *Acta Informatica, 5*, 1975, 257–271.

[50] GILOI, W. K., Interactive Systems—Patterns and Prospects, *Proc. IBM Int. Symp. Interactive Systems*, Bad Homburg, Germany, 1976.

[51] GILOI, W. K., On High-Level Programming Systems for Structured Display Programming, *Computer Graphics 9,*1, Spring 1975, ACM-SIGGRAPH Publ., 61–69.

[52] GILOI, W. K., Special Displays—A Medium for the Physician to Communicate Topological Facts in His Findings, paper presented at *AFIPS NCC 1973*, session: "The Augmented Knowledge Workshop."

[53] GILOI, W., R. BRUEDERS, AND G. TROELLER, A Display Terminal for Interactive Picture Processing, *Proc. Int. Conf. "Man and Information Science,"* Bordeaux (France), 1972, Institut de la Vie.

[54] GILOI, W. K., P. GÜNTHER, AND G. TROELLER, The Evolution of TV Raster Displays for Graphics, in press.

[55] GILOI, W. K., AND J. ENCARNACAO, APLG—An APL Based System for Interactive Computer Graphics, *Proc. AFIPS 1974 NCC*, vol. 43, 521–528.

[56] GILOI, W. K., *Programmieren in APL*, de Gruyter, Berlin, 1976.

[57] GILOI, W. K., J. ENCARNACAO, AND W. KESTNER, APLG—APL Extended for Graphics, *Online 72*, vol. 2, 1972, 579–599.

[58] GILOI, W. K., AND R. HOFFMANN, Adding a Modern Control Structure to APL Without Changing Its Syntax, *Proc. APL Symposium 1976*, 189–194.

[59] GNATZ, R., Sprachliche Aspekte bei Computer Graphics, in *Computer Graphics Symp. 1971*, W. K. Giloi (ed.), GI-Bericht Nr. 2, 1971.

[60] GORDON, W. J., AND R. F. RIESENFELD, Berstein–Bézier Methods for the Computer-Aided Design of Freeform Curves and Surfaces, *JACM 21,*2, Apr. 1974, 293–310.

[61] GORDON, W. J., AND R. F. RIESENFELD, B-spline Curves and Surfaces, in *Computer-Aided Geometric Design*, R. E. Barnhill and R. F. Riesenfeld (eds.), Academic Press, New York, 1974.

[62] GORDON, W. J., Distributive Lattices and the Approximation of Multivariate Functions, in *Approximations with Special Emphasis on Spline Functions*, I. J. Schoenberg (ed.), Academic Press, New York, 1969, 223–277.

[63] GOURAUD, H., Computer Display of Curved Surfaces, University of Utah, Department of Computer Science, *UTEC-CSc-71-113*, June 1971.

[64] GRAUSTEIN, W. C., *Introduction to Higher Geometry*. Macmillan, New York, 1930.

[65] GRAY, J. C., AND C. A. LANG, ASP—A Ring-Implemented Associative Structure Package, *CACM 11*,8, Aug. 1968, 550–555.

[66] GREEN, B. F., Computer Languages for Symbol Manipulation, *IRE Trans. Electronic Computers*, EC-10, Dec. 1961, 729–735.

[67] GREVILLE, T. N. E., Introduction to Spline Functions, in *Theory and Applications of Spline Functions*, T. N. E. Greville (ed.). Academic Press, New York, 1969.

[68] GROSSKOPF, G., Beschreibung, Untersuchung und Erprobung der Visibilitätsverfahren FLAVER und FLAKA, Diploma thesis, Technical University of Berlin, Institute for Information Processing, Sept. 1970.

[69] GUTHERY, S. B., DDA: An Interactive and Extensible Language for Data Display and Analysis, *Computer Graphics 10,* 1, Spring 1976, ACM-SIGGRAPH Publ., 24–31.

[70] HAGAN, T. G., R. J. NIXON, AND L. J. SCHAFER, The ADAGE Graphics Terminal, *Proc. AFIPS FJCC 1968*, vol. 33, pt. 2, 747–755.

[71] HEINDEL, L. E., AND T. J. ROBERTO, *An Interactive Language Design System* American Elsevier, New York, 1975.

[72] HURWITH, A., J. P. CITRON, AND J. B. YEATON, GRAF: Graphic Addition to FORTRAN, *Proc. AFIPS SJCC 1967*, 553–557.

[73] INFORMATION DISPLAYS, INC., IDIIOM *Software Manual*, vol. I. Mount Kisco, N.Y., 1969.

[74] INFORMATION DISPLAYS, INC., IDIIOM *Technical Description*, Mount Kisco, N.Y., 1969.

[75] INTERNATIONAL BUSINESS MACHINES, IBM 1130/2250 Graphic Subroutine Package for Basic FORTRAN IV, *IBM Systems Ref. Library Form C 27-6934-1.*

[76] INTERNATIONAL BUSINESS MACHINES, IBM System/360 Operating System, PL/I (F) Language Reference Manual, *IBM Systems Ref. Library Form C 2723-1.*

[77] JOHNSON, C. I., Principles of Interactive Systems, *IBM Systems J., 7*,3/4, 1968, 147–174.

[78] JONES, B., An Extended ALGOL-60 for Shaded Computer Graphics, *Computer Graphics 10,* 1, Spring 1976, ACM-SIGGRAPH Publ., 18–23.

[79] KAY, A. C., FLEX—A Flexible Extendable Language, University of Utah, Department of Computer Science, *Tech. Rept.* 4-7, June 1968.

[80] KERR, H. D., A Microprogrammed Processor for Interactive Computer Graphics, *Proc. 2nd Ann. Symp. Computer Architecture*, IEEE Comp. Soc., 28–33.

[81] KESTNER, W., Ueber die Erzeugung und Handhabung graphischer Objekte in digitalen Rechananlagen, Ph.D. thesis, Technical University of Berlin, 1974.

[82] KEYDATA, *Computer Display Review*, vols. I–IV, Adams Associates, 1966–1971.

[83] KLOS, W. F., Einheitliche formale Beschreibung, qualitative und quantitative Untersuchung von Visibilitätsverfahren, Diploma thesis, Universität des Saarlands, Fachbereich Informatik, 1975.

[84] KNUTH, D. E., *The Art of Computer Programming, vol. 1*, Chapter 2, Addison-Wesley, Reading, Mass., 1968.

[85] KNUTH, D. E., *The Art of Computer Programming, vol. 3*, Addison-Wesley, Reading, Mass., 1970.

[86] KNUTH, D. E., Structured Programming with GOTO Statements, Stanford University, *Tech. Report STAN-CS-74-416*, May 1974.

[87] KUBERT, B. R., A Computer Method for Perspective Representation of Curves and Surfaces, Aerospace Corp., San Bernardino, Calif., Dec. 1968.

[88] KULSRUD, H. E., A General Purpose Graphic Language, *CACM 11,* 4, Apr. 1968, 247–254.

[89] LAFATA, P., AND J. B. ROSEN, An Interactive Display for Approximation of Linear Programming, *CACM 13,* 11, Nov. 1970, 651–659.

[90] LOUTREL, P. P., A Solution to the Hidden-Line Problem for Computer-Drawn Polyhedra, *IEEE Trans. EC C-19* , 3, Mar. 1970, 205.

[91] LUM, V. Y., P. S. T. Yuen, and M. Dodd, Key-to-Address Transform Techniques, *CACM 14*, 4, Apr. 1971, 228–239.

[92] MAHL, R., Visible Surface Algorithms for Quadric Patches, University of Utah, Department of Computer Science, *UTEC-CSc-70-111,* Dec. 1970.

[93] Mathematical Applications Group, Inc., MAGI, *Datamation 14*, Feb. 1968, 69.

[94] Maxwell, E. A., *General Homogeneous Coordinates in Space of Three Dimensions.* The University Press, Cambridge, England, 1961.

[95] Meads, J., A Terminal Control System, *Proc. IFIP Working Conf. on Graphic Languages*, North Holland, Amsterdam, 1972, 271–290.

[96] Mealy, G. H., Another Look at Data, *Proc. AFIPS FJCC 1967*, 525–534.

[97] Michaels, A. S., B. Mittman, and C. R. Carlson, A Comparison of Relational and CODASYL Approaches to Data-Base Management, *Computing Surveys 8*, 1, Mar. 1976, 125–151.

[98] Myer, J. H., and I. E. Sutherland, On the Design of Display Processors, *Proc. IEEE, 2*, 6, June 1968, 410–414.

[99] Negrete, J., W. K. Giloi, and J. Encarnacao, The Application of Computer Graphics to Automatic Medical Interviews, *Proc. Conf. sobre Sistemas y Computadoras* (Proc. IEEE Conf. on Systems and Computation), Mexico, 1971.

[100] Newman, W. M., and R. F. Sproull, *Principles of Interactive Graphics*. McGraw-Hill, New York, 1973.

[101] Newman, W. M., and R. F. Sproull, An Approach to Graphic System Design, *Proc. IEEE 62,*4, Apr. 1974, 471–483.

[102] Newman, W. M., Display Procedures, *CACM 14,*10, Oct. 1971, 651–660.

[103] Newman, W. M., A Graphical Technique for Computer Input, *Computer J. 11*, May 1968, 63–64.

[104] Newman, W. M., A System for Interactive Graphical Programming, *Proc. AFIPS SJCC 1968*, 47–54.

[105] Newman, W. M., An Experimental Display Programming Language for the PDP-10 Computer, University of Utah, Department of Computer Science, *UTEC-CSc-70-104*, July 1970.

[106] Newell, M. E., R. G. Newell, and T. L. Sancha, A New Approach to the Shaded Picture Problem, *Proc. ACM Nat. Conf. 1972*, 443–450.

[107] Peters, G. J., Interactive Computer Graphics Application of the Bi-Cubic Parametric Surface to Engineering Design Problems, *Proc. AFIPS NCC 1974*, vol. 43, 491–511.

[108] Pfalz, J. L., and A. Rosenfeld, Web grammars, *Proc. 1st Int. Joint Conf. Artificial Intelligence*, Washington, D.C., 1969.

[109] Prenter, P. M., *Splines and Variational Methods*. Wiley, New York 1975.

[110] RECHENBERG, P., *Programmieren für Informatiker mit PL/I.* Oldenbourg Munich, 1974.

[111] ROBERTS, L. G., Homogeneous Matrix Representation and Manipulation of *n*-dimensional Constructs, *Computer Display Rev.*, July 1966.

[112] ROBERTS, L. G., Machine Perception of Three-Dimensional Solids, MIT Lincoln Lab., TR 315, May 1963. Also in Tipper et al. (eds.), *Electro-Optical Information Processing*, MIT Press, Cambridge, Mass., 159.

[113] ROBERTS, L. G., The Lincoln Wand, *Proc. AFIPS FJCC 1966*, 223.

[114] ROMNEY, G. W., Computer Assisted Assembly and Rendering of Solids, University of Utah, Department of Computer Science, *TR-4-20*, 1970.

[115] ROSENFELD, A., AND A. C. KAK, *Digital Picture Processing*. Academic Press, New York, 1976.

[116] ROUGELOT, R. S., AND R. SHOEMAKER, G. E. Realtime Display, General Electric Co., Syracuse, N. Y., *NASA Report NAS 9-3916.*

[117] SAVITT, S., FLAVI Program Description, University of Minnesota, Special Interactive Computing Lab., 1972.

[118] SCHOENBERG, I. J., Contributions to the Problem of Approximation of Equidistant Data by Analytic Functions, *Quart. Appl. Math. 4*, 1946, pt. A, 45–99; pt. B, 112–141.

[119] SCHOENBERG, I. J., On Spline Functions, in *Inequalities*, O. Shisha (ed.), Academic Press, New York, 1967, 255–291.

[120] SCHOENBERG, I. J., On the Variation Diminishing Approximation Methods, in *On Numerical Approximation*, R. E. Lanfer (ed.), University of Wisconsin Press, Madison, 1959, 249–274.

[121] SCHUMACKER, R. A. B. BRAND, M. GILLILAND, AND W. SHARP, Study for Applying Computer-Generated Images to Visual Simulation, *AFHRL-TR-69-14*, U.S. Air Force Human Resources Lab., Sept. 1969.

[122] SCIENCE ACCESSORIES CORP., Graf/Pen Sonic Digitizer, Southport, Conn., 1970.

[123] SCOOP, K., The Design and Use of a PL/I Based Graphic Programming Language, *Online 72*, vol. 2, 1972, 601–615.

[124] SEEGMÜLLER, G., Design Considerations for New Programming Languages, *Proc. IBM Int. Conf. Interactive Systems*, Bad Homburg, Germany, Sept. 1976.

[125] SENKO, M. E., E. B. ALTMAN, M. M. ASTRAHAN, AND P. L. FEHDER, Data Structures and Accessing in Data-Base Systems, *IBM System J. 12,* 7, 1973, 40–93.

[126] SHAW, A. C., A Formal Picture Definition Scheme as a Basis for Picture Processing Systems, *Information & Control 14,* 1, 1969, 938–947.

[127] SIBLEY, E. H., Guest Editor's Introduction: The Development of Data-Base Technology, *Computing Surveys 8,* 1, Mar. 1976, 1–5.

[128] SINGLETON, R. C., Algorithm 347, An Efficient Algorithm for Sorting with Minimal Storage, *CACM 12,* 3, 1969, 185–187.

[129] SPROULL, R. F., Omnigraph: Simple Terminal-Independent Graphics Software, Xerox Palo Alto Research Center, *Tech. Rept. CSL-73-4*, 1973.

[130] STEIN, M., AND I. L. MORISS, The Adage Graphics Terminal—an Integrated Hardware/Software System for the Display and Manipulation of Structured Images, Technical Rept., Adage INC., Boston, 1969.

[131] STRASSER, W., Schnelle Kurven- und Flächendarstellung auf graphischen Sichtgeräten, Ph.D. thesis, Technical University of Berlin, 1974.

[132] SUTHERLAND, I. E., Computer Displays, *Sci. Amer. 222,* 6, June 1970, 3–18.

[133] SUTHERLAND, I. E., AND R. F. SPROULL, A Clipping Divider, *Proc. AFIPS FJCC 1968*, 765.

[134] SUTHERLAND, I. E., R. F. SPROULL, AND R. A. SCHUMACKER, A Characterization of Ten Hidden-Surface Algorithms, *Computing Surveys 8,* 1, Mar. 1974, 1–55.

[135] SUTHERLAND, I. E., Three-Dimensional Data Input by Tablet, *Proc. IEEE 62,* 4, Apr. 1974.

[136] SUTHERLAND, W. R., The CORAL Language and Data Structure, MIT Lincoln Lab., *Tech. Report 405*, Lexington, Mass., May 1966.

[137] TANENBAUM, A. S., *Structured Computer Organization*, Prentice-Hall, Englewood Cliffs, N. J., 1976.

[138] TAYLOR, R. W., AND R. L. FRANK, CODASYL Data-Base Management Systems, *Computing Surveys 8,* 1, Mar. 1976, 67–103.

[139] TEIXEIRA, J. F., AND R. P. SALLEN, The Sylvania Tablet: A New Approach to Graphic Data Input, *Proc. AFIPS SJCC 1968*, 315.

[140] TROELLER, G., Display-Systeme nach dem Fernsehrasterverfahren unter besonderer Berücksichtigung graphischer Darstellungen, Ph.D. thesis, Technical University of Berlin, 1974.

[141] TSICHRITZIS, D. S., AND F. H. LOCHOVSKY, Hierarchical Data-Base Management, *Computing Surveys 8,* 1, Mar. 1976, 105–123.

[142] VAN DAM, A., AND D. EVANS, Data Structure Programming System, *Proc. IFIP Congress 1968*, vol. 1, North-Holland, Amsterdam, 557–564.

[143] VAN DAM, A., G. M. STABLER, AND R. J. HARRINGTON, Intelligent Satellites for Interactive Graphics, *Proc. IEEE 62,* 4, Apr. 1974.

[144] VARGA, R. S., *Matrix iterative analysis*; Prentice-Hall, Englewood Cliffs, N.J., 1962.

[145] VECTOR GENERAL INC., Interactive Graphics for Mini-Computers, Canoga Park, Calif., 1970.

[147] WALLACE V. L., The Semantics of Graphic Input Devices, *Computer Graphics, 10,* 1 Spring 1976, ACM-SIGGRAPH Publ., 61–65.

[147] WARD, J., ET AL., An Integrated Hardware–Software System for Computer Graphics in Time-Sharing, *Technical Report MIT-ESL*, Dec. 1968.

[148] WARNOCK, J. E., A Hidden-Surface Algorithm for Computer-Generated Halftone Pictures, University of Utah, Department of Computer Science, *TR 4-15,* June 1969.

[149] WATKINS, G. S., A Real-Time Visible Surface Algorithm, University of Utah, Department of Computer Science, *UTECH-CSC-70-101,* June 1970.

[150] WEGNER, P., *Programming Languages, Information Structure, and Machine Organization*, McGraw-Hill, London, 1971.

[151] WEISS, R. A., BE VISION, A Package of IBM 7090 FORTRAN Programs To Draw Orthographic Views of Plane and Quadric Surfaces, *JACM 8,* 2, Feb. 1966, 194–204.

[152] WILLIAMS, R. A., A Survey of Data Structures for Computer Graphics Systems, *ACM Computer Surveys 3,* 1, Mar. 1971.

[153] WILLIAMS, R., A General Purpose Graphical Language, in *Graphic Languages*, F. Nake and A. Rosenfeld (eds.), Proc. IFIP Working Conf. on Graphic Languages, North-Holland, Amsterdam, 1972.

[154] WIRTH, N., *Systematic Programming: An Introduction*, Prentice-Hall, Englewood Cliffs, N.J., 1973.

[155] WOODSFORD, P. A., The Design and Implementation of the GINO 3D Graphics Software Package, *Software Practice and Experience 1*, Oct. 1971, 335.

[156] WOODSFORD, P. A., Numerical Methods in Computer Graphics, in *Computer Graphics Symposium 1971*, W. K. Giloi (ed.), *GI-Bericht Nr. 2*, 1971.

[157] ZAHN, C. T., A Control Statement for Natural Top-Down Structured Programming, *Proc. Symp. Programming Language*, Paris, 1974 (see also [85]).

Appendix:

Implementation of the GRIP Concept†

A.1 GENERAL CONSIDERATIONS

In Chapter 10, the general requirements and functions of a graphical programming system are presented and discussed. A hypothetical language, GRIP, is used to illustrate the concepts of structured graphical programming.

Any graphical programming package is simply an interface between the graphical primitive-generating hardware and the person engaged in the use of the interactive graphics system. In this appendix, the internal system aspects of such an interface will be revealed through an examination of how a particular structured graphical programming package, named GRAP (for GRaphical Application Package), has been implemented. This particular programming package provides interactive graphics capabilities in a time-sharing environment. GRAP is a FORTRAN-callable subroutine package that is modeled after GRIP.

Environment: The particular design and implementation of any interactive graphics programming package is determined both by the desired properties of the package as well as by the properties of the computer system and graphics hardware device for which the programming package is to be developed. The desired properties of

†This appendix was written by Steven L. Savitt.

the GRAP system include:

1. Device independence.
2. Portability.
3. A high degree of structuring.
4. Extensibility.

Device Independence: A wide variety of graphical devices are available on the market today, and for this reason it was necessary to design the GRAP system so that several different devices could be supported. This device independency of the GRAP system was provided by the inclusion of an intermediate graphical communications language (see Section 9.3) which serves as the interface between the high-level language routines and the display file in the "intelligent" terminal. The intermediate language contains commands meaningful to all types of graphical devices, such as "draw line" or "draw text." The interpretation of these intermediate language commands is performed by the local intelligence within the particular graphics device in use. Each graphics device must therefore include a local interpreter program to perform the interpretation of the commands of the intermediate language.

Any device that has sufficient local intelligence to perform this interpretation can therefore be used with the GRAP system. Our present terminals use a 16-bit minicomputer to provide the local intelligence necessary for the intermediate language interpretation. Presently available graphics devices that do not possess any local intelligence could be augmented with a low-cost microprocessor to provide for the intermediate language interpretation.

Portability: The GRAP system is coded entirely in standard FORTRAN to ensure a reasonable degree of portability so that the system can be moved to other host computers. The system is presently operational on three different host computers, and the users have no difficulty moving their application software from one machine to another.

High Degree of Structuring: The programming of an application is greatly facilitated by the ability to deal with graphic objects that are more complex than the basic primitives provided by the display hardware (see Section 3.1). The segment–item–primitive hierarchy as described in the text was incorporated into the GRAP system to provide the desired high degree of graphic structure.

Extensibility: Some devices possess unusual and unique features which may not be supported by the existing GRAP system and its intermediate language. For this reason, the commands of the intermediate language as well as the set of available

GRAP routines are not viewed as being either all-inclusive or final, and the system is structured to permit easy insertion of new intermediate language commands and new GRAP routines. The application programmer must, of course, not program a feature which does not exist on the specific device that is being used. For example, an "erase entity" command would be meaningless to a plotter device.

Beyond the desired properties of the GRAP system as discussed above, the constraints imposed by the existing hardware must be considered. The following discussion will describe the host computer and graphics devices with which the GRAP system is being used.

Host Computer: The particular host computer of our graphics system is a Control Data CYBER 74 computer system. The operating system of this computer supports timesharing and batch processing. This computer is physically remote from the site of the graphics terminals, so telephone lines must be employed for communication between the computer and the graphics terminals. A line speed of 2400 bauds has proven to be satisfactory.

Intelligent Graphics Terminal: The "intelligent terminal" is composed of a vector display console and the associated display processor unit. A general-purpose minicomputer with $16K \times 16$-bit words is included to provide the local intelligence. The memory of the minicomputer is shared by the minicomputer and the display processor. This sharing is accomplished by dedicating one section of the memory for the display processor code that is executed by the display processor, and using the remaining memory for the terminal software that is executed by the terminal minicomputer. This memory sharing not only reduces the cost of the overall system but provides the most direct means through which the display file can be accessed and updated by the terminal minicomputer.

The user interaction with the intelligent terminal is supported by three devices: a lightpen, a 32-key function keyboard, and a teletype.

Software: The software for this system falls into three separate categories. The first of these categories comprises the user-callable routines that reside on the host computer. These routines provide the high-level language interface between the application programmer and the graphics system. These routines exist in the form of FORTRAN-callable library subroutines and are collectively referred to as the GRAP package.

The second category of software consists of two programs which are resident in the intelligent terminal and which control the information exchange between the GRAP package routines and the terminal-based display file. The first of these two programs is the L^4 interpreter, which interprets the incoming commands of the intermediate graphical communications language. This L^4 interpreter serves the added role of an L^4 generator when data are sent from the terminal back to the host computer. In this reverse mode of operation, the GRAP package which normally serves as an L^4 generator must now become an L^4 interpreter for the commands generated by the terminal. The second program within the category of terminal software is the terminal DRIVER program, which contains the device drivers for the telephone dataset controller and the terminal teletype controller. The terminal DRIVER program also contains software that falls into the third category, as described below. The third category contains the support software which is used in the development and maintenance of the first two categories of software but is not used during the actual execution of the GRAP package. A more detailed discussion of the three categories will begin with this third category.

Support Software: In order to facilitate the development of the software for the intelligent terminal, a cross-assembler was implemented on the host computer so that source files could be maintained on-line and easily edited using the time-sharing text editor. The main problem with using a cross-assembler, aside from the task of writing it, is that the machine code output from the cross-assembler must be transported somehow to the intelligent terminal for execution. To accomplish this, a binary down-loader was written which takes the output from the cross-assembler and transmits it to the intelligent terminal through the telephone link. One part of the down-loader resides on the host computer and the other part executes on the terminal computer. The portion of the down-loader which executes on the host computer encodes one 16-bit machine word into three 6-bit characters so that the standard time-sharing ASCII character set can be used. The extra two bits are used for error detection. In addition, each block of characters is followed by a checksum character which is useful in detecting a missing character. If an error is detected, the terminal program requests a retransmission of the last character block. The portion of the down-loader that executes in the intelligent terminal decodes the incoming character strings into binary machine words and then stores the data into the proper memory cells.

Another program provided for the terminal computer is a binary debugging routine. This routine allows the inspection and changing of registers and memory cells. In addition, traps can be set to permit partial execution of a program to a breakpoint. A binary load and dump routine for paper tape is provided. This routine allows the remainder of the terminal software to be punched onto paper tape so that it can be locally re-stored. The binary load and dump routine must itself be loaded using a manually entered bootstrap program.

A teletype driver program was written to allow the terminal teletype to operate as a conventional time-sharing terminal. A circular buffer was provided to allow data to be received over the telephone link at 2400 bauds and then be printed at the slower teletype speed of 110 bauds. For long terminal sessions, the low speed and noise of the teletype can be annoying. To overcome this, an alphanumeric CRT terminal simulator was written which takes user input from the teletype keyboard and displays computer output on the graphics display console. This CRT simulator provides the features of scrolling, character editing, and cursor display. In addition, the character size is reduced automatically when a line of text reaches the 73rd character position, so that up to 132 characters of text can be displayed on

Table A.1: TERMINAL DRIVER PROGRAM COMMANDS

Command	Function
CRT	Execute the CRT simulator program
TTY	Execute the TTY time-sharing program
PNCH	Execute the punch binary paper tape program
LOAD	Execute the load binary paper tape program
DEBUG	Execute the binary debug program
RUN	Execute the last down-loaded program (usually the L^4 interpreter)
CLR	Remove the last down-loaded program from the memory
BYE	Sign off the terminal from the host computer

a single line. This feature is useful in displaying output which has been formatted for a line printer rather than a time-sharing terminal. A final feature of the CRT simulator is that the previous 200 lines of display are stored as four 50-line pages, and the user can select the display of a previous page using the system sense switches.

Significant memory is saved in the intelligent terminal by collecting all these routines into one single program which has one common set of device drivers and which serves as a mini-operating system. This single program will be referred to as the terminal driver program. A set of user commands was established that allow the terminal user to invoke various functions of the terminal driver program (Table A.1).

A.2 TERMINAL GRAPHICS SOFTWARE

A.2.1 Data Flow in the Terminal

Data from the host computer enter the intelligent terminal as a string of ASCII coded characters. This character string is composed of two types of data blocks, one that contains graphical data and the other that contains nongraphical data. Each graphical data block begins with an ASCII SOM character and ends with an ASCII ETX character. The nongraphical data block consists of all data that follow the ETX character until the next SOM character is encountered. The terminal DRIVER program routes the nongraphical data blocks to a circular communication queue, where the characters are either delayed until the teletype is ready for them or passed through the queue to the CRT simulator, where they are immediately displayed on the CRT screen. The graphical data blocks, on the other hand, are not printed or displayed but are routed to the L^4 interpreter program. Each graphical data block contains two lists, which are separated by an additional ASCII SOM character. The first list is the message table (MT) and contains the L^4 instructions that are to be interpreted. The second list is the associated data list (ADL), which contains the variable-length data associated with each of the L^4 instructions in the message table (see Section 9.3).

Graphical data blocks enter the L^4 interpreter and are split into the incoming message table list (MT_i) and the incoming data list (ADL_i). When the end of the incoming graphical data block is encountered, control is transferred to the L^4 decoder routine, which consecutively processes each of the L^4 instructions in the MT_i. If an L^4 instruction is a Group I or II instruction (Table 9.2), a display file management routine is called to update either the display processor code list (DPCL) or the correlation table (CT). If a Group III L^4 instruction is encountered, an appropriate interaction-handling routine is called to process the desired user interactions.

User interactions result in the generation of Group IV L^4 statements which are placed in the output message table (MT_o) and associated data list (ADL_o). When the last instruction in the MT_i has been interpreted and all necessary user interactions have been performed, the MT_o and the ADL_o are encoded into a graphical output data block and sent back to the host computer. The teletype keyboard is disabled from the time that the MT_i begins to be decoded to the time that the resulting output graphical data block has been sent. Hence, the terminal communication cycles between a graphical data mode and a nongraphical data mode. Figure A.1 illustrates the data flow in the terminal.

A.2.2. Display File Overview and L^4 Interpreter Data Structures

All information that concerns the pictorial representation of graphic objects as well as all information that is required for the interactive identification of graphic entities is stored in the display file. The complete display file is composed of the name table (NT), the correlation table (CT), and the display processor code list (DPCL) (see Chapter 8). The name table contains the graphical entity names which are generated and utilized within the high-level application program. Since the high-level application program is resident in the host computer, the name table is also located in the host computer. The display processor code list, on the other hand, is interpreted by the display processor to provide the picture refresh and must, therefore, reside in the intelligent terminal. The third list of the display file, the correlation table, provides the two-way mapping between the name table in the host computer and the display processor code list in the intelligent terminal. To facilitate this two-way mapping, part of the correlation table resides in the host computer and the other part resides in the intelligent terminal (Fig. A.2). The name table and the correlation table have the same number of entries (rows), which is equal to the total number of segments, symbols, symbol instances, and items which have been defined. The INDEX parameter is an integer number that gives the row number of one specific entry in the correlation table and name table.

Display File—Name Table: The name table in our FORTRAN implementation of GRIP is an integer vector of 60-bit words. Several data types can be stored in the name table as long as each entry fits within the 60-bit-wide field. For example, a character string composed of ten 6-bit bytes could be used as one name in the

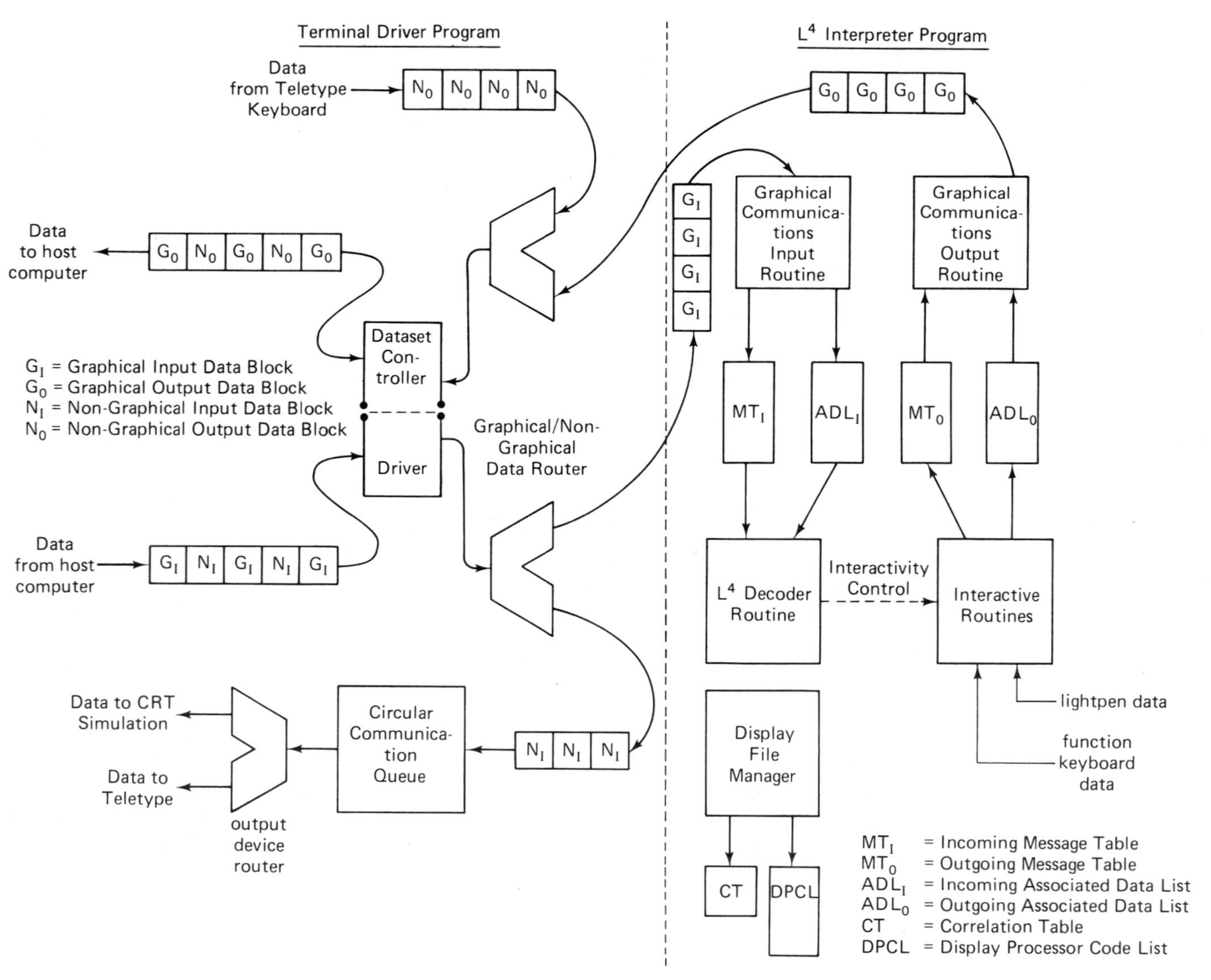

Figure A.1 Intelligent terminal data flow.

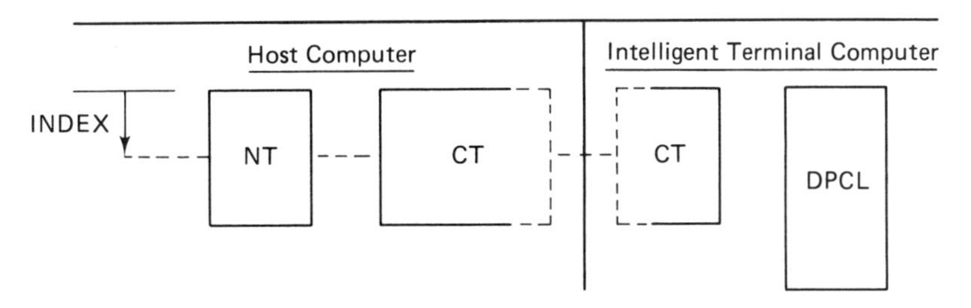

Figure A.2 Location of the display file lists.

table, and at the same time other names in the table could be integers that are generated by a DO-Loop index. The use of character string entries in the name table results in an application program that is self-documenting, whereas the use of ascending or descending integer values permits the simple generation of unique names for graphical entities. The name table in our APL implementation is restricted to character strings, but each entry can be of any desired length, since the row dimension can be dynamically increased to accommodate the longest name.

Display File—Correlation Table: The correlation table contains three types of entries for segments, symbols, and items. The portion of the correlation table that resides in the host computer consists of five data fields per entry (Fig. A.3). The first column of the correlation table is set to one for segment or symbol entries, and is set to the complement of the item type code in the case of item entities.

The second column of the correlation table contains the symbol flag that is used to distinguish symbol entries from segment entries. Columns 3–5 give the global attribute vector (IAV) and the absolute origin (MXA,MYA) for segment entries. These three columns always are set to zero for symbol entries, since the appearance and location of symbols are not defined until the symbols are actually invoked. Columns 3–5 of an item entry contain the number of primitives in the item (N), and the relative origin of the item (MXR,MYR).

The last three columns of the correlation table are resident in the intelligent terminal and contain the start and end pointers (SPTR, EPTR) to the blocks in the

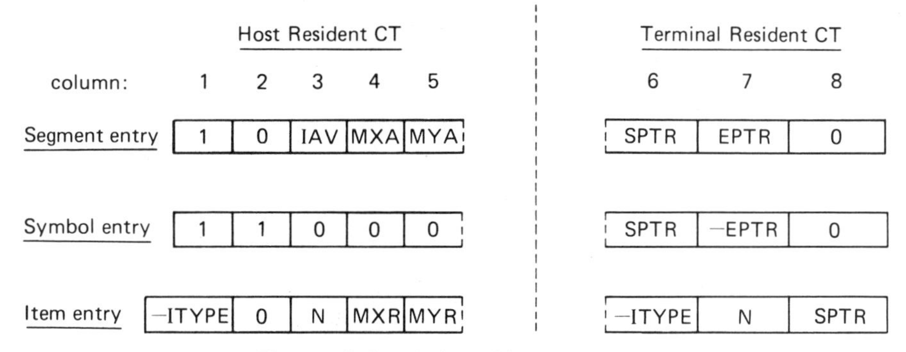

Figure A.3 Correlation table entry formats.

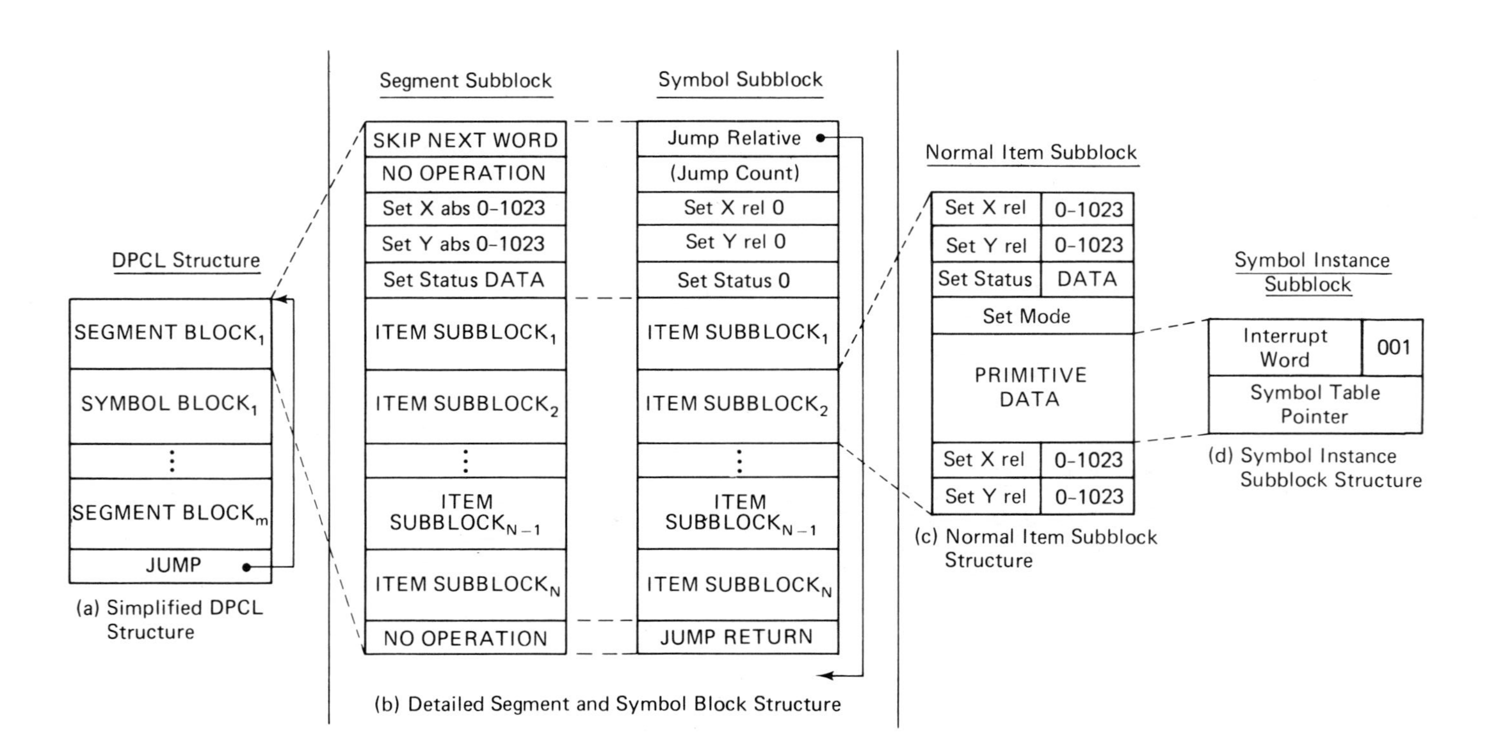

Figure A.4 Display processor code list structure (DPCL).

DPCL for each segment and symbol. In case of an item entry, column 6 of the correlation table contains the complement of the item type code, column 7 contains the number of DPCL words (N) consumed by the item, and column 8 contains a pointer to the beginning of the DPCL subblock for the item. The sign bits of fields 6 and 7 are used to identify the three types of entries that are used.

Display File—Display Processor Code List (DPCL): The display processor instructions, executed by the terminal display processor in the course of the picture refresh, are stored in a linear list called the display processor code list (DPCL). This list contains one block of instructions for each segment and symbol. Each of these blocks is subdivided into subblocks for the various items and/or symbol instances that comprise each of the segments and symbols (Fig. A.4).

Segment and symbol blocks both contain a five-word header preceding the first item subblock and a single-word trailing entry succeeding the last item subblock. In the case of a segment block, the first two words and the last word of the block are No Op instructions, whose only purpose is to pad up the segment block so that it has the same format as the symbol block. The third through fifth words of the segment block define the absolute segment origin and the global attributes (i.e., line style, intensity, and blink mode).

The first two words of the symbol block cause the remainder of the symbol block to be skipped during the scan of the DPCL for picture refresh. This permits the symbol blocks to be interspersed with the segment blocks in the DPCL, thus eliminating the need for a separate DPCL symbol list. Words 3–5 of the symbol block are No Op instructions, since the origin and display status of the symbol are specified in the calling sequence for a symbol instance rather than within the heading of the symbol block. The last word in the symbol block is a JUMP RETURN instruction and is described below. The item or symbol instance subblocks begin with a four-word initialization heading. The first two words of this initialization specify the desired item/instance origin. This origin is relative to the absolute origin defined in the segment block heading. At the very end of the subblock are two instructions that move the beam back to the segment origin so that the following item/instance origin will again be relative to its segment origin. The third word of the item heading defines the attributes of the item primitives for those elements of IAV which differ from the default segment attributes.

The fourth word of the item subblock defines the mode with which the subsequent primitives data are interpreted. For example, if the character mode is specified, the item data are interpreted as 8-bit ASCII character codes packed two per word. Since an item contains primitives of the same type, only one mode instruction is needed in each item subblock. Further, by having only one mode of data interpretation within a specific item subblock, the index number of a picked primitive within an item block can be determined more easily.

Symbols: A multilevel (nested) symbol capability is provided by the programming system. The display processor, however, only provides a single return address saving register, hence nested symbol-calling sequences are not directly supported by the display processor. To support nested symbol calls, a return address stack was implemented using two interrupt-driven routines, one of which pushes a return address onto the stack upon entry to a symbol, and the other which pops a return address off the stack upon exit from a symbol. The interrupts that invoke these two routines are generated by two special display processor instructions.

A detailed description of the symbol linkage implementation will now be presented. A many-to-one relationship can exist between symbol instances and a given symbol If the instance invocation contained the actual entry address of the desired symbol block, every invocation would have to be modified whenever the called symbol block changed locations within the DPCL. This problem was solved through the inclusion of a symbol jump table (SJT) which contains pointers to all the symbol entries in the correlation table (Fig. A.5). Once an entry is placed in the symbol jump table, the location of the entry is never changed; only its value is changed to keep the entry always pointing to the desired correlation table entry.

The calling sequence for a symbol instance consists of two words, the first of which is an interrupt-generating instruction, and the second of which is the address of the symbol jump table entry that is associated with the desired symbol block. Upon execution of the interrupt instruction, an interrupt service routine pushes the return address onto the symbol address stack and then follows the pointers from the second word of the symbol-calling sequence through the symbol jump table, onward to the correlation table, and finally back to the DPCL to arrive at the entry point of the desired symbol block. The entry point address is then placed in the

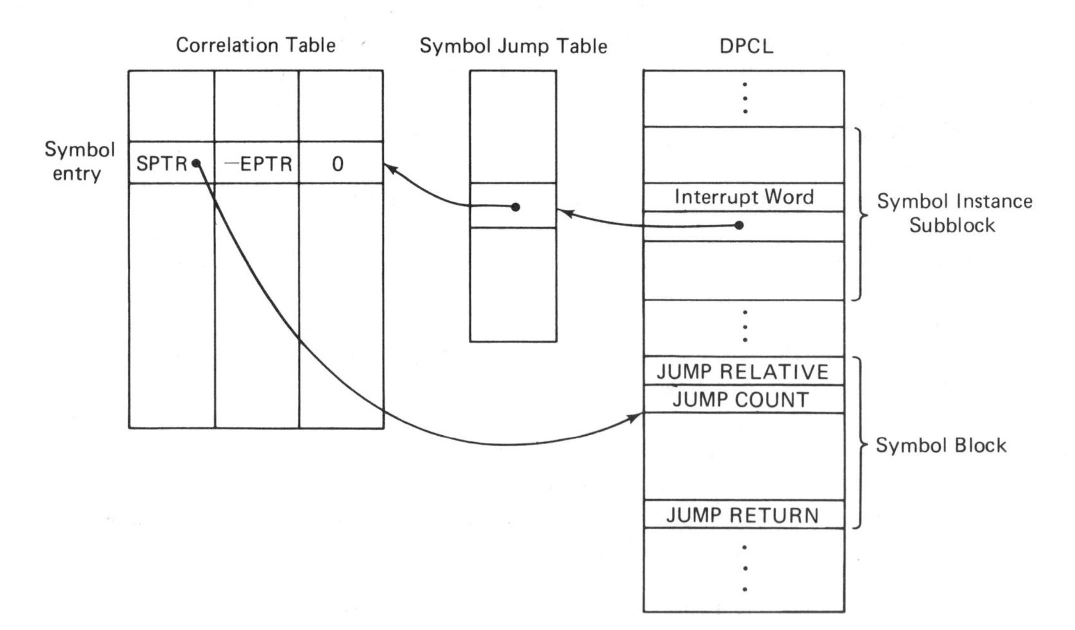

Figure A.5 Symbol linkage structure.

program address register of the display processor, and the display processor is restarted to execute the symbol. The last instruction of the symbol block is a JUMP RETURN interrupt instruction. Execution of this instruction invokes an interrupt service routine which removes the top element from the return address stack and places it in the display processor address register to accomplish the proper return from the symbol block back to the calling sequence. Both the symbol table and the symbol address stack can hold up to 64 entries.

A.2.3 L^4 INTERPRETER

A.2.3.1 L^4 Interpreter—Utility Routines

A number of utility routines are required to create and manage the internal data of the L^4 programs and their interpreter. These are basically the following.

Block Storage Memory Manager: The limited size of the memory of the terminal processor calls for its efficient utilization. Therefore, a block storage manager dynamically allocates the available memory space to the four communication lists ADL_i, ADL_o, MT_i, and MT_o. The display file lists are not dynamically allocated because of the overhead and complexity associated with performing random accessing within block storage lists.

Error-Processing Facilities: A generalized error-handling procedure is included as a programming aid during the debugging of the L^4 interpreter. Each routine in the L^4 interpreter performs checks on data that are passed to it to determine the presence of invalid data as well as the underflow or overflow of internal data structures. If an error is encountered, an error message is printed, listing an error number, an error code, the location of the error, and the entry point address of the routine in which the error occurred.

Correlation Table and Symbol Jump Table Manager: A series of routines are provided to manage the correlation table (CT) in the intelligent terminal. There is a routine for initializing the CT, for retrieving an entry, for creating a new entry, for replacing an existing entry by a new one, or for deleting an entry from the CT. Other important routines search the CT for the entry that is associated with a given DPCL address or update the CT whenever a change was made in DPCL. In the discussion of the symbol implementation, a symbol jump table was described whose entries are pointers to entries in the CT. Therefore, the symbol jump table must be updated whenever the CT is modified. Two more routines handle the symbol return address stack that allows for the nesting of symbol invocations.

Display Processor Code List (DPCL) Management Routines: A series of routines are provided to manipulate the DPCL. There is a routine for initializing DPCL, for inserting a data word, for deleting a data word, and for adding a primitive to an existing item. As it is not desirable to change or relocate display processor code during picture refresh, two routines are provided to start and stop the display processor, respectively.

Communications and Input/Output Routines: A set of device drivers are provided to control the exchange of information between the L^4 interpreter and the peripheral devices (teletype and telephone data set). Two general input/output routines may be employed to read characters from a specified device driver or to write a number of characters to a device driver. An input routine is activated whenever the beginning of a graphical data block is recognized, coming from the host computer. This routine decodes the incoming character string into L^4 instructions and places them into MT_i or ADL_i, respectively. Furthermore, a data block is checked for transmission errors (such a check is made possible through the addition of a "checksum" character to each block). The routine exits at the end of a data block.

An output routine sends the message table MT_o and the associated data list ADL_o from the terminal to the host computer. The output routine may repeat a transmission several times if it receives retransmission requests from the host computer (in the case that a transmission error was detected). Finally, two routines are provided to insert entries into MT_o and ADL_o, respectively.

Tracking Cross Routine: A tracking cross routine provides the means with which the lightpen can be used to interactively supply position data.

A.2.3.2 L^4 Interpreter—Entry Points

The L^4 interpreter has two separate entry points. The first one is used for the initialization call to the interpreter. This call is made immediately after the interpreter is loaded into the memory and anytime it becomes necessary to manually restart the interpreter. When this call is processed, all tables and lists as well as all global variables are initialized. The second entry to the interpreter is used every time the beginning of a graphical data block is sensed by the terminal driver program. This entry causes the input routine discussed above to be executed.

A.2.3.3 L^4 Interpreter—Instruction Interpretation

Up to this point, we have discussed the various data structures that are part of the L^4 interpreter program, as well as several sets of routines which are provided to manipulate these data structures. All the discussion above involved portions of the L^4 interpreter that are only secondary to its major task, which is the interpretation of L^4 instructions. We will now discuss this primary aspect of the L^4 interpreter.

After the MT_i and the ADL_i are received from the host computer, a message table decoder routine is called to sequentially interpret the L^4 instructions that are held in the MT_i, beginning with the first instruction and continuing until an ENDDATA opcode is encountered. As a result of the instruction decoding, the appropriate display file management or interaction-handling routines are invoked.

Each MT_i entry is broken into three fields; the OPCODE, the PARAMETER field, and the REFERENCE field (see Section 9.3). The PARAMETER and REFERENCE field values are placed in global variables so that their values can be easily accessed by other routines. The 6-bit OPCODE is used as an index to a 64-element jump table which contains the entry points to the various *command execution procedures* (CEPs). These CEPs perform the actual tasks required by the various L^4 instructions. Additional L^4 instructions can be implemented by adding an appropriate CEP to the interpreter.

The L^4 instruction set is divided into four major groups (Table 9.2). CEPs are provided for the first three groups of instructions. The Group I CEPs are those which generate display processor instructions and insert them into the DPCL. The Group II CEPs encompass those L^4 instructions that require changes to be made to the correlation table as well as modification to be made to the existing DPCL. The CEPs in Group III provide the interaction-handling capability of the system. Any data that are generated as a result of executing instructions from Group III CEPs will be returned to the host computer in the form of L^4 instructions from Group IV. The Group IV instructions therefore reverse the roles of the terminal and the host computers by providing for L^4 instruction generation in the terminal computer followed by L^4 interpretation in the host computer. Since the Group IV commands are not interpreted within the intelligent terminal, no Group IV CEPs exist.

The Group I CEPs provide the interface between the L^4 instructions and the display processor instructions. A detailed description of the Group I CEPs would not be of general value as the construction of the CEPs is dictated by the specific display processor that is being used. One illustrative example of a group I CEP will be given for the SET ORIGIN command. This L^4 instruction references two associated data list entries, X and Y. The display processor of our terminal has one instruction to set the X position and another instruction to set the Y position of the beam. Therefore, the CEP that interprets the single L^4 SET ORIGIN instruction must generate two display processor instructions. Further, the correlation table must be interrogated using the PARAMETER field of the L^4 instruction to determine whether the origin instruction is being used for a segment initialization or an item initialization. This differentiation between a segment and item initialization is necessary because the segment origin is an absolute screen position, whereas the item origin is a relative screen position. The SET ORIGIN CEP must set or clear a bit in the display processor origin instructions to indicate whether an absolute or relative positioning is to be carried out.

The group II instructions manipulate the display file. Consequently, the CEPs of this group reflect, more than those of other groups, the characteristics of our picture structure and the display file organization suggested in Chapter 8.

Therefore, it may be worthwhile to discuss the CEPs of group II in a little more detail.

DELETE: The beginning (FWA) and ending (LWA) addresses of the DPCL section to be deleted are first computed. Then an appropriate routine is called to move the section of the DPCL between LWA + 1 and the end of the DPCL up to the FWA location, thus effectively removing the unwanted segment or item. Another routine is then called to update the correlation table pointers so that they continue to point correctly to the remaining segments and items in the DPCL, and finally the unwanted segment or item entry is removed from the correlation table.

INSERT SEGMENT: This CEP is called to insert a new segment entry into the correlation table. The segment start and end pointers (SPTR, EPTR) of the new entry point to the current end of the DPCL. Then the first two and the last words of the segment block are added to the DPCL.

INSERT SYMBOL: Same as INSERT SEGMENT, except that the first two and last words require different display processor instructions in a symbol block.

INSERT ITEM: This CEP searches the correlation table backward from the entry index for the new item to the beginning of the correlation table, until the associated segment entry is found. The segment end pointer field (EPTR) then provides the DPCL address where the item can be inserted. The two trailing origin commands are added to the DPCL, and finally the new item entry is inserted into the correlation table.

NEWSTATUS: The correlation table entry for the specified segment or item is fetched and used to compute the DPCL address of the display processor SET CONTROL instruction. The new SET CONTROL instruction with the desired display attribute values is then substituted for the old one.

NEWORIGIN: The correlation table entry for the specified segment or item is fetched and used to compute the DPCL address for the two display processor origin commands. The new commands are then generated by this CEP and substituted for the old ones. In the case of an item origin change, the trailing origin commands are also modified so that the CRT beam still returns to the segment origin after the item is generated.

WAIT FOR PICK: This CEP will wait for the execution of the display processor code to be interrupted by a lightpen strike. When this interrupt occurs, the symbol return address stack is checked to see if it is empty. If it is, then an item within a segment was picked, and the address of the display processor instruction which was executing when the interrupt occurred is used to search the correlation table to determine which segment and item was picked. If the symbol return address stack is not empty upon occurrence of a lightpen strike, then a symbol was picked and the return address at the top of the stack is used to determine which instance was picked. Finally, this CEP generates an L^4 instruction (PICK RETURN) which is used to pass the index of the picked item back to the host computer for use by the high-level application program.

WAIT FOR FUNCTION KEYBOARD HIT: This CEP will illuminate one or more keys of the 32-key function keyboard to indicate which keys are enabled. When the terminal user pushes one of the illuminated keys, an L^4 instruction (HIT RETURN) is generated and placed in MT_o and ADL_o for subsequent transmission of the key number back to the host computer and the application program.

WAIT FOR POINTING: This routine displays the tracking cross and ACCEPT lightbutton on the screen. The user can move the tracking cross to any desired screen position. When the ACCEPT lightbutton is picked, an L^4 instruction (POINTING RETURN) is generated and placed in MT_o and ADL_o so that the X, Y-coordinates of the selected screen position can be returned to the host computer.

WAIT FOR RUBBERBANDING: This routine provides four interactive object generation capabilities. The first capability is a generalized line-drawing procedure. The generalized line drawing is accomplished by first moving the displayed tracking cross to one end point of a line and then moving the tracking cross a second time to the other desired end point of a line. As the cross is moved the second time, a vector is generated that dynamically connects the two endpoints as the tracking cross is moved. By releasing the lightpen switch, the tracking cross can be alternately moved from one end point of the line to the other. When the ACCEPT light button is picked, an L^4 instruction is generated that returns the X, Y-coordinates of the two end points to the host computer.

The second rubberbanding capability allows only the second end point of a vector to be moved with the tracking cross and requires that the initial starting point of the vector be specified in the invoking L^4 instruction. This second mode of rubberbanding a vector is used for continuing a line drawing from the end point of a previously drawn vector.

The third interactive rubberbanding capability involves the generation of a circle by using the tracking cross to initially specify the center and then finally a point on the circumference of a circle. Since mathematical functions are not available in the minicomputer of the intelligent terminal, the radius R of the circle is approximated from the position of the tracking cross (X_2, Y_2) and the center point (X_1, Y_1) by first computing an initial estimate R^*,

$$R^* = \frac{\max(|X_1 - X_2|, |Y_1 - Y_2|) + \min(|X_1 - X_2|, |Y_1 - Y_2|)}{2}, \qquad \text{(A.1)}$$

and then refining the estimate with a single Newton iteration:

$$R = \frac{1}{2}\left[R^* + \frac{1}{R^*}\left((X_1 - X_2)^2 + (Y_1 - Y_2)^2\right)\right]. \tag{A.2}$$

The fourth interactive rubberbanding capability involves the generation of a rectangle by specifying two opposing vertices with two moves of the tracking cross. In all four modes of rubberbanding, the picking of the ACCEPT light button causes the locally generated object to disappear and an L^4 instruction to be generated that sends the two user-selected coordinate pairs back to the host computer so that the application program can generate the object as a permanent part of the display file.

A.3 HIGH-LEVEL GRAPHIC PROGRAMMING PACKAGES

A.3.1 Introduction

Up to this point, we have discussed the general concepts of the graphics programming system and have presented a detailed description of the support software for the intelligent terminals. Although it is certainly feasible for a user of the terminal to directly communicate with it, using L^4 commands entered through the terminal keyboard, such a low-level-language approach would suffer the same disadvantages as exhibited by the use of a machine assembly language when higher-level languages are readily available. To capitalize on the advantages offered by high-level languages, two user-callable graphics application programming packages were written. In each case, the GRIP language model (see Section 10.5) was followed as closely as possible within the confines of the syntax of the utilized language.

A.3.2. FORTRAN Graphics Package (GRAP)

GRAP (GRaphics Application Package) is our FORTRAN-callable subroutine package for interactive computer graphics. It exists as a library of subroutines that may be linked to a user-written application program. Although FORTRAN is not a block-structured language, the structuring of the display file as dictated by the segment–item–primitive hierarchy, is realized through the use of subroutine calls which signal the beginning and ending of each new block. For example, a CALL SEGMENT (SEGNAME, MX, MY, IAV) will open a new block, which remains open until a CALL END is encountered to close the block. Such blocks only effects graphical entities and do not limit the scope of any variables within the FORTRAN source coding.

The routines that comprise the GRAP package are divided between those that are user-callable and those that are invoked solely by other GRAP routines and are therefore referred to as internal routines. These internal routines will be discussed first.

SUBROUTINE L4STATS (IAV): This routine generates four associated data list (ADL) entries from the four-digit-decimal attribute vector IAV. If this routine is called as a result of an item-generating procedure, then any zero (default) elements in the attribute vector cause the segment attribute to be used. If this routine is called as a result of a segment initialization, then any zero elements in the attribute vector are replaced with the system-defined default values. The global segment attributes are placed in a labeled common block, so that their values are readily available in subsequent item calls to this procedure.

SUBROUTINE L4GNMT (OPCODE,INDEX): This routine generates one L^4 instruction and places it in the message table. The instruction consists of the desired L^4 instruction opcode as given by the OPCODE parameter, the INDEX or data field of the L^4 instruction, and the pointer to the associated data area in the ADL whenever the instruction includes associated data. This ADL pointer is maintained in a labeled common block.

SUBROUTINE L4GNADL (DATA): This routine inserts one DATA word into the associated data list (ADL). A global ADL pointer that is located in a labeled common block is used to indicate the next available position in the ADL. Once the DATA word is placed there, the pointer is increased by one.

SUBROUTINE L4CNVRT (WORD, NUMCHR, ARRAY, INDEX): This routine converts WORD into NUMCHR number of characters and places the characters into ARRAY at the position pointed to by INDEX. The routine further provides the conversion between the binary data in the MT and ADL, respectively, and the ASCII character strings that are needed to send the MT and ADL to the intelligent terminal through the time-sharing interface. In our implementation, the MT and ADL are actually stored in their ASCII character representation rather than in their binary format because of the economy of putting 10 characters in one 60-bit machine word instead of only one binary value.

FUNCTION L4INSR (NAME, ITYPE, NUM, MXR, MYR): This function creates a new item entry in the host computer name table (NT) and correlation table (CT) by placing the NAME parameter in the name table and the ITYPE, NUM, MXR, MYR parameters into the correlation table. The new entries are made at the beginning of the current segment area as pointed to by an entry in a labeled common block. Everything below this point in the NT/CT is moved down one notch. The ITYPE parameter gives the item type code, the NUM parameter gives the number of primitives in the item, and the MXR, MYR parameters give the relative item origin. The function value upon return is the CT index where the item entry was actually made.

SUBROUTINE L4CTDL (INDX1, INDX2): This routine deletes the NT/CT entries between the positions pointed to by the INDX1 and INDX2 parameters. All entries beyond the INDX2 position are moved up to the INDX1 position.

FUNCTION L4INDX (NAME, IFLAG): This function searches the NT/CT in three different modes, as determined by the IFLAG parameter:

IFLAG = −1 searches the current segment area for an item that matches the NAME parameter
= 0 searches the NT/CT for the next segment entry or the end of the NT/CT
= 1 searches the entire NT/CT for a segment that matches the NAME parameter

The function value returns the NT/CT index to the desired entry.

SUBROUTINE L4GTMT (I, IOPCOD, INDEX, IREF): This routine decodes the Ith entry in the MT that was received from the intelligent terminal and returns the value of the opcode in IOPCOD, the value of the data field in INDEX, and the value of the ADL reference pointer field in IREF.

SUBROUTINE L4INPUT: This routine reads in the MT and the ADL that is sent back to the host computer from the intelligent terminal. The MT and the ADL are stored in their binary representation. The conversion between the incoming ASCII character strings and the binary data format of the MT and ADL is performed in the next subroutine.

SUBROUTINE L4INPCK (IBUF, JBUF, ICHK): This routine will convert a string of ASCII characters from the intelligent terminal into an array of 16-bit binary numbers packed one per each 60-bit machine word. The incoming character string is delimited by either a space byte or a zero byte. Three characters of the string are used to form each 16-bit binary word using the lowest-order 4+6+6 bits of each character code. IBUF is the incoming character string. JBUF is the output binary data array. A checkword is received at the end of each character block and the ICHK parameter is returned as zero if no transmission errors were detected.

The remaining routines in the GRAP package are those that are called directly by the user application program. They are basically the GRIP commands (see Section 10.4) which are preceded by CALL statements to conform to FORTRAN syntax.

A3.3 APL Graphic Package (APL-G)

An APL graphics package was also programmed for our system. The internal procedures are practically the same as were discussed above for the FORTRAN package, except that they are now given the syntactic form of APL defined functions. The user-callable defined functions in APL-G are absolutely identical with the GRIP commands listed in Section 10.4. From the viewpoint of the APL syntax, the parameter list of a GRIP command is declared as a character string (thus the quotes). Each APL-G function representing a GRIP command includes a string analysis part which looks for the occurrences of the list separators (;) and, accordingly, breaks the input string into substrings representing the various list elements. If a substring represents a name, it is directly assigned to a local variable for further reference. If a substring represents an expression, it is evaluated (by the use of the EXECUTE operator), and the obtained value is then assigned to a local variable for further use. Values are exchanged between functions by using global variables as carriers. Hence, the defined functions that represent GRIP statements are all monadic and without explicit result. The names of the internal procedures and global variables all begin with the character combination L4. Therefore, the user must never use names which begin with these characters.

Procedures for the various picture transformations are existent in the form of mnemonically named defined functions. Here, the regular parameter passing scheme of APL (by value) is employed. Furthermore, a special package of defined functions provide all desirable modern control constructs [58]. As a result, APL-G is a mnemonic "end user" language for graphical programming that can be used without any knowledge of APL. On the other hand, if a user is proficient in APL, he/she may use APL-G procedures within regular APL programs.

Environment Idiosyncracies: In that the full APL character set is not supported by our intelligent terminal, a reduced character set is used. This results in the encoding of each 16-bit binary word of the MT/ADL into four rather than three ASCII characters, using the four lowest order bits of each of the characters. This results in a one-third slowdown in transmission time for a graphical data block relative to our FORTRAN graphics package. Further, the execution speed of the APL interpreter is noticably slower than an equivalent task performed using a FORTRAN implementation. Finally, the amount of memory consumed by the APL workspace is about 50% more than that consumed by FORTRAN for the same application.

APL advantages: Despite the disadvantages of the APL package mentioned above, definite advantages emerge when we compare the APL package to its FORTRAN counterpart. The source code of the APL routines is approximately one-third as lengthy as that of their FORTRAN equivalents. Many of the tedious programming tasks involved in the FORTRAN implementation, such as character string manipulation and array processing, become trivial in the APL implementation. Finally, the construction of APL graphics application programs is greatly facilitated by the ability to immediately execute any partially completed section of the total program and to receive comprehensive error information as to where the execution was terminated as well as why the program aborted. At present, the FORTRAN programming package is more widely used than the APL package, primarily because of the greater number of programmers who have had FORTRAN experience. This author's opinion is that the advantages of APL far outweigh the disadvantages, and that in time more and more users will be making the switch to APL.

Index